Probability and Mechanics of Ship Collision and Grounding

Probability and Mechanics of Ship Collision and Grounding

Shengming Zhang
Lloyd's Register, United Kingdom

Preben Terndrup Pedersen
Technical University of Denmark, Denmark

Richard Villavicencio
Lloyd's Register, United Kingdom

ELSEVIER

Butterworth-Heinemann
An imprint of Elsevier

Butterworth-Heinemann is an imprint of Elsevier
The Boulevard, Langford Lane, Kidlington, Oxford OX5 1GB, United Kingdom
50 Hampshire Street, 5th Floor, Cambridge, MA 02139, United States

Notices

Knowledge and best practice in this field are constantly changing. As new research and experience broaden our understanding, changes in research methods, professional practices, or medical treatment may become necessary.

Practitioners and researchers must always rely on their own experience and knowledge in evaluating and using any information, methods, compounds, or experiments described herein. In using such information or methods they should be mindful of their own safety and the safety of others, including parties for whom they have a professional responsibility.

To the fullest extent of the law, neither the Publisher nor the authors, contributors, or editors, assume any liability for any injury and/or damage to persons or property as a matter of products liability, negligence or otherwise, or from any use or operation of any methods, products, instructions, or ideas contained in the material herein.

Library of Congress Cataloging-in-Publication Data
A catalog record for this book is available from the Library of Congress

British Library Cataloguing-in-Publication Data
A catalogue record for this book is available from the British Library

ISBN: 978-0-12-815022-1

For information on all Butterworth-Heinemann publications
visit our website at https://www.elsevier.com/books-and-journals

Publisher: Matthew Deans
Acquisition Editor: Carrie Bolger
Editorial Project Manager:
Thomas Van Der Ploeg
Production Project Manager:
Nirmala Arumugam
Cover Designer: Matthew Limbert

Typeset by SPi Global, India

Contents

Preface

Ship collision and grounding events continue to constitute a major hazard for shipping and operation of offshore installations. These events can result in the loss of human lives and severe environmental damages. This has led the maritime community to further increase the efforts to mitigate the probability and consequences of ship collision and grounding accidents.

To develop effective Rules and procedures to reduce the risk associated with ship collision and grounding events requires firstly a procedure to evaluate the probability of ship collision and grounding in specific sea traffic conditions. Secondly, given a ship collision or grounding, a procedure is needed to estimate the energy that will be released for structural damage of the involved ship structures. Knowing the energy to be absorbed by structural damage, the final step is to estimate the structural damage to the striking ship and the struck ship or the struck offshore installation.

In this book, mathematical models for the assessment of ship collisions and grounding events are presented, including methods for determining the probability of ship collisions, collision energy losses, collision forces, and the extent of the structural damage. The analysis procedures are divided into three main parts: the collision probability, the external dynamics, and the internal mechanics. Methods relating the absorbed energy and the destroyed material volume are also developed. All these analysis methods have been extensively validated by both experimental results and numerical simulation results to provide confidence in their use for full-scale collision assessments. A special chapter is dedicated to the finite element analysis techniques used for simulating structural damage due to a ship collision.

The ship collision and grounding probability model in Chapter 1 is based on a division of the problem into different ship collision and grounding scenarios and subsequent application of mathematical models to quantify the risk from human and technical errors. The mathematical models for accidents related to human errors are mainly based on calculation of the number of collision candidates in the case of blind navigation and the use of causation factors to model the action of the navigators on board the ship. The presented model is applied to determine the probability of ship–ship collisions, grounding events, ship collisions against bridges over navigational channels, and ship collisions against fixed and floating offshore installations including offshore wind turbine parks.

In Chapter 2 on external dynamics, analytical methods are developed for the collision energy loss and the impact impulse in arbitrary ship–ship collisions, ship collisions with offshore platforms, ship collisions with rigid walls, and ship grounding. The method can be used for any kind of vessels with no limitations in size, impact velocity, impact location, or striking angle. Collisions involving glancing blows are also included. The energy loss to be dissipated by destroying the ship structure is given in closed-form expressions. Thus, the calculation of the collision energy is simple and fast, and it can be used for probabilistic analyses of ship collisions where a large number of cases are simulated. Comparisons of the analytical method with reported numerical simulation results and experimental results show good agreement.

In Chapter 3 on internal mechanics, material properties for ship collision and grounding analysis are reviewed and discussed. The fundamental upper-bound theorem is used for ship collision damage estimation and applied to a series of analyses for basic structural elements. The analyses include large plastic deformation modes of the shell plating subjected to various loadings; folding and crushing of web frames; denting and crushing of web intersections; and crushing of L, T, and X elements. Ship–bridge collision damage analysis, ship–ship collision damage, and ship–offshore collision damage analysis are presented. In addition, theoretical models for tearing of plates in ship grounding events are established. These basic analysis methods are verified by existing experimental data and numerical solutions.

Closed-form expressions relating the absorbed energy and the destroyed material volume during collisions and grounding are developed in Chapter 4. This approach overcomes a major drawback of Minorsky's well-known method since it takes into account the structural arrangement, the material properties, and the damage modes. The validity of the method is verified against a large number of experimental results and numerical simulations. Further, the procedure is illustrated by analyses of full-scale ship collision and grounding accidents.

Finite element analysis techniques for ship collision analysis are summarized in Chapter 5. Here, the emphasis is on the nonlinear material characterization, in particular the material failure criteria available for assessments of full-scale collision scenarios, and some of the present limitations of the finite element method.

The objective of the book is to present theories and tools for analysis and assessment of ship collisions and grounding events. Illustrative calculation examples are included in each chapter to facilitate the industrial applications and further academic research.

The authors would like to thank publishers, companies, and individuals who granted the permissions to reuse or reproduce some of the figures printed in the present book. They are all acknowledged in the figure captions by indicating 'copy from (cf.)' or 'based on' followed by the respective reference. Special thanks to the individuals who have provided the original copies of the figures and supports.

Thanks to the publishers/editors, Cambridge University Press, Elsevier, IOS Press, Springer, Taylor & Francis Group, The National Academy of Sciences, The Society of Naval Architects and Marine Engineers (SNAME), Standards Norway, and ISO Standards, and the companies Lloyd's Register, Ramboll Group A/S, Fehmarn Link, and GateHouse Maritime A/S.

Thanks to the following individuals: Professor Jørgen Amdahl, Professor Sören Ehlers, Professor Norman Jones, Mr Tim Kent, Professor Bin Liu, Professor Kun Liu, Dr Marie Lützen, Dr Ingo Martens, Dr Jörg Peschmann, Professor Jeom Paik, Professor Jonas Ringsberg, Professor Carlos Guedes Soares, Professor Yousheng Wu, Dr Yasuhiro Yamada and Professor Ling Zhu.

The contribution from the first and the last author was primarily developed when the authors worked at universities before joining in Lloyd's Register. Lloyd's Register Group Limited; its affiliates and subsidiaries; and their respective officers, employees, or agents are, individually and collectively, referred to in this clause as the 'Lloyd's Register'. Lloyd's Register assumes no responsibility and shall not be liable to any person for any loss, damage, or expense caused by reliance on the information or advice in this document or howsoever provided, unless that person has signed a contract with the relevant Lloyd's Register entity for the provision of this information or advice, and in that case, any responsibility or liability is exclusively on the terms and conditions set out in that contract.

Shengming Zhang
Lloyd's Register, United Kingdom

Preben Terndrup Pedersen
Technical University of Denmark, Denmark

Richard Villavicencio
Lloyd's Register, United Kingdom

Probability of ship collision and grounding

1

1.1 Introduction

Every year, ship–ship collisions, ship collisions to offshore structures and bridges, and grounding events cause the loss of lives, economic losses, damages to the environment, and other unwanted events. It is important that the design and operation of a ship or an offshore installation meet risk levels that can be estimated to be so low that risks are of little concern to users and third parties. On the other hand, the required risk levels should still allow construction and operation of the maritime structure at feasible cost levels. Therefore, one of the many performance goals during the design phase of ships and offshore structures should be to ensure that serious accidents and service disruptions are low enough to be acceptable to all stakeholders, that is, owners, the public, and those responsible for public safety. To demonstrate such safety levels, risk analyses have to be performed.

Risk evaluation procedures are based on an estimation of the probability of the considered hazard followed by an estimate of the consequences.

The probability of the occurrence of collision and grounding events can be estimated from historical data, expert opinions, and predictive calculations. Historical data can in some highly trafficked navigational areas provide realistic figures. However, these data are difficult to use for future predictions since they are relevant to ship structures constructed in the past that differ from those used today and they do not take into account new operational procedures and new navigational equipment. Expert opinions can be useful and have the advantage of avoiding highly complex mathematical models, but for quantitative predictions, they are rarely very accurate. For these reasons, it is important to develop and apply rational mathematical models for the prediction of the frequency of hazard occurrence that also allow for selection of the most effective risk control options.

The most cost-effective way to reduce risk is normally by reducing the probability that the associated adverse events take place.

The focus of this book is on ship collision and grounding hazard evaluation, and the first chapter is specifically devoted to mathematical tools to determine the ship and route specific probability that collision and grounding events will take place.

1.1.1 Risk based analyses

Formally, the risk related to a given activity can be defined as a function of the hazards, the probability of occurrence, and all the related consequences:

$$Risk = f(hazards, probabilities, consequences) \tag{1.1}$$

Probability and Mechanics of Ship Collision and Grounding. https://doi.org/10.1016/B978-0-12-815022-1.00001-3

The risks associated with ship collisions and grounding events are primarily the loss of lives and injuries, environmental damages, and economic losses, but risk analyses can also be useful planning tools to form the basis for the development of emergency services for contamination at sea. Consistent risk-based models can predict the effect of specific geographical locations and configurations of the emergency services for combating oil and chemical pollution at sea.

The functional relationship (Eq. 1.1) can be a challenge related to risk evaluation, since it may require that consequences to humans, environment, and economical values to be related to the same scale.

In order to solve this problem and to introduce evaluation criteria for making decisions related to risk reduction, a common measure or scale is often introduced. Typically, the scale to which the consequences are related is selected as the monetary scale. Thus, a risk definition could have the form:

$$Risk = SUM(P_i(H_i; C_i) \cdot U(C_i)) \tag{1.2}$$

where the sum is over all the consequences C_i related to the hazard H_i and P_i is the probability of the ith consequence. The function U is a utility function that expresses the consequence in some common measure, most often as the monetary value. This translation of the different types of consequences into some other measure not only depends on the type of consequence of the hazard but also should depend on to whom it is a consequence. The utility function U must be different for those who have a direct interest in the activity, that is, a gain from a successful operation, and for instance a third party such as the public who may only have the risk from the activity. The present chapter describes procedures to estimate the probability P_i for ship collision and grounding events and Chapters 2–5 deal with tools for analysis of the consequences C_i.

1.1.2 Risk acceptance criteria

For a rational reduction of risk related to hazards, such as ship collisions and grounding events, it is necessary to establish a risk acceptance criterion. Without a generally agreed acceptance criterion, it is not possible to find the balance between safety in terms of risk reduction and the cost to the stakeholders.

Usually, risk acceptance criteria must be established for three main types of risks:

- Fatalities
- Pollution of the environment
- The loss of property or financial exposure

The acceptance criteria for fatalities related to shipping accidents are normally based on two principles:

- The *individual fatality risk* shall be approximately the same as typical for other occupational hazards.
- The frequency of accidents with several fatalities, that is, the *societal fatality risk*, shall not exceed a level defined as unconditionally intolerable, and moreover, the general as low as

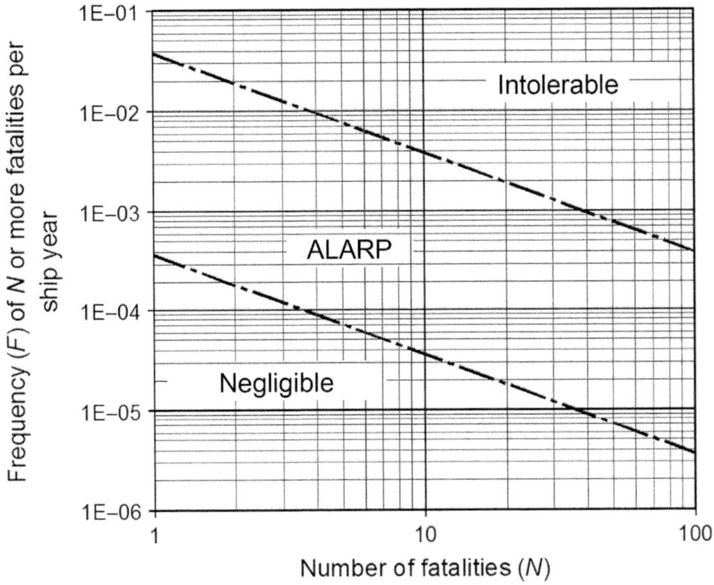

Fig. 1.1 Typical risk acceptance criterion, F-N diagram.

reasonably practicable (ALARP) risk management shall be applied. Fig. 1.1 illustrates the principle of the ALARP criterion.

The latter criterion is introduced because society is often more concerned about single accidents with many fatalities than many accidents with few fatalities per accident.

If the risk is estimated to be in the 'intolerable' region in Fig. 1.1, then the activity should not be allowed. For risks that are not in the negligible region, the general ALARP risk management shall be applied. That is, in principle for all nonnegligible risks, it is required that measures for risk reduction are to be identified and analysed and their societal value assessed. In this ALARP region, an economic criterion can be applied to consider the effectiveness of safety measures or risk control options; see Eq. (1.2).

Similarly, ALARP criteria have been used by authorities to reduce accidental oil spills from tankers. Here, the cost for preventing an oil spill accident must be based on the cost of oil, the clean-up cost, the environmental damage cost, etc. Shipping and grounding accident costs are often dominated by clean-up costs due to oil spillage.

Authorities like International Maritime Organization (IMO), International Association of Classification Societies (IACS), and national administrations normally focus on these types of risks one after the other. That is, so far, the ALARP principle has been applied separately for fatalities and for environmental impacts when considering new rules. The general economic costs associated with severe accidents have not been considered very often.

With a significant percentage of all ships involved annually in a serious and costly accident, the economic loss will have a significant influence on the outcome of expressions such as Eq. (1.2). This indicates that it is often possible to introduce risk mitigation measures that can be cost-effective. One example is improved navigational safety measures that influence all three risk categories at the same time.

Ideally, the additional cost of risk-reducing measures in the form of construction cost plus present value of operational costs is evaluated against the effect of the risk in the ALARP region. The condition for a decision to introduce risk-reducing measures could be.

$$C < X + I + E + D$$

where

C is the cost of the considered risk-reducing measure
X is the direct costs concerned with the accident
I is the accumulated change of cost of all 'individual' user risks
E is the quantified change in cost of 'environmental' impact
D is the change in cost due to economic loss caused by 'disruption'

where all costs are converted to present value costs.

1.1.3 Rules and regulations

The ship collision risk is addressed by different governing standards that include regulatory standards, classification society standards or guidelines, industry standards such as American Petroleum Institute (API) and International Organization for Standardization (ISO), and corporate standards.

To ensure that serious ship accidents and service disruptions are low enough to be acceptable to all stakeholders and to enhance safety, IMO Maritime Safety Committee (2013) established a procedure for Formal Safety Assessment (FSA) as a proactive approach to evaluate risk associated with different hazards. The procedure adopted by IMO is an approach to find the likelihood of scenarios, which may develop from hazards such as ship collisions and groundings, and the severity of their consequences in order to calculate the risk. The FSA method must be applied together with some risk acceptance criteria.

The same procedure is also well suited for the evaluation of ship collision risk for offshore installations and for bridges crossing shipping lanes. For the evaluation of the risk for ship collisions for offshore structures, the procedure is depicted graphically in Fig. 1.2.

1.2 Probability of ship–ship collisions

Reported causes for ship collisions and grounding events show that approximately 80% of collision and grounding accidents involve at least some human errors or questionable judgments and approximately 20% are due to technical errors. Unfortunately, the human error mechanisms are not yet well understood.

Pioneering work on risk models for ship collisions and grounding events has been made in Japan by Fujii and co-workers (1971, 1974, 1998) and by Kaneko

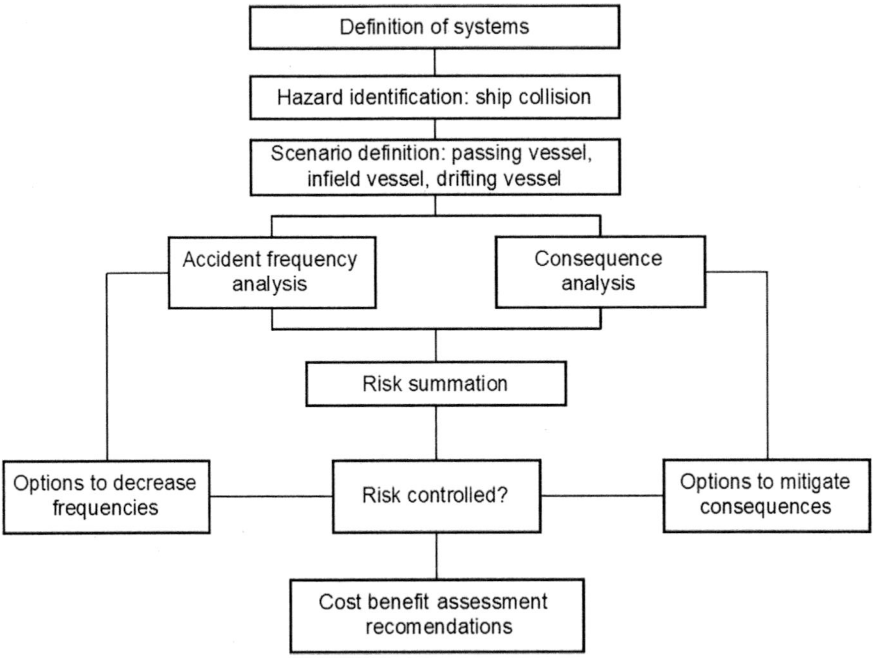

Fig. 1.2 Illustration of a Formal Safety Assessment (FSA) process for ship collisions against offshore installations.

and Hara (2007). As a result of this and subsequent work, a model for ship collisions and grounding has been established on the basis of observations of ship track distributions and estimations of human and technical failure probabilities; see Pedersen (1995). This ship–ship collision model is based on knowledge on shipping routes, number of ships, ship track distributions, ship size distributions, and distributions of ship speeds. A review of ship collision probability models has been given in Pedersen (2010) and Li et al. (2012).

The main principle behind the risk model is to determine the number of possible ship accidents N_a, that is, the number of collisions in case of blind navigation and no aversive manoeuvres. This number N_a of possible accidents is in principle determined as:

$$N_a = N_{ships} \cdot P_{geometrical} \tag{1.3}$$

where N_{ships} is the number of ships on the route in a given time interval and $P_{geometrical}$ is the geometrical probability that a ship on this route will be on collision or grounding course. The number of possible collision candidates N_a is then multiplied by a causation probability P_c in order to find the actual number of ship collisions F_{col} during the considered time interval:

$$F_{col} = P_c \cdot N_a \tag{1.4}$$

The causation probability P_c is the fraction of the accident candidates that results in an accident. This ratio is also often termed aberrant vessels.

The causation probability P_c can be estimated on the basis of available accident data collected at various locations and then transformed to the area of interest. Another approach is to analyse the cause leading to human inaction or external failures and set up a fault-tree procedure or apply a Bayesian network procedure for the calculation of P_c.

1.2.1 Ship traffic distribution

The traffic intensity on sailing routes is key information when estimating the frequency of ship accidents; see Silveira et al. (2013). The more ship movements, the more accidents must be expected.

Information about the annual number of ship movements and the volume of traffic on each sailing route in a given geographical area and the distribution of ship types and sizes are basic parameters when estimating the frequencies of ship collisions. For the consequence estimation of ship accidents, the size and the type of ships involved are also important factors. A large oil tanker collision accident can lead to large environmental consequences in case of oil spill, and a collision involving a large passenger vessel can give severe consequences in the form of number of fatalities if subsequently the passenger vessel capsizes or sinks.

Detailed analysis of the existing ship traffic can be based on analysis of automatic identification system (AIS) data. AIS is a system that makes it possible to exchange data between ships and between ships and land-based stations. According to a decision by the IMO, all ships with a gross tonnage of 300 and upwards engaged in international navigation, cargo ships of more than 500 tons gross tonnage not engaged in international voyages, and all passenger ships irrespective of size are required to be fitted with AIS. Navy vessels are not obliged to transmit AIS signals.

AIS data can be received by an entity free of charge (see Fig. 1.3). A ship that is equipped with AIS will continuously transmit information on, for example, course and position. The information is normally transmitted by radio to other ships equipped with AIS and to land-based AIS stations. An AIS receiver usually receives messages from vessels, base stations, or aids to navigation at a maximum range of 20–25 nautical miles. If there is a need for vessel data outside of the range of a terrestrial AIS receiver or network of receivers, satellite AIS is needed.

Each part of the information in the AIS data is categorized as either dynamic data or static data. The static parts are generated manually by the navigator, whereas the dynamic parts are generated automatically.

A ship database can be imported to the registered AIS data to identify the specific ship type and specific physical properties. These combined data can include vessel dimensions, displacement, gross tonnage, dead weight tonnage, vessel type, speed, age, and constructive details such as bulbous bow and ice class.

From an analysis of AIS data, the annual number of movements of ships with a gross tonnage of 300 and upwards can be determined and distributed to a number of main routes in the considered geographical area. See Fig. 1.4.

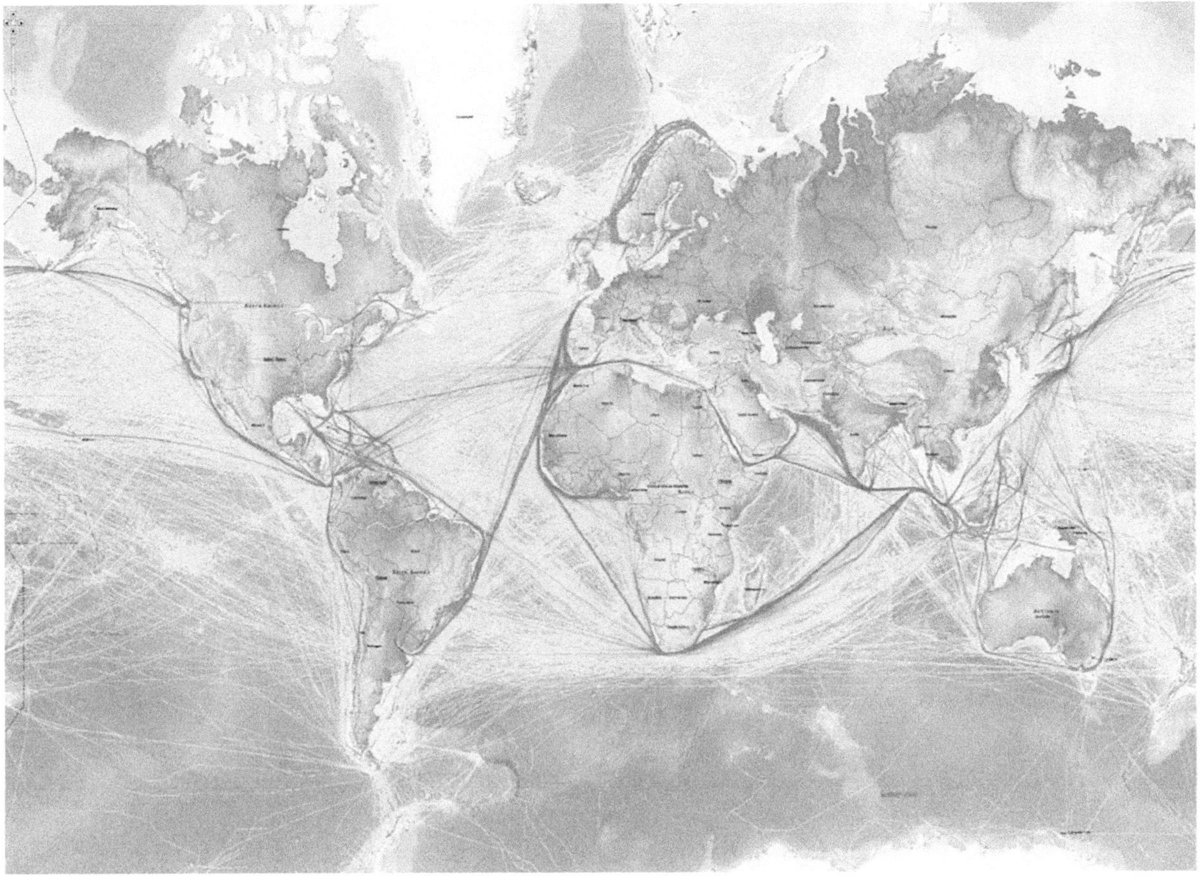

Fig. 1.3 AIS tracks showing the global ship traffic pattern based on satellite receivers. Reproduced with permission from GateHouse Maritime A/S.

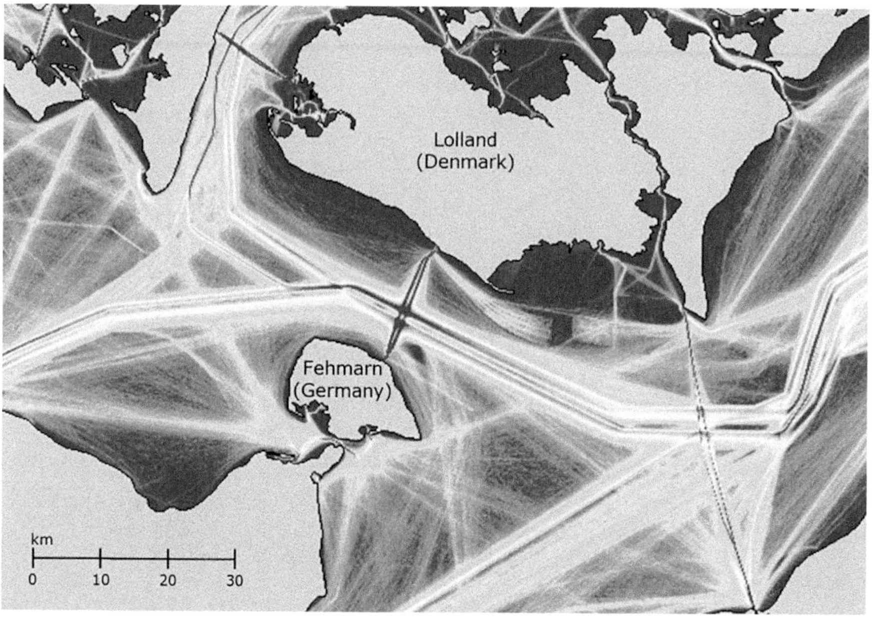

Fig. 1.4 Ship traffic intensity plots in Fehmarn Belt between Denmark and Germany. By Rambøll and reproduced with permission from Femern A/S.

An estimate of the traffic segment in a given geographical area with a gross tonnage below 300 and naval vessels can be provided on the basis of radar data. The traffic segment with a gross tonnage below 300 can consist of small commercial ships, workboats, small tugs, fishing vessels, and pleasure crafts. The majority of ships in this segment does not use the main navigational routes and is often better modelled as background traffic with constant intensity throughout the area.

When a detailed risk analysis is needed, the ship traffic can be grouped in categories. One example could be as follows:

- Bulk carriers
- Chemical tankers
- Container vessels
- Gas tankers
- Oil tankers
- Passenger vessels
- Ro-Ro vessels
- Other vessels

Each of the categories contains a number of ship classes. Each ship class is defined by the following properties:

- Length
- Breadth
- Depth

and, for both loaded and ballast conditions,

- speed,
- draught,
- displacement,
- height of deck,
- height of bulb.

For an overview of the ship collision risks in a specific geographical area, all these ship categories can be collapsed into just one group containing a limited number of ship size classes. See Table 1.1. An example of observed ship traffic within two shipping lanes is given in Table 1.2.

The distribution of the ship traffic transverse to each lane is important for the estimation of the probability of ship collisions. For example, the average distance

Table 1.1 Example on simplified data set for ship properties.

Class	Gross tonnage	Mean length (m)	Mean breadth (m)	Mean draught (m)	Mean speed (m/s)
1	0–500	35.0	8.9	3.2	3.8
2	500–1000	56.1	10.4	3.7	4.5
3	1000–2000	77.5	11.9	4.3	5.0
4	2000–3000	90.8	13.7	5.2	5.5
5	3000–5000	104.9	16.1	6.0	6.0
6	5000–10,000	127.6	19.4	7.6	7.0
7	10,000–20,000	162.4	23.9	9.5	7.5
8	20,000–50,000	199.4	30.6	12.0	7.5
9	50,000–200,000	255.7	42.9	14.6	8.0

Table 1.2 Annual ship traffic on two shipping routes.

Class	Route 1		Route 2	
	North	South	East	West
1	897	786	1388	1221
2	567	530	704	745
3	639	628	688	698
4	745	823	780	836
5	338	307	361	327
6	229	271	269	269
7	357	360	352	351
8	45	20	35	10
9	20	50	10	20
Total	3830	3775	4586	4479

Table 1.3 Parameters for observed spatial distribution of ship traffic on two shipping routes.

Route	Direction	Standard deviation (m)	Uniform part	Uniform width (m)	Mean separation μ (m)
1	North	290	3%	2400	200
1	South	310	2%	2425	200
2	East	315	2%	2100	190
2	West	160	2%	2220	190

and the lateral distribution of the traffic on two lanes with opposite sailing directions, belonging to the same sailing route, will be needed when estimating the probability that ships with opposite sailing directions are on collision course. See Table 1.3 and Fig. 1.5. These distributions of ship traffic transverse to the sailing routes can also be based on analyses of AIS data. It is often convenient to fit the observed lateral distribution of the ships in the various traffic lanes into analytical expressions. If a combination of normal and uniform distribution is used for the transverse distribution, $f(z)$, then the ship location distributions are determined as

$$f(z) = \alpha U(z; a, b) + (1 - \alpha)N(z; \mu, \alpha) \tag{1.5}$$

where

α is the ratio of ships following the uniform distribution
$U(z;a,b)$ is the uniform distribution at z with lower bound a and upper bound b
$N(z;\mu,\sigma)$ is the normal distribution at z with mean μ and standard deviation σ.

Fig. 1.5 Lateral distribution of ship traffic: Histogram and fitted Gaussian distribution.

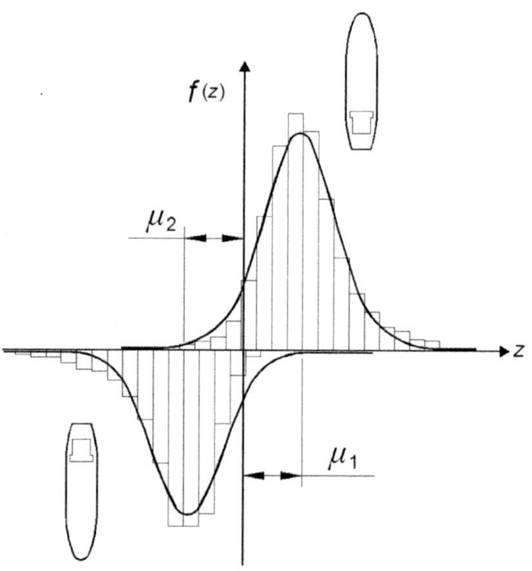

The normal distribution represents the ordinary deviations from the ideal route, whereas the uniform distribution represents ships under control but due to gross navigational errors or intentionally making a short cut and therefore deviating from the ideal route. The standard deviation σ of the ships relative to the ideal route can depend on the size of the ships and the width of the navigational channel. If the navigational channel is well defined by buoys and lights, the standard deviation is smaller than on navigational channels in the open sea. The mean value μ models the eccentricity induced by navigational markings for opposite directions in the same navigational channel.

Instead of curves, fitting the ship traffic distribution based on normal or Gaussian distributions as well as log-normal distributions, Gumbel distributions, and Weibull distributions can be used.

1.2.2 Ship collision candidates

For calculation of the number of possible ship collision candidates N_a (see Eq. 1.3), firstly two crossing waterways are considered in Fig. 1.6. Here, the ship traffic is assumed to be known and has been grouped into a number of different ship classes according to the vessel type, deadweight tonnage or length, loaded or ballasted, etc.

1.2.2.1 Collisions in crossing waterways

In the following, the calculation model based on Pedersen (1995) will be presented for the number N_a of possible events where two ships will collide in the overlapping hatched area in Fig. 1.6, if no aversive manoeuvres are made. The crossing routes

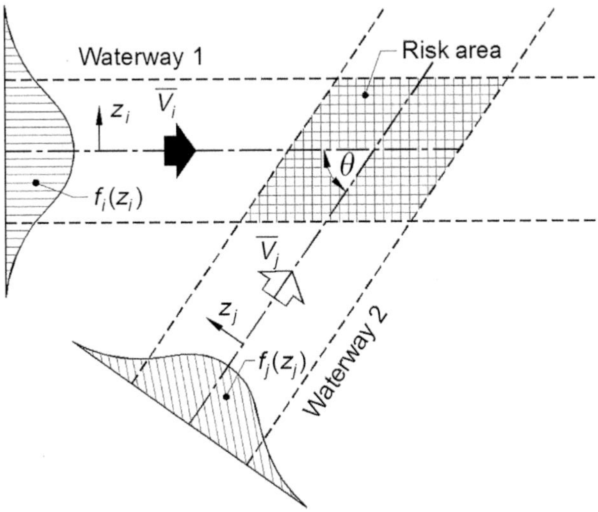

Fig. 1.6 Crossing waterways with risk area of ship–ship collision.
Based on Pedersen, P.T., 1995. Collision and grounding mechanics. In: Proc. WEMT'95. Danish Society of Naval Architecture and Marine Engineering, Copenhagen, pp. 125–157.

are defined to have one-way traffic only. If two-way traffic exists in the crossing route, the route must be included twice, one for each direction of the traffic. This strategy is adopted to allow for different traffic distributions in the two directions.

From geometric analysis, it is seen that in a segment dz_j of waterway 2, the number of class '*j*' ships on collision course with one '*i*' ship during the time dt is

$$\frac{Q_j^{(2)}}{V_j^{(2)}} f_j^{(2)}(z_j) D_{ij} V_{ij} dz_j dt \tag{1.6}$$

where $Q_j^{(2)} = Q_j^{(2)}(t)$ is the traffic flow, that is, the number of ships per unit of time of ships class j in waterway no. 2 (which could be loaded bulk carriers with length between 140 and 160m; see also Table 1.1), and $V_j^{(2)}$ is the speed. The lateral distribution of this ship traffic in waterway 2 is denoted $f_j^{(2)}(z_j)$. If the lateral ship traffic distribution is approximated by a Gaussian distribution, then

$$f_j^{(2)}(z_j) = \frac{1}{\sigma_j^{(2)}\sqrt{2\pi}} \exp\left(\frac{-\left(z_j - \mu_j^{(2)}\right)^2}{2\left(\sigma_j^{(2)}\right)^2}\right) \tag{1.7}$$

where z_j is the distance from the centreline of navigation route 2, $\mu_j^{(2)}$ is the mean value of the spatial distribution, and $\sigma_j^{(2)}$ is the standard deviation. D_{ij} is the geometrical collision diameter defined in Fig. 1.7.

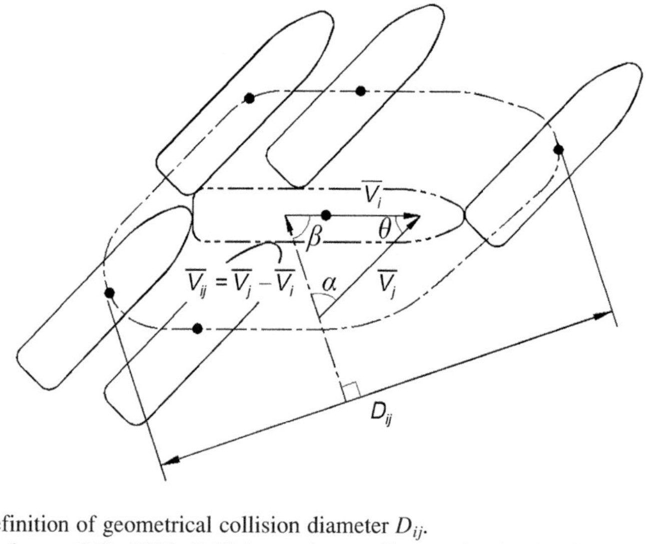

Fig. 1.7 Definition of geometrical collision diameter D_{ij}.
Based on Pedersen, P.T., 1995. Collision and grounding mechanics. In: Proc. WEMT'95. Danish Society of Naval Architecture and Marine Engineering, Copenhagen, pp. 125–157.

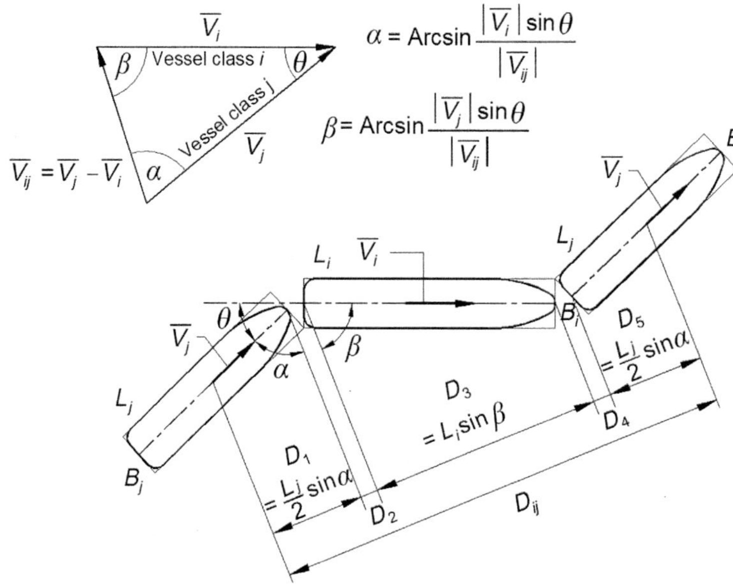

Fig. 1.8 Calculation of the geometrical collision diameter D_{ij}.
Based on Pedersen, P.T., 1995. Collision and grounding mechanics. In: Proc. WEMT'95. Danish Society of Naval Architecture and Marine Engineering, Copenhagen, pp. 125–157.

Assuming that the waterline area of the ships can be approximated by rectangular shapes, then the geometry depicted in Fig. 1.8 shows that

$$D_{ij} = \frac{L_i^{(1)} V_j^{(2)} + L_j^{(2)} V_i^{(1)}}{V_{ij}} \sin\theta + B_j^{(2)} \left\{ 1 - \left(\sin\theta \frac{V_i^{(1)}}{V_{ij}} \right)^2 \right\}^{1/2} + B_i^{(1)} \left\{ 1 - \left(\sin\theta \frac{V_j^{(2)}}{V_{ij}} \right)^2 \right\}^{1/2} \tag{1.8}$$

where the relative velocity V_{ij} is determined as

$$V_{ij} = \sqrt{\left(V_i^{(1)}\right)^2 + \left(V_j^{(2)}\right)^2 - 2V_i^{(1)} V_j^{(2)} \cos\theta} \tag{1.9}$$

and where B_i is the width and L_i is the length of ship i.

Expression (1.6) is for one ship in waterway 1. Multiplication by the number of class i ships $Q_i^{(1)}$ per unit time in the considered segment in waterway 1

$$\frac{Q_i^{(1)}}{V_i^{(1)}} f_i^{(1)}(z_i) dz_i \tag{1.10}$$

followed by a summation over the ship classes i and j and integration over the considered risk area $\Omega(z_i, z_j)$ and the considered time period Δt gives the number of possible ship–ship collisions:

$$N_a = \sum_i \sum_j \int_0^{\Delta t} \int_{\Omega(z_i z_j)} \frac{Q_i^{(1)} Q_j^{(2)}}{V_i^{(1)} V_j^{(2)}} f_i^{(1)}(z_i) f_j^{(2)}(z_j) V_{ij} D_{ij} dA dt \tag{1.11}$$

The expected number of ship–ship collisions is then determined by multiplication with the causation factor P_c (see Eq. 1.4) as

$$F_{ship-ship} = P_c \cdot N_a \tag{1.12}$$

where the empirical causation factor P_c can be estimated as described in Section 1.2.3.

Given the expected number of annual collisions $F_{ship-ship}$, the probability of having a collision can be estimated on the assumption of arrivals of the collision events as points in a Poisson process:

$$F[\text{Collision}] = 1.0 - \exp\left[-F_{ship-ship}\right] \approx F_{ship-ship} \text{ for } F_{ship-ship} \rightarrow 0 \tag{1.13}$$

provided that the rate of collision frequency is time-invariant.

In the special case where the considered accident area is extended to infinity, the crossing angle $\theta \neq 0$ and the traffic intensity are constant; then, the area integral in Eq. (1.11) can be evaluated analytically. In this case, the number of possible ship–ship collisions during the time period Δt is independent of the spatial distributions $f(z)$:

$$N_a = \sum_i \sum_j \frac{Q_i^{(1)} Q_j^{(2)}}{V_i^{(1)} V_j^{(2)}} D_{ij} V_{ij} \frac{\Delta t}{\sin\theta} \tag{1.14}$$

When the two waterways are perpendicular to each other, that is, $\theta = \pi/2$, we find from Eq. (1.14) that

$$N_a^{cross} = \sum_i \sum_j \frac{Q_i^{(1)} Q_j^{(2)}}{V_i^{(1)} V_j^{(2)}} \left[L_i^{(1)} V_j^{(2)} + L_j^{(2)} V_i^{(1)} + B_j^{(2)} V_j^{(2)} + B_i^{(1)} V_i^{(1)}\right] \Delta t \tag{1.15}$$

To determine how often a specific vessel is the struck vessel, the case where vessel i in waterway 1 is the striking vessel, vessel j in waterway 2 is considered. Seen from vessel i in waterway 1, vessel j in waterway 2 will move in the direction $\overline{V}_{ij} = \overline{V}_i^{(1)} - \overline{V}_j^{(2)}$ with speed $V_{ij} = |\overline{V}_{ij}|$ given by Eq. (1.9); see Fig. 1.9. The apparent collision diameter seen from vessel i, $D_i^{(1)}$, can be determined as

$$D_i^{(1)} = \frac{L_j^{(2)} V_i^{(1)} \sin\theta}{V_{ij}} + \frac{1}{2} B_j^{(2)} \left\{1 - \left(\sin\theta \frac{V_i^{(1)}}{V_{ij}}\right)^2\right\}^{1/2} + \frac{1}{2} B_i^{(1)} \left\{1 - \left(\sin\theta \frac{V_j^{(2)}}{V_{ij}}\right)^2\right\}^{1/2} \tag{1.16}$$

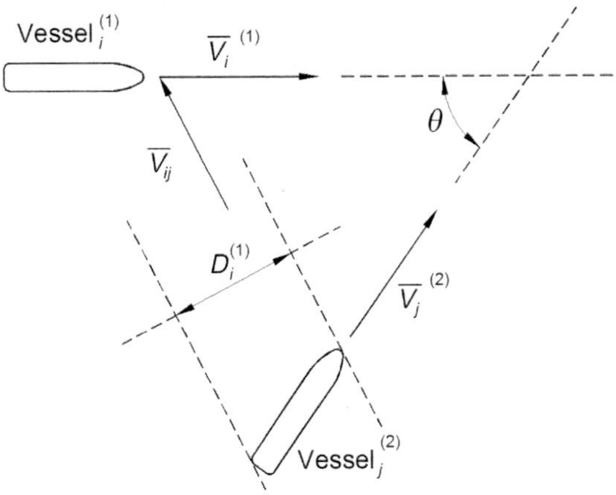

Fig. 1.9 Illustration of the apparent diameter $D_i^{(1)}$ for the struck vessel 1 and the striking vessel 2.

Similarly, for the case where vessel j in waterway 2 is the striking vessel i in waterway 1, the apparent collision diameter is

$$D_j^{(2)} = \frac{L_i^{(1)} V_j^{(2)} \sin\theta}{V_{ij}} + \frac{1}{2} B_i^{(1)} \left\{ 1 - \left(\sin\theta \frac{V_j^{(2)}}{V_{ij}} \right)^2 \right\}^{1/2} + \frac{1}{2} B_j^{(2)} \left\{ 1 - \left(\sin\theta \frac{V_i^{(1)}}{V_{ij}} \right)^2 \right\}^{1/2} \tag{1.17}$$

As expected, it is seen that the total collision diameter D_{ij} is the sum of the two apparent collision diameters, that is,

$$D_{ij} = D_i^{(1)} + D_j^{(2)}$$

The probability that vessel i in waterway 1 is the striking vessel j in waterway 2 given a collision is then found as

$$P[\text{vessel } i \rightarrow \text{vessel } j | \text{collision}] = \frac{D_i^{(1)}}{D_{ij}} \tag{1.18}$$

Similarly, the probability that vessel j in waterway 2 is the striking vessel i in waterway 1 is found as

$$P[\text{vessel } j \rightarrow \text{vessel } i | \text{collision}] = \frac{D_j^{(2)}}{D_{ij}} \tag{1.19}$$

From these expressions, it is in general found that in case of collisions between fast vessels and ordinary merchant vessels, then, the probability that the high speed vessel will be the striking vessel is significantly increased.

1.2.2.2 Head-on and overtaking collisions

Besides ship–ship collision candidates in *crossing waterways*, it may also be of interest to consider *overtaking situations* where two ships in the same direction pass each other within the considered area and *head-on situations* where two ships in opposite directions pass each other; see Fig. 1.10. Collision models for meeting and for overtaking in parallel waterways and for contacts with floating objects are just special cases of the procedure described by Eqs (1.11), (1.12).

For parallel waterways of length L_w separated by a distance μ and where $\theta = 180$ degrees, Eq. (1.11) gives the number of possible ship collisions during the time interval Δt as

$$N_a = L_w \sum_i \sum_j \frac{Q_i^{(1)} Q_j^{(2)}}{V_i^{(1)} V_j^{(2)}} \left(V_i^{(1)} + V_j^{(2)}\right) \int_{-\infty}^{\infty} \int_{z_j = z_i - B_{ij}}^{-z_i + B_{ij}} f_i(z_i) f_j(z_j)\, dz_i dz_j \Delta t$$

or

$$N_a = L_w \sum_i \sum_j \frac{Q_i^{(1)} \; Q_j^{(2)}}{V_i^{(1)} \; V_j^{(2)}} \left(V_i^{(1)} + V_j^{(2)}\right) \int_{-\infty}^{\infty} f_i(z_i) \left[F_j(-z_i + B_{ij}) - F_j(-z_i - B_{ij})\right] dz_i \Delta t$$

where $B_{ij} = (B_i^{(1)} + B_j^{(2)})/2$ is the average ship breadth.

Integration of the standard normal distribution (see Eq. 1.7) gives the cumulative distribution function (CDF):

$$F(z) = \Phi\left(\frac{z-\mu}{\sigma}\right) \left(= \frac{1}{2}\left[1 + \left(\operatorname{erf}\left(\frac{z-\mu}{\sigma\sqrt{2}}\right)\right)\right]\right)$$

Therefore, when $f_i(z)$ and $f_j(z)$ both follow normal distributions with mean values μ_i and μ_j and standard deviations σ_i and σ_j, then this equation can be expressed as

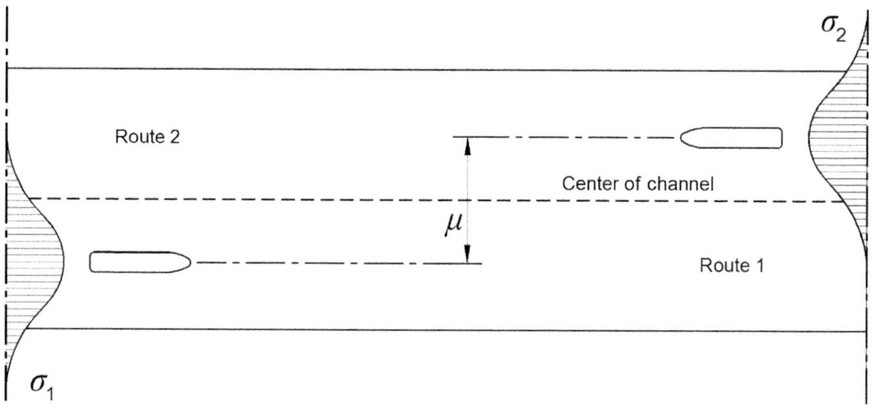

Fig. 1.10 Definition of mean distance μ and lateral traffic distribution σ.

$$N_a = L_w \sum_i \sum_j \frac{Q_i^{(1)} Q_j^{(2)}}{V_i^{(1)} V_j^{(2)}} \left(V_i^{(1)} + V_j^{(2)} \right) \left[\Phi \left\{ \frac{B_{ij} - \mu}{\sigma_{ij}} \right\} - \Phi \left\{ \frac{B_{ij} - \mu}{\sigma_{ij}} \right\} \right] \Delta t \tag{1.20}$$

where

$\mu = \mu_i^{(1)} + \mu_i^{(2)}$ is the mean sailing distance between the two lines and

$\sigma_{ij} = \sqrt{\left(\sigma_i^{(1)}\right)^2 + \left(\sigma_j^{(2)}\right)^2}$ is the standard deviation of the joint distribution.

Provided that the variation of the Gaussian distribution functions $f_i(z)$ over the width of the meeting ships can be neglected, then Eq. (1.11) yields

$$N_a = \frac{1}{\sqrt{2\pi}} L_w \sum_i \sum_j \frac{Q_i^{(1)} Q_j^{(2)}}{V_i^{(1)} V_j^{(2)}} \left(V_i^{(1)} + V_j^{(2)} \right) \left(B_i^{(1)} + B_j^{(2)} \right) \frac{1}{\sqrt{\left(\sigma_i^{(1)}\right)^2 + \left(\sigma_j^{(2)}\right)^2}} \exp\left(\frac{-\mu^2}{2\left(\left(\sigma_i^{(1)}\right)^2 + \left(\sigma_j^{(2)}\right)^2\right)} \right) \Delta t \tag{1.21}$$

1.2.2.3 Collisions in bends of the seaway

Another type of collision on waterways occurs at bends where two straight route segments intersect. See Fig. 1.11. At such an intersection, a ship can become a collision candidate if the course is not changed at the bend of the route. Omission of this change of course is associated with a human error that may be corrected after some time.

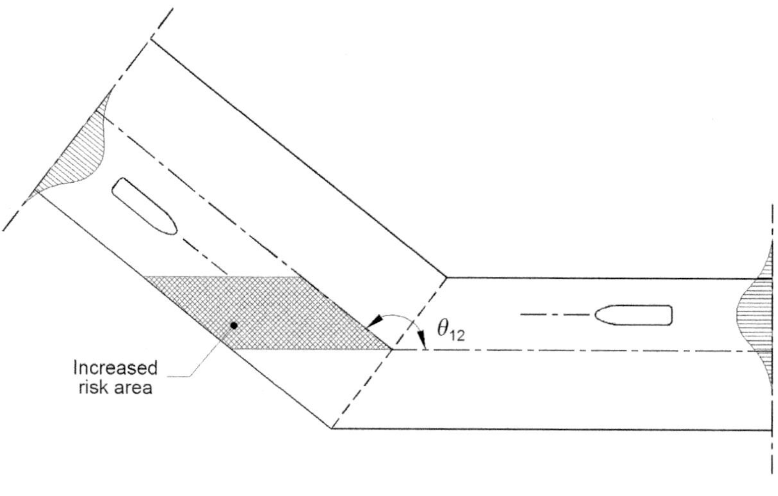

Fig. 1.11 Collision between ships in opposite directions at an intersection between two straight waterway segments.

That is, the probability of omission P_0 is a function of the distance d sailed from the missed waypoint and the distance a sailed between position checks. The probability P_0 of omission of route change can be taken as

$$P_0 = P_t + (1 - P_t) \cdot P_d^{\frac{d-a}{a}} \tag{1.22}$$

where values for P_t and P_d depend on a number of factors such as whether there is a vessel traffic system (VTS) in the considered area or the ship has a pilot on board. The distance, a, sailed between position checks depends on the navigational conditions. Often the distance, a, is taken to be six to eight times the length of the ship. The distance a is related to the check frequency δ and the ship speed V through

$$a = V\delta$$

In the open sea, the position may be checked once every half hour, and in more difficult navigational areas, it may be every 60 s or even more often. The expression (1.22) takes into account that observations show that shipping accidents have happened far away from shipping lanes due to total inactivity of the navigators. See Section 1.5.

An alternative expression with fewer parameters for the probability of omitting to change course at a given turning point has been proposed by Fujii and Mizuki (1998):

$$P_0 = e^{-d/a} \tag{1.23}$$

There are large uncertainties related to the probabilities to be entered into Eqs (1.22), (1.23) and therefore to the results that can be obtained.

1.2.3 Estimation of the causation factor P_c

The causation factor P_c is the fraction of collision candidates N_a that results in an accident; see Eq. (1.4). That is, the causation factor represents the conditional distribution of no action from own vessel, given that the other vessel does not act. The reason for no action depends on those technical, environmental, and human factors that have an effect on the avoidance of a collision. The numerical value of P_c is not a unified value but varies for different geographical locations. The causation probability P_c can be estimated on the basis of available accident data collected at various locations and then transformed to the area of interest.

Among the parameters in such a transformation are the following:

- Environmental factors (visibility, darkness, wind, current, sea state, and ice)
- Ship and crew characteristics (ship type, ship size, ship speed, standard of the ship, mechanical failure rates, radar status, Electronic Chart Display and Information System (ECDIS), standard of crew, pilot on board, etc.)
- Navigational conditions (VTS system and marking of the shipping route)

Table 1.4 Causation probabilities from observations in Japan by Fujii and Mizuki (1998).

	Log P_c	+/−	P_c
Head-on	−4.31	0.35	$4.90 \cdot 10^{-5}$
Overtaking	−3.96	0.36	$1.10 \cdot 10^{-4}$
Crossings	−3.89	0.34	$1.29 \cdot 10^{-4}$
Grounding	−3.80	0.26	$1.59 \cdot 10^{-4}$
Object	−3.73	0.36	$1.86 \cdot 10^{-4}$

For ordinary ship traffic, the causation factor P_c is in the range of $0.5 \cdot 10^{-4}$ to $3.0 \cdot 10^{-4}$. Table 1.4 shows causation factor probabilities based on observations by Fujii and Mizuki (1998) in Japan. These values are estimated for conditions where the frequency of a low visibility (1.0km or less) is smaller than 3.0%. General observations show that the causation factor is proportional to the inverse of the visibility. If the frequency of visibility of 1.0km or less is between 3.0% and 10%, the values for the causation factor should be multiplied by a factor of two and so on.

VTS are typically established in areas of high navigational complexity where an accurate monitoring or guidance of the vessels is of importance. Often, VTS is established where large bridges span the navigational routes or in highly trafficked areas. The main effects of a VTS system will be an added awareness of the navigator due to the contact and advice from the VTS centre and the information on the distribution of ships along the routes. Based on a study of the VTS system in the Great Belt in Denmark, Lehn-Schiøler et al. (2013) found that the effect of the presence of a VTS system may reduce the causation probability for ship–bridge collisions by a factor of 2–3. Similarly, the effect of having a pilot on board is also often modelled by a reduction of the causation factor to about one-half to one-third. Since part of the reduction of the causation factor P_c from VTS and pilotage is caused by additional awareness of navigators on board the ship, then the combined effect will be less than the product of the two reduction factors.

Another approach to determine the causation factor P_c is to analyse the cause leading to human inaction or external failures and set up a fault-tree procedure for the estimation of the causation factor.

At present, the most comprehensive tool to estimate the causation factor is based on a Bayesian network method. By this method, it is possible to include the effect of the actual navigational and environmental conditions and factors such as ship–bridge layout and manning. See Friis-Hansen and Pedersen (1998).

By the use of Bayesian networks, it is possible to quantify the effect of improved manoeuvrability, two navigators on the bridge, AIS coupled to the radar, and the like. To demonstrate the parameters included in such an analysis, Fig. 1.12 shows a sketch of the starting point of a calculation model. A vessel with speed V_2 is on collision course with a ship with speed V_1. With a given weather condition, it is possible to estimate when the other vessel can be seen, either by vision or by radar. Based on the speed of the vessels, a calculation can be made of the time available for detection

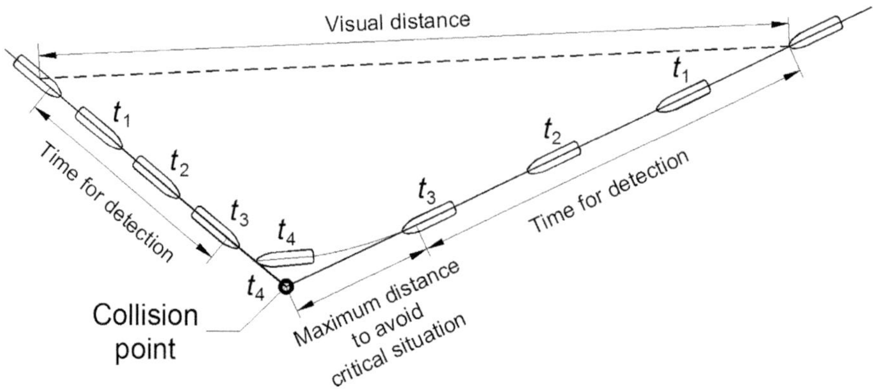

Fig. 1.12 Probabilistic space–time model used as a basis for derivation of causation probabilities using Bayesian networks.
Based on Pedersen, P.T., 2010. Review and application of ship collision and grounding analysis procedures. Mar. Struct. 23, 241–262.

of the other vessel such that an aversive action can be taken. This time for detection is narrowed by the time needed for manoeuvring and for planning the change of course. The visibility is a stochastic quantity, and the probability of detection of the approaching vessel during the time available depends on a number of factors such as manning, bridge layout, and AIS. These interdependencies of stochastic elements make modelling of the causation probability suitable by Bayesian networks. A Bayesian network is a network with nodes and directed arcs. The nodes represent random variables and decisions. Arcs into random variables indicate probabilistic dependence, while arcs into decisions specify the information available at the time of the decision. As an example, one node in the network may represent the weather, whereas another node may represent the visibility. An arc from weather to visibility indicates that visibility is conditionally dependent on weather; see Fig. 1.13.

With the aid of mathematically based analytical tools such as fault-tree or Bayesian network-based approaches, it is possible to analyse the effect of *risk control options* such as new navigational systems, vessel traffic systems, the effect of having a pilot on board, the effect of having two navigators on the bridge, and improved manoeuvrability.

The Bayesian network shown in Fig. 1.13 is taken from Friis-Hansen and Pedersen (1998) where a comparative risk evaluation of traditional watch keeping and one-watch keeping has been modelled. The results of the modelling were compared with observations, and good agreement was obtained. Later, the modelling has been extended to also cover ship–ship collisions. See, for example, Lützen and Friis-Hansen (2003), Hänninen and Kujala (2012), and Sotiralis et al. (2016).

It should be noted that the causation probability as described above cannot be directly used with other geometrical models for the estimation of the number of collision candidates N_a, and the values presented in Table 1.4 should not be compared with other causation probabilities without paying attention to the geometrical models from which they originates.

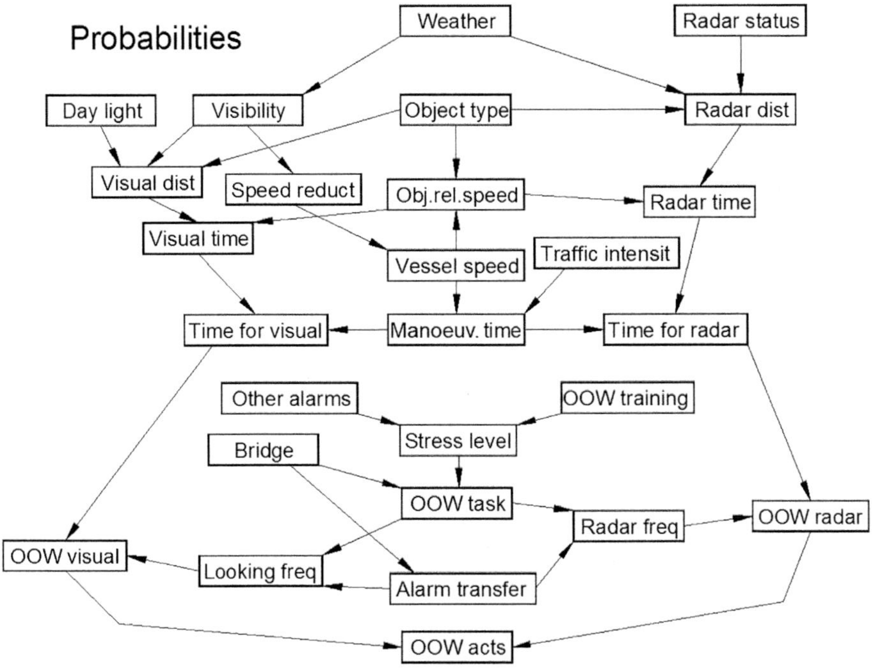

Fig. 1.13 Example of Bayesian network for navigator to react in the event of being on collision course with an object.
Based on Friis-Hansen, P., Pedersen, P.T., 1998. Risk Analysis of Conventional and Solo Watch Keeping. Submitted to International Maritime Organization (IMO) Maritime Safety Committee by Denmark at 69th Session.

The value of the causation factor P_c is typically adjusted by calibration to registered accident data in the considered region using the mathematical model described in Section 1.2.1.

1.2.4 Effect of aversive manoeuvres

Having estimated the probability of ship–ship collisions using the expressions (1.3) and (1.4), further data are needed to estimate the distributions of collision speeds, angles, and collision locations in order to derive collision energy distributions. These collision variables depend on the aversion procedures taken by the involved navigators during the last minutes up to the event, and since so far no mathematical models have been proposed for these quantities, such collision variables have to be based on empirical statistical data.

Based on recorded data related to ship collision damages, Lützen (2001) found that in case of ship–ship collisions, the probability density functions for the velocity of the struck vessel at the moment of impact can be approximated by a triangular distribution with the most likely value equal to zero speed, and the velocity of the striking ship can

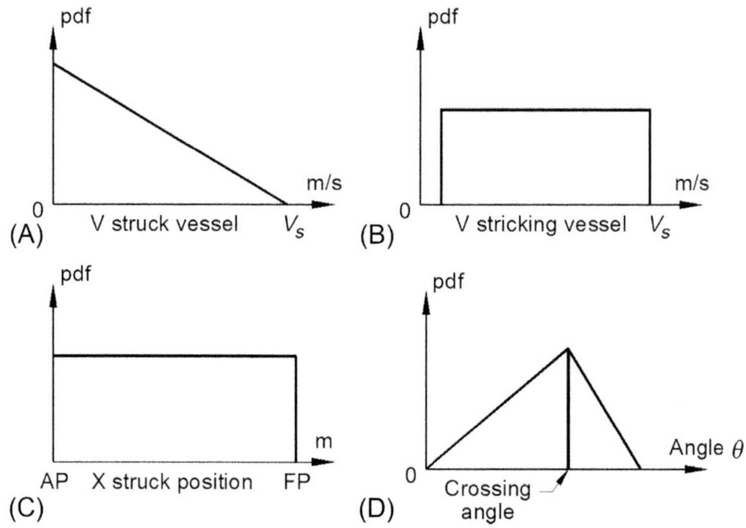

Fig. 1.14 Proposed density functions for impact velocities for (A) the struck vessel, (B) the striking vessel, (C) the longitudinal position of the impact on the struck vessel, and (D) the striking angle.

be modelled by a uniform distribution for velocities between almost zero and the service speed. The impact location along the length of the struck vessel can be assumed to be uniformly distributed. Finally, for a given route, the collision angle may be approximated by a triangular distribution with the most likely value equal to the crossing angle between the two routes. See Fig. 1.14.

1.2.5 Calculation examples

Example 1.1: Crossing ship collisions

As a simplified example of the application of the risk model, we can consider an average year in a strait with a crossing international north–south ship traffic and east–west ferry traffic. The north–south bound international traffic is approximately 20,000 passages, and the east–west ferries have approximately 50,000 voyages.

Find the return period for collisions between these crossing ships when the average vessel length is equal to 130 m, the breadth is equal to 16 m, and the average speed is 6 m/s. Observations have shown three accidents in 17 years.

Solution to Example 1.1

From Eq. (1.15),

$$N_a^{cross} = \sum_i \sum_j \frac{Q_i^{(1)} Q_j^{(2)}}{V_i^{(1)} V_j^{(2)}} \left[L_i^{(1)} V_j^{(2)} + L_j^{(2)} V_i^{(1)} + B_j^{(2)} V_j^{(2)} + B_i^{(1)} V_i^{(1)} \right] \Delta t$$

The annual number of collision candidates N_a is estimated as

$$N_a^{cross} = \frac{50{,}000 \cdot 20{,}000}{6 \cdot 6 \cdot (365 \cdot 24 \cdot 60 \cdot 60)^2} \cdot (130 \cdot 6 \cdot 2 + 16 \cdot 6 \cdot 2) \cdot (365 \cdot 24 \cdot 60 \cdot 60) \cdot 1$$

or

$$N_a^{cross} = 1543$$

According to Table 1.4, a suggested causation probability for crossing ship traffic is $P_c = 1.29 \cdot 10^{-4}$. However, in this example where two experienced navigators always are present on the bridge on the ferries, a reduction of the causation factor should be introduced similar to the effect of having a pilot on board. That is, a reduction of the causation factor P_c up to around one-half to one-third can be expected. A mean value could be a reduction factor of 0.4.

Since only the ferries have improved navigational conditions, it is reasonable to use a combined factor for the causation factor for collisions between ordinary merchant vessels and ferries. Such a combination could be

$$P_{c,combined} = 0.63 P_c$$

With this assumption, the estimated annual number of ship–ship collisions at this traffic intersection is determined as

$$F_{ship-ship} = 0.63 P_c N_a^{cross}$$

$$F_{ship-ship} = 0.63 \cdot 1.29 \cdot 10^{-4} \cdot 1543$$

$$F_{ship-ship} = 0.13$$

That is, the calculated return period for collisions between the ordinary merchant vessels and the crossing ferries is estimated as 8 years. This result is comparable with the observed three collisions in 17 years.

Example 1.2: Meeting collisions in straight channel

Consider a 10 km-long straight channel segment as shown in Fig. 1.10, the total number of ship movements in each direction is 10,000, the mean ship length is 100 m, the mean breadth is 15 m, and the average ship speed is 6 m/s. The standard deviation σ for the two ship track distributions can be estimated to be 150 m, and the mean distance μ between the two shipping lanes is 100 m.

Calculate the expected annual frequency of head-on ship–ship collisions between two meeting ships.

Solution to Example 1.2

The number of ship collision candidates can be estimated from Eq. (1.21)

$$N_a = \frac{1}{\sqrt{2\pi}} L_w \sum_i \sum_j \frac{Q_i^{(1)} Q_j^{(2)}}{V_i^{(1)} V_j^{(2)}} \left(V_i^{(1)} + V_j^{(2)}\right)\left(B_i^{(1)} + B_j^{(2)}\right) \frac{1}{\sqrt{\left(\sigma_i^{(1)}\right)^2 + \left(\sigma_j^{(2)}\right)^2}}$$

$$\exp\left(\frac{-\mu^2}{2\left(\left(\sigma_i^{(1)}\right)^2 + \left(\sigma_j^{(2)}\right)^2\right)}\right)\Delta t$$

With figures inserted,

$$N_a^{head-on} = \frac{1}{2\sqrt{\pi}} \cdot 10 \cdot 10^3 \cdot \frac{10{,}000^2}{(6 \cdot (365 \cdot 24 \cdot 60 \cdot 60))^2} \cdot (2 \cdot 6) \cdot (2 \cdot 15) \cdot \frac{1}{150} \cdot \exp\left(-\frac{100^2}{4 \cdot 150^2}\right) \cdot (365 \cdot 24 \cdot 60 \cdot 60) \cdot 1$$

or

$$N_a^{head-on} = 534$$

Using the proposed value of the causation factor for meeting ships given in Table 1.4, $P_c = 4.9 \cdot 10^{-5}$, the annual number of estimated head-on ship–ship collisions is.

$$F_{ship-ship} = N_a^{head-on} P_c$$

$$F_{ship-ship} = 534 \cdot 4.9 \cdot 10^{-5}$$

$$F_{ship-ship} = 0.026$$

This value corresponds to a return period of 38 years.

Example 1.3: Overtaking ship collisions

Consider one lane of the 10km-long straight channel segment as shown in Fig. 1.10. Two different ship classes, i and j, are sailing within this lane both with an intensity of 10,000 ships annually. The mean length L_i of class i ships is 180m, the breadth B_i is 25m, and the mean speed is 7m/s. The class j ships have a mean length of 240m, a breadth of 35m, and a mean speed of 13m/s. The standard deviation σ for the ship track distributions can be estimated to be 190m, and the mean position of the lane is 200m from the centreline.

Calculate the expected annual frequency of overtaking ship–ship collisions.

Solution to Example 1.3

The only difference to the head-on meeting calculation illustrated in Example 1.2 is the sign of the two speeds. Thus, Eq. (1.21) can be applied

$$N_a^{Overtaking} = \frac{1}{\sqrt{2\pi}} L_w \sum_i \sum_j \frac{Q_i^{(1)} Q_j^{(2)}}{V_i^{(1)} V_j^{(2)}} \left(V_i^{(1)} + V_j^{(2)}\right) \left(B_i^{(1)} + B_j^{(2)}\right) \frac{1}{\sqrt{\left(\sigma_i^{(1)}\right)^2 + \left(\sigma_j^{(2)}\right)^2}} \exp\left(\frac{-\mu^2}{2\left(\left(\sigma_i^{(1)}\right)^2 + \left(\sigma_j^{(2)}\right)^2\right)}\right) \Delta t$$

With figures inserted, the annual number of possible overtaking ship–ship collisions is determined as

$$N_a^{Overtaking} = \frac{1}{2\sqrt{\pi}} \cdot 10 \cdot 10^3 \cdot \frac{10,000^2}{7 \cdot 13 \cdot (365 \cdot 24 \cdot 60 \cdot 60)^2} \cdot (13-7) \cdot (25+35) \cdot \frac{1}{190} \cdot \exp\left(-\frac{0^2}{4 \cdot 190^2}\right) \cdot (365 \cdot 24 \cdot 60 \cdot 60) \cdot 1$$

or

$$N_a^{Overtaking} = 186$$

By the application of the proposed value $1.10 \cdot 10^{-4}$ for the causation factor given in Table 1.4, the annual number of estimated overtaking ship–ship collisions is found as

$$F_{ship-ship}^{Overtaking} = P_c N_a^{Overtaking}$$

$$F_{ship-ship}^{Overtaking} = 1.10 \cdot 10^{-4} \cdot 186$$

$$F_{ship-ship}^{Overtaking} = 0.020$$

that is, a return period of about 50 years.

1.3 Probability of ship collisions against offshore structures

Offshore installations in the form of fixed and floating offshore oil exploration and production platforms and wind energy farms involve many types of risks. Collision or allision (allision is defined as violent striking with a fixed object such as a stationary offshore installation or a bridge structure) between ships and platforms has been pointed out as one of the major risks. An offshore installation may be hit by bypassing ships; by drifting ships; and by infield vessels such as supply ships, tankers involved in offshore loading, and other operations close to the installation.

The world statistics show that the annual collision risk per offshore oil platform is around 10^{-3}. See International Association of Oil and Gas Producers (2010). This probability depends very much on the location of the installation. Statistics from the North Sea show accident rates that are one magnitude higher.

Collisions with infield vessels are the most frequent types of collisions, and for these collisions, the consequences for the structural integrity of the platform should be small. Most rules and regulations (for instance, NORSOK Standard N-004 (2004)) specify that for structural design purposes, the mass of the colliding supply ships should not be taken to be less than 5000 tons and the speed not less than 0.5 m/s for the ultimate limit state (ULS) design and 2 m/s for accidental limit state (ALS) design checks. Supply vessels are growing in size, and several supply ships have displacements around 7500 tons. An accidental collision with a large supply vessel with considerable speed may therefore lead to an emergency case that can have large economic and environmental implications.

Oil tankers for offshore loading will, in the loaded case, have displacements in the order of 100,000 tons, so also in this case can the collision energy be quite large even for small impact velocities.

The most severe collision accidents against oil installations are caused by impact from passing vessels sailing with service speeds. To minimize the collision risk from passing vessels, safety zones are established extending 500 m from any part of an offshore installation. To protect both the safety of the people working on or in the vicinity of the installation and the installation itself, it is critical for the installation to monitor and manage their 500m zones.

Offshore installations are not all oil-related. A large number of offshore windmill farms have been or are being built close to areas with considerable ship traffic. Also wind turbines require regular visits from service vessels.

Ship impact analyses for offshore installations are not only required during the design phase. Around Europe, a safety case has to be prepared for oil offshore installations at regular intervals, for example, every 5 years, in order to demonstrate that the installation continues to meet safety and legal requirements. Part of this safety case is a quantitative assessment of the collision risk. In the following, a mathematically based tool for the evaluation of the collision probability shall be presented; see Fig. 1.2.

1.3.1 Quantitative assessment of ship collision probability

At present, the frequency of impacts from infield vessels such as standby boats, supply vessels, working vessels, and offshore tankers have been based on historical data. Much effort has been made to reduce the frequency of collisions with infield vessels, and the extensive experience with accidental impacts has led to a number of risk-reducing operational measures. In practice, the design procedure against infield vessel impacts against offshore installations has been based on a reduction of the consequences of such impacts. Rules and regulations seek to establish design guidelines that ensure sufficient crashworthiness of offshore installations such that the risk associated with impacts from infield vessels can be considered satisfactory. See Moan et al. (2017).

Collisions with bypassing ships and large drifting vessels should be events with very little probability since the energy released for damage of the installation and/or the ship can be very significant.

To perform a quantitative risk analysis for offshore installations exposed to ship collisions from passing ships, a first step is to collect basic background information such as installation geometry and position; estimation of present and future ship traffic characteristics and volume; and wind, current, and ice conditions.

Information about the present characteristic ship traffic includes information on sailing routes, the traffic on these sailing routes, and ship traffic particulars. A sailing route for line traffic is defined as the centreline between predicted route points. A sailing route normally has two lanes used by traffic in opposite directions, that is, one lane on each side of the route centreline; see Fig. 1.10. The frequency that ships are on collision course with the installation depends on the geographical location of the sailing route, the distribution of traffic transverse to the lanes, and the location of the installation. The volume of the traffic on each sailing route is important for the estimation of the frequency of ship collisions from passing vessels. The volume of the ship traffic and the spatial distribution of this traffic can be obtained from the compulsory AIS and from radar observations; see Section 1.2.1.

The observed ship traffic must afterwards be modified to account for the development in ship traffic during the lifetime of the installation. The spatial distribution of the ship traffic determined from the AIS data can typically be approximated by a combination of a normal distribution or log-normal distribution and a small uniform part. See Eq. (1.5) and Fig. 1.5.

One problem concerned with a collision risk estimate for a planned installation is that the presence of new offshore installations has an effect on the traffic pattern and location of shipping lanes. For offshore petroleum installations in open waters, the change in the spatial ship traffic distribution has been studied by Hassel et al. (2017). Their observations show that in general, ships will alter course such that the traffic moves away from the installations to achieve a passing distance of one to three nautical miles. However, a small but important volume of the ships will be very close to the guard zone around the installations.

1.3.2 Probability of ship collision events

Knowing the composition of the ships in the considered waterway area and the spatial distribution of these vessels, the next step will be to estimate the probability of ship collisions. Most of the existing mathematical models that describe the probability of ship impacts against offshore installations are based on the calculation of the number of collision candidates in the case of blind navigation and a subsequent use of causation factors to model the action of the navigators on board the ships; see Section 1.2. That is, the collision frequencies F_{col} are estimated as the number of collision candidates N_a multiplied by causation factors P_c (Eq. 1.4):

$$F_{col} = P_c \cdot N_a$$

The model for the estimation of the number of collisions in the case of blind navigations, N_a, can be based on a division of the ship collision problem into a number of different phenomena and subsequent application of mathematical models for the quantification of the risk from each category; see Pedersen (2014).

Category 1:

This category includes ships following the ordinary, straight route at a normal speed and with a normal ship track distribution. See Fig. 1.15. For this category, the expected number $N_{Cat,1}$ of blind collisions per unit time Δt can be determined from the following expression:

$$N_{Cat,1} = \frac{1}{\Delta t} \sum_{\substack{Ship \\ class}}^{n} \int_{z=A}^{B} \int_{t=0}^{\Delta t} Q_i f(z)\,dt\,dz \tag{1.24}$$

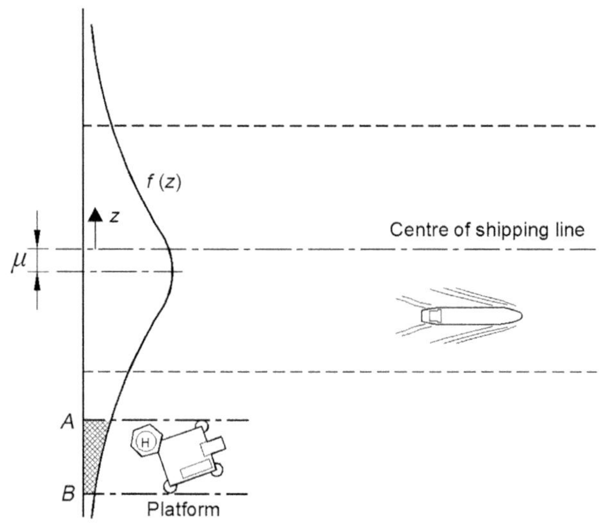

Fig. 1.15 Collision candidates from Category 1 (Pedersen, 1995).

where $Q_i = Q_i(t)$ is the traffic density function of class i ships given by the number of ships per time unit, $f(z)$ is the spatial distribution (see Eq. 1.5), and $z = A$ and $z = B$ are the transverse coordinates of the offshore installation boundaries minus and plus the half breadth of the ship.

Category 2:

This category includes ships that fail to make a proper change of course at a turning point in the vicinity of the installation. To determine the added probability for ship collisions in the vicinity of bends of the seaway, see Fig. 1.16, where the vessels may not make the turn, a procedure similar to the one used for Category 1 can be used provided that the procedure is augmented with the probability that the ships fail to act on the changing route direction. The probability P_0 that a ship fails to change course at a turning point can, for example, be modelled by Eq. (1.22) or (1.23) from Section 1.2.2:

$$P_0 = P_t + (1 - P_t) \cdot P_d^{\frac{d-a}{a}} \quad \text{or}$$

$$P_0 = e^{-d/a}$$

where P_t is the probability that the navigators continue on the wrong course and completely neglect the needed course change, P_d is the probability of omission to check the position of the ship, d is the distance sailed on wrong track, and a is the average sailing length between position checks by the navigator. Instead of using the distance sailed d as measure for the probability of updating the course, then, another possibility is to use the time needed for detection of a wrong course.

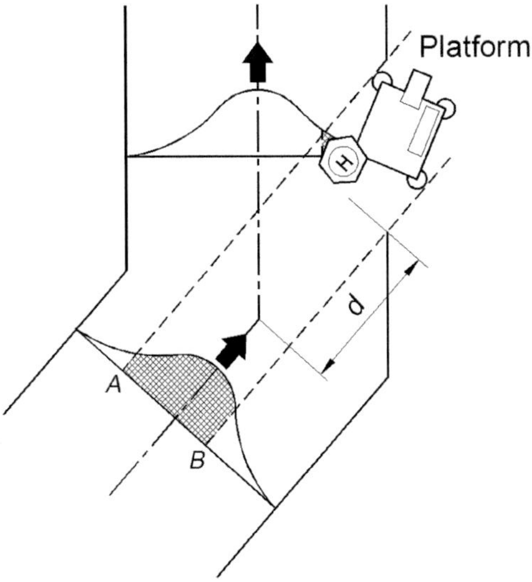

Fig. 1.16 Collision candidates from Category 2.
Based on Pedersen, P.T., 2014. Risk assessment for ship collisions against offshore structures. In: Guedes Soares, Santos (Eds.), Invited Lecture, Maritime Technology and Engineering. Taylor & Francis Group, London, ISBN: 978-1-138-02727-5.

For this category, the expected number $N_{Cat,2}$ of blind collisions per unit time Δt can be determined from the following expression:

$$N_{Cat,2} = \frac{1}{\Delta t} \sum_{\substack{Ship \\ class}}^{n} P_0 \int_{z=A}^{B} \int_{t=0}^{\Delta t} Q_i f(z) dt dz \tag{1.25}$$

Category 3:

This category includes ships drifting due to mooring or anchoring failures, drifting ships at work such as infield vessels and fishing vessels, and ships drifting due to loss of propulsion or control. For drifting passing ships or ships with control failure, it is normally assumed that the ships at the time of loss of propulsion are located on the idealized sailing route.

Complete loss of propulsion can be caused by failure in power supply and tripping of generators known as blackout. If propulsion is lost, the ship will lose speed and finally also lose steering ability and become adrift under influence of current and wind. Drifting ships normally have their shipside perpendicular to the wind direction and the speed and drifting direction governed by wind speed and current. For a typical commercial ship, the frequency of loss of propulsion has been estimated to be (Rasmussen et al., 2012)

$$P_{\text{Loss of propulsion}} = 1.5 \cdot 10^{-4} \text{ failures per hour per ship}$$

With an assumed average speed of 12 knots, this corresponds to a failure rate of

$$\begin{aligned} P_{\text{Loss of propulsion}} &= 1.5 \cdot 10^{-4} / (12 \text{ knots} \cdot 1852 \text{ m/n.m.}) \\ &= 6.7 \cdot 10^{-9} \text{ failures per metre} \end{aligned}$$

Technical failures in the rudder control often lock the rudder in the present position or turn the rudder to one of its extreme positions. Loss of steering may lead to powered collision accidents. The speed will decrease somewhat due to the turn itself, and further speed reduction may be achieved by full astern or crash stop of the engine. Rasmussen et al. (2012) estimated

$$P_{\text{Rudder failure}} = 6.5 \cdot 10^{-5} \text{ per vessel and hour}$$

In a blackout event, there may still be control of rudder, but a total blackout may also disable rudder control and lead to the loss of steering. That is, rudder failure and blackout are not necessarily independent events. If it is estimated that the loss of steering occurs in 10% of all blackout events, the following probabilities for technical failures are achieved:

$$P_{\text{Loss of propulsion}} = 1.5 \cdot 10^{-4} - 1.5 \cdot 10^{-5} = 1.35 \cdot 10^{-4} \text{ per vessel and hour}$$

$$P_{\text{Rudder failure}} = 6.3 \cdot 10^{-5} - 1.5 \cdot 10^{-5} = 4.8 \cdot 10^{-5} \text{ per vessel and hour} \quad (1.26)$$

$$P_{\text{Loss of propulsion and rudder failure}} = 1.5 \cdot 10^{-5} \text{ per vessel and hour}$$

These failure rates are observed values for all types of ships, although differences in reserve power and backup systems will be present and change these probabilities. Due to the lack of data, the frequency is assumed constant in time even if the probabilities must be expected to be higher during intense manoeuvring.

When a failure in the propulsion system or control system is detected, then the crew will initiate repair and in most cases be able to regain propulsion within a certain timeframe and avoid accidental events like collision or groundings. According to Rasmussen et al. (2012), the probability of having repaired the error is given by a cumulative distribution as a function of time. This distribution function for self-repair $P_{repair}(t)$ can be estimated for rudder failure and for blackout events with a Weibull function with $k=0.5$ (shape) and $\lambda=0.605$ (scale):

$$P_{repair}(t) = 1 - e^{-\left(\frac{t}{\lambda}\right)^k} \text{ for } t > 0$$

The probability that a ship is still drifting at time t after loss of propulsion $P_{drift}(t)$ is then given by the complementary distribution function of the repair time distribution:

$$P_{drift}(t) = 1 - P_{repair}(t) = e^{-\left(\frac{t}{\lambda}\right)^k} \quad (1.27)$$

Fig. 1.17 shows $P_{drift}(t)$ as a function of time.

The time t_{col} to the possible collision is

$$t_{col} = d_{col}/V_{drift}$$

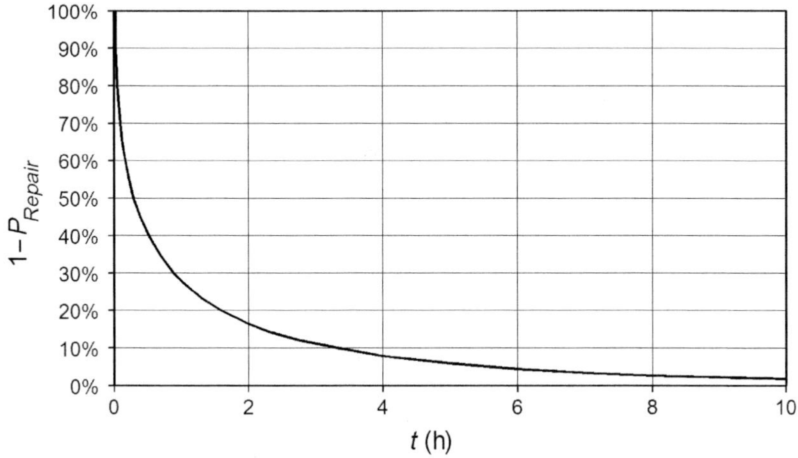

Fig. 1.17 Probability $P_{drift}(t)$ after loss of propulsion or rudder failure based on Weibull function for repair time with $k=0.5$ and $\lambda=0.605$.

in which V_{drift} is the drifting speed and $d_{col}(x)$ is the distance from the leg segment to the installation. The drifting speeds are typically between 1.0 and 3.0 m/s. Depending on the water depth, the navigator may decide to drop the anchor to avoid the vessel drifting into the installation. This may reduce the probability for drifting impacts.

If blackouts and loss of control can be taken to occur as points in a Poisson process, then the probability of a technical error along a segment of length L_w is given by

$$P_{Technical\ error}(L_w) = 1 - \exp\left(-P_f \frac{L_w}{V}\right) \tag{1.28}$$

where P_f represents the probability of a technical failure per time unit such as given by Eq. (1.26) and V is the operational vessel speed.

The number of collisions against an offshore installation due to drifting depends on the wind and current directions and velocities. As an example, the collisions due to drifting can be estimated as

$$\begin{aligned} F_{drift} = N_{ship} \int_{\theta=0}^{360} P_{wind,cur}(\theta) \sum_{segments} P_{Technical\ error}(L_w) \\ \int_{z=0}^{L_w} \int_{v_{drift}} P_{drift}(t_{col}|X) f(v_{drift}) dv_{drift} dz d\theta \end{aligned} \tag{1.29}$$

where $P_{wind,cur}$ represents the probability of having the drifting direction θ. The drifting speed is given by $f(v_{drift})$, and X defines the parameters that give the time to collision.

The total number of ship collisions against the offshore installation from passing ships must be calculated separately for the three situations, and the total is found by adding these contributions:

$$F_{col,offshore} = P_c \cdot N_{Cat,1} + P_c \cdot N_{Cat,2} + F_{drift} + F_{rudder} \tag{1.30}$$

1.3.3 Calculation examples

Example 1.4: Passing vessel collision with offshore installation

A shipping lane and an offshore installation are shown in Fig. 1.18. A bend in the shipping lane is arranged $d = 2500$ m before an offshore installation. Without this bend, the centre of the shipping lane would pass through the centre of the offshore installation. After the bend, the centre of the shipping lane is placed $H = 380$ m from the centre of the offshore installation. The offshore installation has a projected width of $D = 70$ m perpendicular to the two sailing directions. The transverse distribution of the ships in this region has been found to be given by a normal or Gaussian distribution with a mean value $\mu = 0$ m and a standard deviation $\sigma = 230$ m.

(a) Determine the Category 1 probability that one ship from size class 8 in Table 1.1 with a rounded up length $L = 200$ m and a breadth $B = 30$ m on this route will strike the offshore installation. For this calculation, it is assumed that the ship has successfully made the course change.

(b) Assume now that the same ship from size class 8 in Table 1.1 is on a position before the bend and determine the Category 2 probability for this one ship to strike the offshore installation.

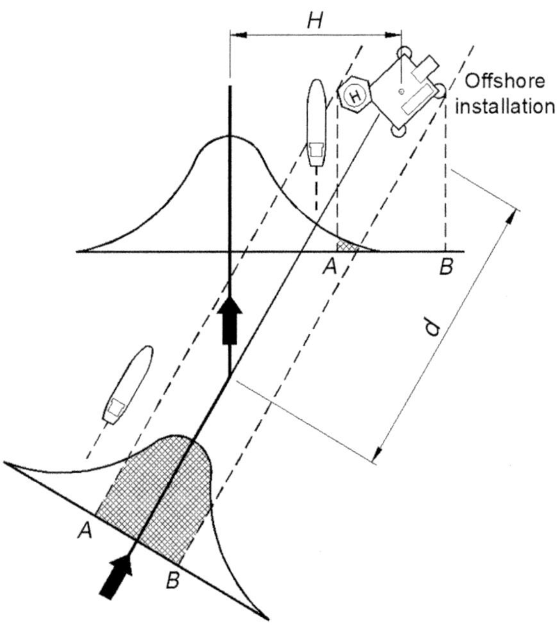

Fig. 1.18 Shipping lane arrangement in the vicinity of an offshore installation.

Solution to Example 1.4

Question (a)

The first task will be to use Eq. (1.24) to determine the geometrical probability, $P_{geometrical}$, that the ship will be placed between $z_1 = H-(B/2+D/2) = 330\,\text{m}$ and $z_2 = H+(B/2+D/2) = 430\,\text{m}$.

The transverse distribution is given as a Gaussian distribution; see Eq. (1.7):

$$f_j(z_j) = \frac{1}{\sigma_j\sqrt{2\pi}}\exp\left(\frac{-(z_j-\mu_j)^2}{2(\sigma_j)^2}\right)$$

Integration of this standard normal distribution gives the CDF as

$$F(z) = \Phi\left(\frac{z-\mu}{\sigma}\right)\left(=\frac{1}{2}\left[1+\left(\text{erf}\left(\frac{z-\mu}{\sigma\sqrt{2}}\right)\right)\right]\right)$$

where the normal distribution $\Phi(x)$ is tabulated in standard textbooks on probability.

For $z = 330\,\text{m}$, then, $F(z<330) = \Phi\left(\frac{330}{230}\right) = 0.925$

For $z = 430$, then, $F(z<430) = \Phi\left(\frac{430}{230}\right) = 0.969.$

So, the geometrical probability $P_{geometrical}(330<z<430) = 0.969-0.925 = 0.044$.

To determine the number of collisions, the geometrical probability must be multiplied with a causation factor; see Eq. (1.4). Table 1.4 suggests $P_c = 1.86\cdot 10^{-4}$. Using Eqs (1.4), (1.13), the Category 1 collision probability is determined as

$$F[Collision\ I] = 1.0 - \exp\left(-F_{ship-ship}\right)$$

or

$$F[Collision\ I] = 1.0 - \exp\left(-0.044\cdot 1.86\cdot 10^{-4}\right)$$

That is, the Category 1 collision probability for one ship of size class 8 is estimated as

$$F[Collision\ I] = 8.18\cdot 10^{-6}$$

Question (b)

To determine the Category 2 collision probability firstly, the probability that the ship forgets to make the course change has to be estimated. In the lack of further data, Eq. (1.23) can be applied:

$$P_0 = e^{-d/a_i}$$

With an estimated position check every eight ships lengths, the probability to be on the wrong course at offshore installation is determined as

$$P_0 = e^{-\frac{2500}{8\cdot 200}} = 0.21$$

With the offshore installation placed in the middle of the erroneous shipping lane ($z=0$), the geometrical probability is determined from the normal distribution function with coordinates:

$z_1=-(B/2+D/2)=-50\text{m}$ and $z_2=(B/2+D/2)=50\text{m}$

for $z=50\text{m}$, then $F(z<50)=\Phi\left(\frac{50}{230}\right)=0.586$

since $\Phi(-u)=1-\Phi(u)$, then for $z=-50\text{m}$, $F(z<-50)=0.414$.

Thus, the geometrical probability $P_{geometrical}(-50<z<50)=0.586-0.414=0.172$.

To determine the number of Category 2 collisions with the installation from Eqs (1.25), (1.4), the probability of omission to change course P_0 must be multiplied by geometrical probability $P_{geometrical}$ and the causation factor P_c. For Category 2 collisions, the same causation factor as for Category 1 can be assumed.

The probability of Category 2 collisions with the considered ship is then found from Eq. (1.13) as

$$F[Collision\ 2]=1.0-\exp\left(-0.21\cdot 0.172\cdot 1.86\cdot 10^{-4}\right)$$

$$F[Collision\ 2]=6.72\cdot 10^{-6}$$

The annual collision probability for one ship in this ship class 8 due to human errors is the sum of the Category 1 and the Category 2 probabilities:

$$F[Collision\ 1+2]=8.18\cdot 10^{-6}+6.72\cdot 10^{-6}=1.49\cdot 10^{-5}$$

To this probability, the collision probability due technical errors such as machinery and/or control failures must be added.

1.4 Probability of ship collisions against bridges

A large number of bridges crossing inland waterways and straits are being constructed all over the world. There are several hazards related to bridges over waterways with significant ship traffic. These hazards include ship collisions (or allisions) to the bridge structure, see Pedersen et al. (1993), increased probability for ship–ship collisions or groundings in the vicinity of the bridge, and environmental damages due to shipping accidents (Fig. 1.19).

History shows numerous accidents where ship collision accidents against bridge piers or ship deckhouse impacts against bridge girders cause bridges to collapse and people travelling on the bridge to be killed. In addition ship–bridge collisions often lead to disruption of the traffic on the bridge for extended periods. Such traffic disruptions can have severe economic consequences for the society.

The introduction of a bridge crossing a major waterway may also influence the ship traffic pattern by introducing new navigational obstacles, for example, pillars, pontoons, and low span sections with restricted air draught. For passing ships, the navigational conditions can be more difficult, and denser ship traffic in dedicated ship lanes and bridge passage span will lead to a higher probability of ship–ship collisions and ship grounding accidents; see Sections 1.2 and 1.5. For this reason, an important design criterion for the bridge will be that the opening of the navigational span of the bridge is so large that ship collisions with the structural elements of the bridge occur

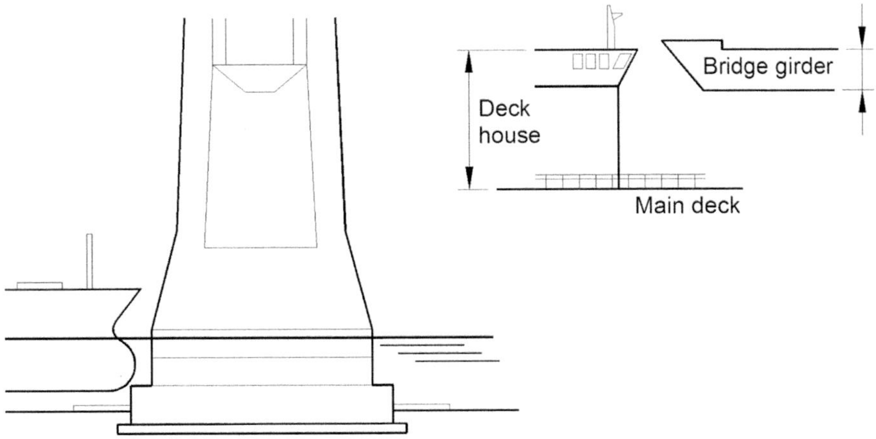

Fig. 1.19 Bow collision against a bridge pylon and deckhouse collision against bridge girder. Based on Pedersen, P.T., Valsgaard, S., Olsen, D., Spangenberg, S., 1993. Ship impacts: bow collisions. Int. J. Impact Eng. 13 (2), 163–187.

only as a result of navigational errors and technical failures on board the passing vessels, and not because of increased navigational difficulties due to the bridge (Pedersen, 1998, 2002).

With these hazards in mind, the following types of operational risks related to bridge–ship collisions should be considered and quantified during the design phase and also at regular time intervals during the operation of bridges:

- Fatalities and injuries to users of the bridge caused by structural failure due to ship impact
- Disruption of service due to failure of the bridge or reduction in capacity due to ship impacts
- Third-party risks related to accidental impact to ships passing the bridge, for example, damage to ships and persons on ships due to collision with the bridge and ship collisions or grounding due to increased navigational complexity
- Environmental impacts from accidents due to releases from a ship impacting the bridge

Acceptance criteria and the general risk management methodology for major bridges will typically be based on the ALARP principle described in Section 1.1.2. This means that risk shall be reduced as long as the cost of the risk-reducing measures is not disproportionate to the effect of the measure. In addition, the risk to users of the bridge should not exceed a level that is unconditionally intolerable chosen such that the level reflects the safety standards elsewhere in the surrounding society. Another safety goal should be that the probability of long-term disruption of the bridge traffic due to accidental loads shall be below prescribed limits. In this section, analytical tools will be presented that can constitute elements in a probabilistic procedure for the estimation of these operational risks.

Standards for modelling ship collisions against bridges have been published by International Association for Bridge and Structural Engineering (IABSE) (Larsen, 1993) in connection with the bridges over the Great Belt in Denmark, by the American

Association of State Highway Transportation Officials (AASHTO) (2009), which was developed for US highway bridges over rivers, and by Eurocode (2006).

1.4.1 Basic information and navigational studies

To perform a consistent quantitative risk analysis of new bridge structures exposed to ship collisions, the first step is to collect basic background information such as sea bed bathymetry; bridge geometry; bridge alignment; future road and railway traffic; wind, current, and ice conditions; navigational routes; future ship traffic characteristics and volume; and spatial distribution of the ship traffic.

The future ship traffic forecast estimate can be based on the following:

- A recording of the present traffic in the region by AIS (see Section 1.2.1)
- A description of the existing ship types, size distribution, and speed (see Section 1.2.1)
- A hearing process with stakeholders in the area about future plans
- A general knowledge of the future global trends in shipping

To reduce the ship–bridge collision risk, it is most often necessary to improve the navigational channel layout in the vicinity of the bridge such that the direction of the navigational channel is close to perpendicular to the bridge line and arrange the location of bends in the navigational channel before the bridge such that ships that omit to change course at these bends have a low probability for collision with the bridge.

In international waterways, the right of navigation requires bridges to be constructed such that the navigation is not interfered with. To estimate the necessary horizontal clearance of the navigation span, information on the number of potentially critical meeting situations in the vicinity of the bridge is needed.

1.4.2 Critical meeting situations

The frequency of ship collision accidents near the bridge is influenced by the number of ships that have to take evasive actions in the vicinity of the bridge. Evasive actions often occur in situations where two or more large ships meet in a confined navigational channel. In this section, a model shall firstly be described for an empirical estimation of the minimum safety domain in which the navigators feel comfortable when passing other ships or a bridge pylon. Based on the safety domain concept, a procedure will then be presented for the calculation of the frequency of critical meeting situations on a certain length interval L_w on a two-way shipping route.

In order to achieve safe navigational conditions, a navigator will try to keep a certain distance between his ship and other vessels and obstacles. This safety area or ship domain has been studied by marine researchers in Europe and in Japan; see Szlapczynski and Szlapczynski (2017). Ship domain theory is useful for the estimation of the fraction of ships that are able to sail through a traffic lane at constant speed without getting too close to fixed objects or other ships.

For free navigation in a channel at service speed with no obstructions, Fujii and Mizuki (1998) have by observation identified the average ship domain as an ellipse with the following:

$$\begin{aligned}&\text{Length in the course direction} : \beta L = 8.0L \\ &\text{Length in the side direction} : \alpha L = 3.2L\end{aligned} \tag{1.31}$$

where L is the ship length.

In narrow channels or in harbours where the vessels travel at reduced speed and where no overtaking or head-on encounters occur, the ship domain is reduced to

$$\begin{aligned}&\text{Length in the course direction} : \beta L = 6.0L \\ &\text{Length in the side direction} : \alpha L = 1.6L\end{aligned} \tag{1.32}$$

In case of one-way ship traffic, the empirical domain theory suggests a minimum navigational bridge span opening equal to the width of the domain for a typical large ship passing the bridge, that is, either 3.2 or 1.6 times the length L of the ship. Ships with lengths above this design ship size will experience a lack of free navigation since their ship domain ellipse will be wider that the lane width. See Fig. 1.20.

For two-way traffic without ship traffic separation, Fig. 1.21 shows conditions related to a critical meeting situation. In this case, the ship domains should not touch the bridge pylons or the meeting ships, and the navigational span should include a transverse separation zone L_a. From Fig. 1.21, it is seen that the width required when two or more ships can be present near the bridge can be determined as

$$W_N = \alpha L_1 + \sum_{i=2}^{N} \left(\frac{\alpha}{2} L_i + \frac{1}{2} B_i \right) + L_a \tag{1.33}$$

where $N \geq 2$ is the number of ships passing each other within a certain distance in the channel, L_1 is the length of the longest ship, and L_i and B_i are the lengths and the breadth of the smaller ships.

For waterways with traffic separation, the ships are not permitted to leave their zone, and therefore, the ships are not allowed to sail in the same area. This situation

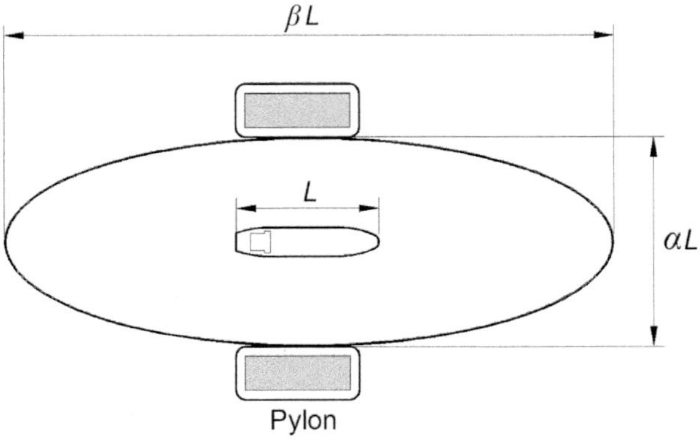

Fig. 1.20 Schematic illustration of a ship domain that ships will try to keep free for other ships and obstacles.

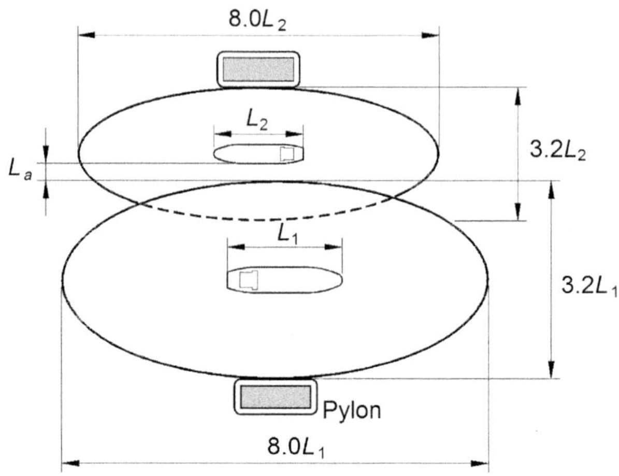

Fig. 1.21 Critical meeting situations for free navigational conditions and without a ship traffic separation zone.

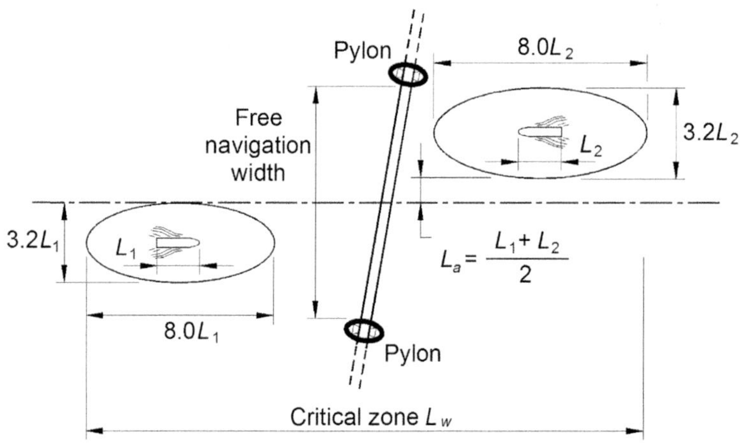

Fig. 1.22 Critical meeting situations for free navigation with ship traffic separation and two-ship meeting.

is shown in Fig. 1.22. With a transverse separation zone, L_a, between the two ship domain areas taken as the mean ship length, Larsen (1993) proposed that the free navigational width of the bridge should be

$$W_2 = (\alpha + 0.5)(L_1 + L_2) \tag{1.34}$$

It may be too costly to provide a bridge span opening that in all cases can fulfil the width criteria expressed by Eqs (1.31)–(1.34). Therefore, these criteria should be augmented by the probability that the largest vessels meet in the vicinity of the bridge.

This will allow the width of the bridge span opening be related to the annual probability of critical meeting situations.

For free navigational conditions, the empirical dimensions, Eq. (1.31), for the ship domain have been verified by studies of AIS tracks. Hansen et al. (2013) used observations from Danish waters of ship–ship encounters and ship–bridge pylon distances to visualize the distances D between all ships in the considered areas in special intensity plots; see Fig. 1.23. The registrations were displaced to the centre of the ships taking into account the actual location of the Global Positioning System (GPS) signal on the ships. All distances D were measured from each ship to all other ships and were then normalized with the length of the current ship. By this method, they could include ships of all sizes in their analysis. Fig. 1.23 shows observed normalized transverse ship-ship distances in the Great Belt, ship–ship distances in the deep water route T through the Great Belt, and distances between ship centreline and bridge pylons. These observed data give credibility to the ship domain dimensions given by Eq. (1.31).

Ship domain models have also been applied to design security zones for safe navigation of liquefied natural gas (LNG) ships and estimation of the capacity of restricted water channels when ship safety has to be maintained. See Liu et al. (2016).

The fraction of ships that experience critical meeting situations can be computed by a simulation of ships through a traffic segment or by analytical analysis when the ship

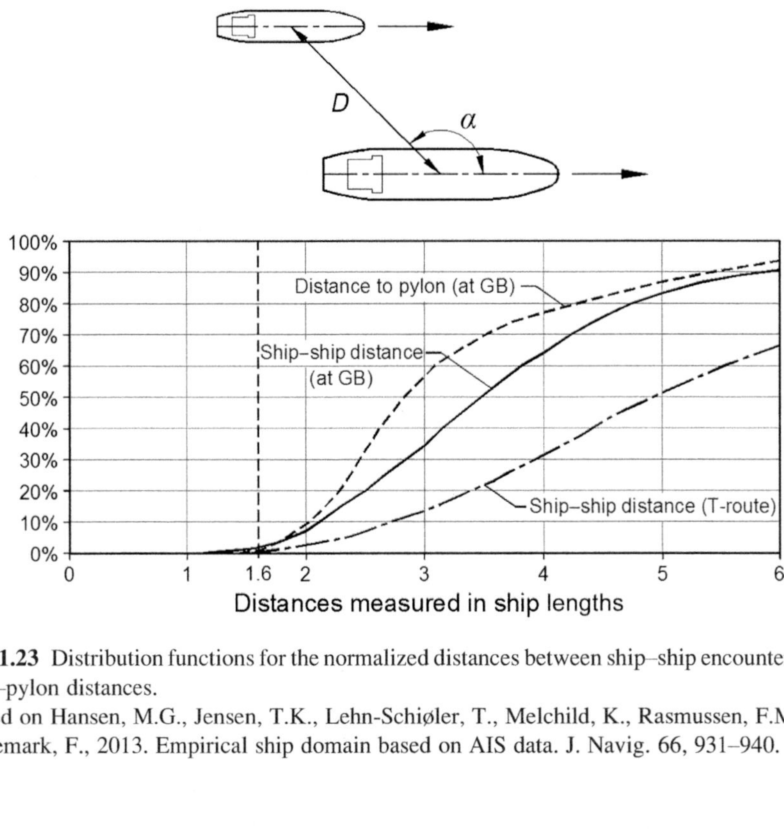

Fig. 1.23 Distribution functions for the normalized distances between ship–ship encounters and ship–pylon distances.
Based on Hansen, M.G., Jensen, T.K., Lehn-Schiøler, T., Melchild, K., Rasmussen, F.M., Ennemark, F., 2013. Empirical ship domain based on AIS data. J. Navig. 66, 931–940.

traffic intensity can be assumed to be constant in time without correlation to the time of the year or day. For bridges with one span for both sailing directions, the number of critical or close encounters can be determined by the use of the safety domains, Eqs (1.33), (1.34), together with a calculation of the frequency of meeting situations involving two or more large ships near the bridge.

In order to determine the frequency of a two-ship meeting situation within a distance of L_w in a two-way traffic situation, Fig. 1.24 can be considered.

Fig. 1.24 shows an east–west shipping lane where the westbound traffic consists of Q_W movements per time unit with a velocity V_W. Similarly, the eastbound traffic is described by Q_E and V_E.

To determine the number of meetings over the distance L_w between location A and B, the time needed for an eastbound ship to cover the distance L_w is first determined:

$$T_{AB} = \frac{L_w}{V_E}$$

If this particular eastbound ship shall be engaged in a meeting situation with a westbound ship, the westbound ship must be situated between point A and point C at the time the eastbound ship passes point A where

$$L_{AC} = L_w + T_{AB} \cdot V_W$$

or

$$L_{AC} = L_w \left(1 + \frac{V_W}{V_E}\right)$$

The average distance between the westbound ships is

$$l_w = \frac{V_W}{Q_W}$$

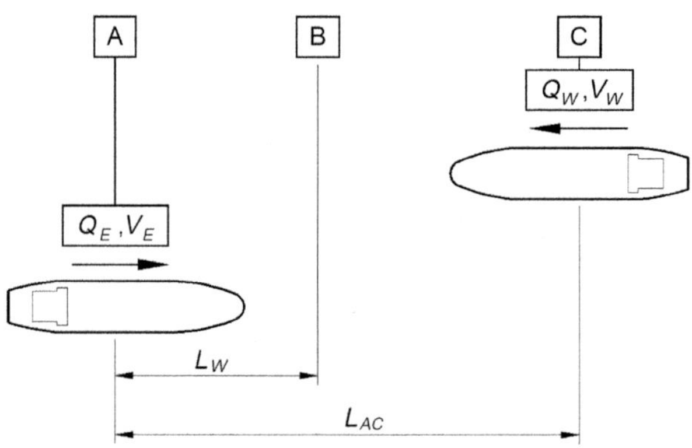

Fig. 1.24 Basic parameters in calculation model for meeting frequencies between two ships.

The average number of westbound $N_{W\,AC}$ ships over the distance L_{AC} is equal to the number of meeting situations with one eastbound ship:

$$N_{W\,AC} = \frac{L_{AC}}{l_w} = \frac{Q_W}{V_W}\left(1 + \frac{V_W}{V_E}\right)L_w \tag{1.35}$$

Thus, the average number of meetings, α_{EW} during one time unit will be given as

$$\alpha_{EW} = N_{W\,AC} \cdot Q_E$$

$$\alpha_{EW} = L_w Q_W Q_E \left(\frac{1}{V_W} + \frac{1}{V_E}\right) \tag{1.36}$$

where

α_{EW} is the average number of meeting situations per time unit
L_w is the distance considered
Q_W and Q_E are west and east ship movements per time unit
V_W and V_E are the average speed

Eq. (1.36) is an approximation when two or more ships with the same direction are in the zone simultaneously. With more than one ship in the zone with the same direction at the same time, Eq. (1.36) will overestimate the expected number of meetings.

Assuming that the meeting ships travel at a common velocity V (m/s), the annual frequency of meeting situations is given by the expression

$$E[N_{\text{annual}}] = 2L_w Q_W Q_E/(365 \cdot 24 \cdot 60 \cdot 60 \cdot V) \tag{1.37}$$

where Q_W and Q_E denote the number of movements in the two opposite directions per year and where the normalizing product, which is equal to the number of seconds in a year, is used to convert the duration to years.

The length L_w is the critical meeting zone along the navigational channel close to the bridge. See Fig. 1.22. This length could be based on the capability of evasive actions for the specific size category. Since the stopping distance for ships can be of the order of 10 times the ship length L and the turning radius is between 4 and 5 times the ship length, the length L_w has been estimated to be around

$$L_w = 16\max(L_1, L_2) \tag{1.38}$$

Thus, for a situation with ship traffic separation, the boundaries to be considered for meetings are estimated as a rectangle with length L_w given by Eq. (1.38) and a width W_2 given by Eq. (1.33) or (1.34).

Let us consider a simple example: for two-way traffic where the largest ship has a length around 200 m, that is, an estimated $L_w = 3200$ m according to Eq. (1.38), and with 12,000 ships annually in each direction having a common speed of 8.0 m/s, the expected number of meetings in the considered meeting zone per hour is

$$E[N_{\text{hour}}] = 2 \cdot 3200 \cdot 12{,}000 \cdot 12{,}000/(365 \cdot 24 \cdot 60 \cdot 60 \cdot 8 \cdot 365 \cdot 24) = 0.42$$

Assuming that the ship traffic in each direction can be described by mutually independent Poisson processes, then the simultaneous occurrence, that is, meetings of two counter-directional ships in the time interval t, can also be approximated by a Poisson process with

$$E[n_m] = \alpha_{EW} t$$

$$Var[n_m] = \sigma_{N_m}^2 = \alpha_{EW} t$$

Since a proper description of the occurrence of vessels in the considered zone, L_w * W_2, requires a statistical description of the volume, the speed, and the size distribution of the ship traffic, a numerical simulation procedure must be applied to determine the expected number of critical meeting situations per time unit. With such a numerical procedure, it is possible to determine the frequency of meeting situations where ship domains are violated as a function of the bridge opening.

Fig. 1.25 from Frandsen et al. (1991) shows an example of such a calculation of the number of critical meeting situations as a function of the bridge navigational span. In this case, it is seen that if the navigational span is chosen to be 1200 m, approximately one critical meeting will occur per day. A span of 1600 m will lead to approximately 17 critical meetings per year, and an 1800 m navigational span will completely avoid critical meeting situations.

For a heavily trafficked two nautical mile-long one-way shipping lane in Fehmarn Belt, Jensen et al. (2013) have presented a similar calculation of the percentage of

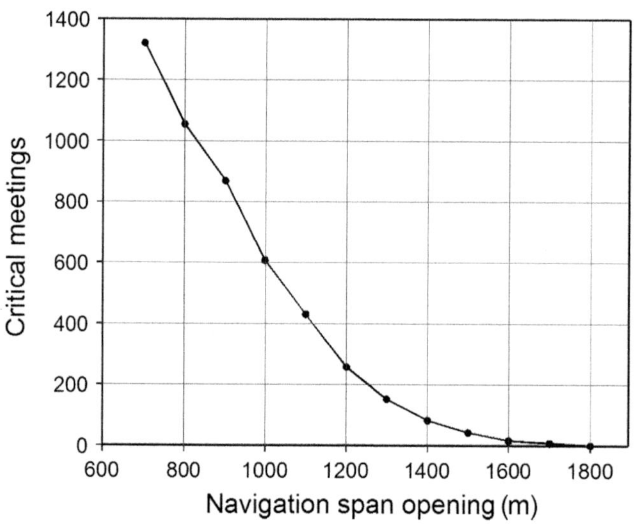

Fig. 1.25 Number of annual critical meeting situations as function of navigational span opening for the Great Belt.
Based on Frandsen, A.G., Olsen, D., Lund, H.T., Bach, P.E., Gerwick, B.C., 1991. Evaluation of minimum bridge span openings applying ship domain theory. Transp. Res. Rec. 1313, 83–90.

ships with free flow, that is, without violation of the ship domain, as function of the width of the traffic lane.

Estimates of the frequency of close encounters given a specific width of the navigational span are somewhat uncertain. Whether a navigator feels that he is too close to the bridge or another vessel depends on many factors such as the weather conditions, the manoeuvrability of the vessel, and the training of the navigator. Therefore, these theoretical calculations are most often supplemented by real-time simulations using computer-based manoeuvring simulations where professional navigators or pilots operate the simulator ships under adverse environmental and traffic situations. Especially situations where encounter analyses have found the bridge span opening is too narrow for free-flow conditions should be studied by real-time simulation in adverse weather conditions. See Sand and Petersen (1998).

1.4.3 Ship–bridge collision probability analysis

In this section, we shall briefly describe a procedure for the calculation of the probability of ship impacts against bridge piers or bridge girders. As before, the model is based on a division of the ship–bridge collision problem into a number of different phenomena and subsequent application of mathematical models to quantify the risk from each category. Accidents caused by human factors and technical errors shall be discussed.

As discussed in Section 1.2, the mathematical models for human errors are mainly based on the calculation of the number of collision candidates in the case of blind navigation, N_a (see Eq. 1.3), and a subsequent use of a causation factor to model the action of the navigators on board the ships; see Eq. (1.4). The causation factor P_c depends on a number of causes such as the annual distribution of visibility in the considered area, the presence of pilots, and the presence of VTS in the region. Causation factors in the range $P_c = 0.35 \cdot 10^{-4}$–$3.2 \cdot 10^{-4}$ have been used. See Section 1.2.3.

As for the analysis procedure presented in Section 1.3 for ship collisions against offshore structures, the collision frequency estimation or a probabilistic measure of the consequences can be based on the evaluation of four accident categories:

$$F_{total} = F_{Cat,1} + F_{Cat,2} + F_{Cat,3} + F_{Cat,4} \tag{1.39}$$

Again, these categories are characterized as follows:

Category 1 includes human errors associated with ships following the ordinary, straight route with a normal speed, see Fig. 1.26, where the spatial distribution can be modelled as a Gaussian or a log-normal distribution; see Eq. (1.5).

The parameters in the transverse distribution can be estimated from AIS observations. The results of these observations can be compared with published empirical expressions for the distribution parameters. Karlson et al. (1998) proposed the standard deviation to be taken as 40% of the navigational channel, and Gluver and Olsen (1998) proposed to apply a standard deviation equal to the ship length. Other proposals are given in Larsen (1993).

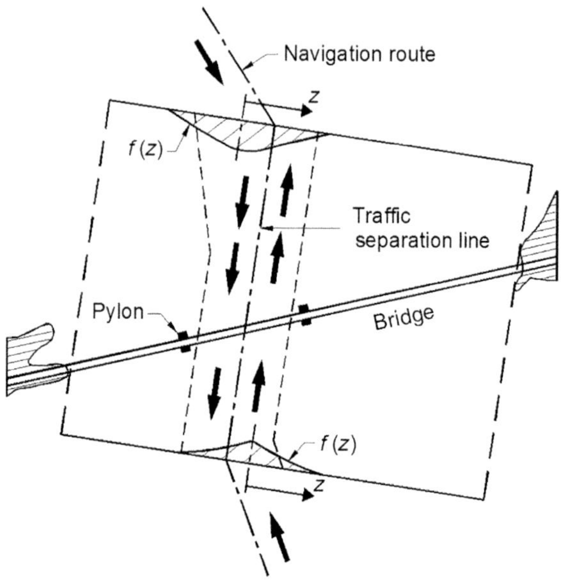

Fig. 1.26 Category 1: Ship–bridge collision candidates due to human errors.
Based on Pedersen, P.T., Valsgaard, S., Olsen, D., Spangenberg, S., 1993. Ship impacts: bow collisions. Int. J. Impact Eng. 13 (2), 163–187.

By the use of these empirical transverse distributions, there would hardly be any ship traffic some distance away from the bridge opening. Therefore, to take into account unusual events and to obtain a certain crashworthiness of the approach spans, it is often assumed that between 1.0% and 2.0% of the ship traffic is uniformly distributed across the channel. See Eq. (1.5). This will take care of the very rare ship collision accidents that can happen away from the main shipping routes due to gross human errors.

All in all, little guidance can be found in the literature on the lateral distribution of ship traffic for a specific regional area, and therefore, AIS observations should be carried out if at all possible.

Given the geometrical distribution $f_i(z)$ of the ship traffic on a specific route, the position of candidates for a collision with a pylon or a pier can be identified by the projection of the pylon or pier on a line perpendicular to the route. See Fig. 1.27. The methodology for determining the candidates for collisions with the ship deckhouse against the bridge girder is similar to the procedure for collisions against pylons and piers, except ship categories with an air draught that do not exceed the vertical clearance at the position of the considered girder element are eliminated from the deckhouse collision candidates. The calculation of the expected number $F_{Cat,1}$ of serious collisions during a time interval Δt for a bridge segment between points A and point B can be based on the following expression:

$$F_{Cat,1} = \frac{1}{\Delta t} \sum_{\substack{Ship \\ class\ i}}^{n} \int_{z=z_a}^{z_b} \int_{0}^{\Delta t} P_{c,i}\, Q_i(t) f(z) R_i dt dz \tag{1.40}$$

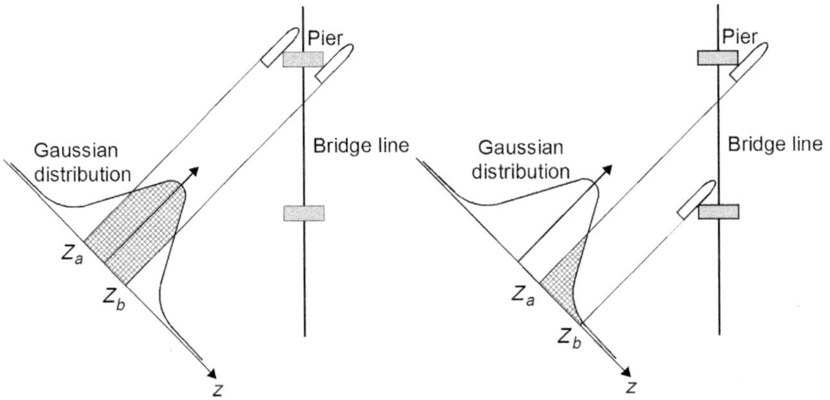

Fig. 1.27 Ships positioned between A $= z_a$ and B $= z_b$ can collide with the considered bridge pier (left) or bridge girder segment (right) provided there is an overlap between ship deckhouse height and bridge girder.

where

i is the ship class as discussed in Section 1.2.1

Δt denotes the time

z is an axis transverse to shipping route

z_a and z_b denote integration limits; see Fig. 1.27

$P_{c,i}$ is the causation factor, that is, the probability of a collision if the ship is on collision course; see Section 1.2.3

$Q_i(t)$ is the number of ships of class i per time unit

$f(z)$ is the lateral ship distribution

R_i is a factor that can express the severity when the consequences to the bridge of the ship collisions are analysed. This factor is often called 'Heinrich factor' (Heinrich, 1959).

The Heinrich factor depends on the ship collision crushing force, that is, ship type, ship size, ship velocity at the moment of impact, the strength of the bridge structure, and a number of other parameters. This factor will be discussed in more detail in Section 1.4.4 of the book. When R_i is given the value 1, then $F_{Cat,1}$ will simply reflect the ship collision frequency for Category 1 collisions.

To reduce the probability of Category 1 accidents, it is necessary to arrange fairways and the bridge layout such that there are ample margins for minor errors and faults on board the passing ships and to ensure that the ship domains are not violated too often due to insufficient navigational spans.

Category 2 includes those ships that fail to make a proper change of course at a turning point in the vicinity of the bridge. See Fig. 1.28. In this case, the ship will continue on the old course and will continue until the ship navigator corrects the course, the ship goes aground, or the ship reaches the bridge line and may collide with the bridge.

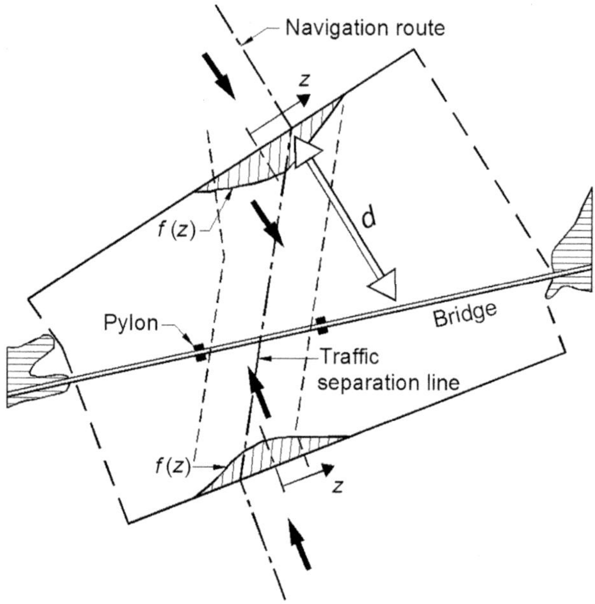

Fig. 1.28 Category 2: Ship–bridge collision candidates due to failure to change course. Based on Pedersen, P.T., Valsgaard, S., Olsen, D., Spangenberg, S., 1993. Ship impacts: bow collisions. Int. J. Impact Eng. 13 (2), 163–187.

The Category 2 event can be modelled by inclusion of the probability for forgetting to change course into Eq. (1.40):

$$F_{Cat,2} = \frac{1}{\Delta t} \sum_{\substack{Ship \\ class\ i}}^{n} \int_{A}^{B} \int_{0}^{\Delta t} P_0 P_{c,i} Q_i(t) f(z) R_i dt dz \tag{1.41}$$

where P_0 is the probability for omission of a course change. See Eq. (1.22) or (1.23).

The Category 2 hazard implicates that it is necessary to carefully consider the layout of sailing routes near the bridge in order to reduce as far as possible the probability that ships that neglect a turn will collide with the bridge.

Category 3 includes ships that do not follow the ordinary route recommended in charts due to technical errors such as ships drifting due to mooring or anchoring failures, drifting ships at work such as fishing vessels, ships drifting due to the loss of propulsion, ships in pack ice drifting against the bridge, and ships with a steering gear failure.

For ships drifting due to engine failure, it can be assumed that the ships at the time of loss of propulsion are located on the idealized sailing route. The moving direction of a drifting ship depends on wind and current direction. Typically, the drifting ship will be oriented such that the longitudinal axis of the ship is transverse to the moving direction. See Fig. 1.29. Suggested probabilities for loss of propulsion and for loss of

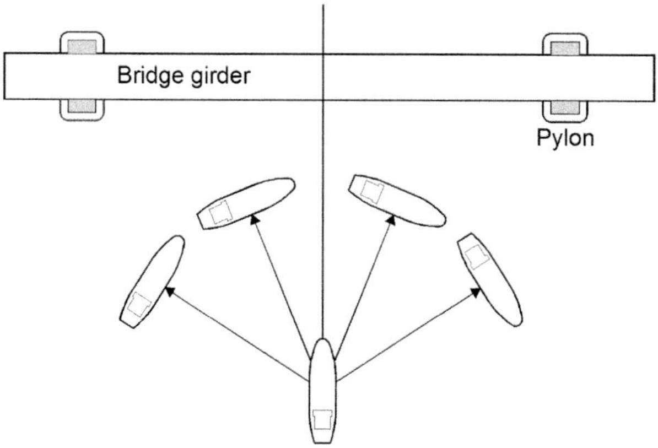

Fig. 1.29 Ship drifting due to loss of propulsion and driven by wind and current.

control are given by Eq. (1.26). Loss of propulsion or loss of control is normally repaired by the crew within a relatively short time span. An empirical estimation of the probability of a drifting vessel and the subsequent probability of drifting as function of time is given by Eq. (1.27) and depicted in Fig. 1.17.

Drifting ships and ships without control can collide with the bridge structure at any point along the bridge line, although excluding areas with low water depth, where grounding can occur before hitting the bridge. If the water depth is not too high, then the navigator may be able to stop the drifting by the use of the ship's anchoring system.

The number of collisions per unit time due to drifting depends on the wind and current directions and velocities. As an example, the collisions due to drifting can be determined by the expression similar to Eq. (1.29) but including the probability for successful anchoring:

$$F_{drift}^{bridge} = \frac{1}{\Delta t} \int_0^{\Delta t} \sum_{ship\ class\ i} Q^i(t)(1 - P_{anchor}) \int_{\theta=0}^{360} P_{wind,cur}(\theta) \sum_{segments} P_{Technical\ error}(L_w) \int_{z=0}^{L_w} \int_{v_{drift}} P_{drift}(t_{col}|X) f(v_{drift}) dv_{drift} dz d\theta dt \tag{1.42}$$

Here, P_{anchor} is the probability of a successful anchoring in a propulsion failure situation, and $P_{wind,cur}$ represents the probability of having the drifting direction θ. The drifting speed is given by $f\ (v_{drift})$, and X defines the parameters that give the time to collision.

When a steering system fails, the rudder will be locked, and the ship may start sailing in circular path. The diameter depends on the locked position of the rudder and the underkeel clearance. Low water depth will increase the turning diameter. Often, the rudder will be locked in an extreme position, and provided a large underkeel

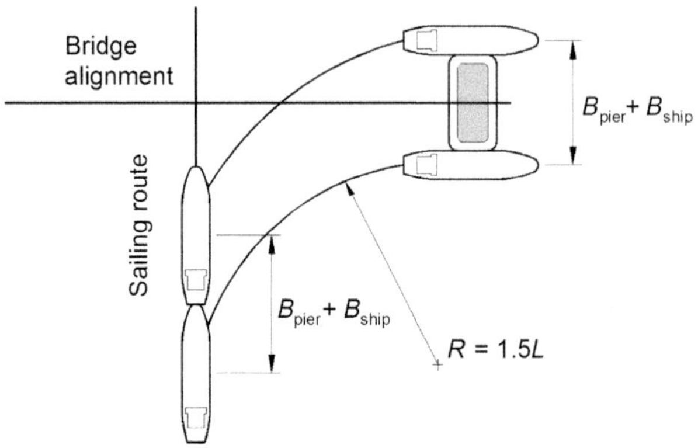

Fig. 1.30 Ship collision with bridge pier due to steering gear failure.

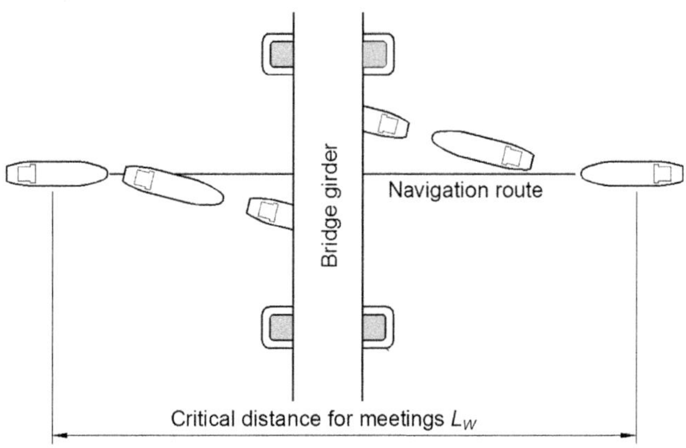

Fig. 1.31 Evasive manoeuvres near the bridge line.

clearance, the ship may go into a circular path with a diameter equal to two to three times the ship length. See Fig. 1.30.

Category 4 includes ships, which make evasive manoeuvres in the vicinity of the bridge due to the presence of other vessels and, because of such manoeuvres, collide with the bridge. See Fig. 1.31. The number of critical encounters can be estimated on the basis of the ship domain theory described in Section 1.4.2.

If the bridge span opening is sufficiently wide with good margins for evasive manoeuvres and if the numbers of critical encounters near the bridge are infrequent, then this risk contribution is normally relatively insignificant.

The procedure presented here for the estimation of the probability of ship impacts depends on a number of empirical and uncertain factors such as the lateral distribution $f_i(z)$ of the ships and the causation factor P_c. For this reason, the mathematical model should be verified by a comparison between a prediction of the number of collisions and groundings predicted by this model and the number of actual accidents over a period in the region. Also, the number of collisions against lighthouses in the region can be used to verify the model.

A typical set of results for the annual collision frequency for impacts by different ship classes (see Table 1.1) against some bridge structural elements are shown in Table 1.5.

1.4.4 Acceptance criteria and Heinrich factor

A typical acceptance criterion for the bridge structure related to the availability of the bridge for traffic could be that the annual frequency of collapse of the bridge or disruption of the bridge traffic for an extended period shall be less than 10^{-4}. Such an acceptance criterion is global and shall be distributed among the bridge structural elements. The criterion shall normally also include other types of hazards than ship collisions.

With given impact strengths of the bridge structural elements and with estimated collision loads related to the individual ship classes, it is possible from impact probability results such as those indicated in Table 1.5 to choose a ship class or design ship (a typical ship size and velocity) that the individual bridge structural elements shall be able to withstand, in order to fulfil the global disruption criterion. The distributions of the global acceptance criteria to the individual bridge structural elements should be done such that the results are optimal from a bridge design point of view.

The design ship for an individual bridge element is defined as a vessel striking a bridge structural element that the structural element must withstand without critical failure in order for the total bridge line to fulfil the design criteria. That is, in order for the acceptance criterion to be met, impacts from all ships equal to or smaller than the size of the design vessels should not lead to critical failure of any of the bridge elements.

In Eqs (1.40), (1.41), the consequence or severity of the collision is modelled by the factor R_i, often called a Heinrich factor, Heinrich (1959). This factor originates from an observation of how accidents in industry can result in different degrees of severity from (a) none to (b) minor or to (c) serious consequences, depending on the cause. In connection with ship collisions, the Heinrich factor can be used to give the ratio between impact incidents that cause critical failure and all incidents that have the potential to cause critical failure. The American AASHTO Guideline (2009) refers to the Heinrich factor as 'probability of collapse'.

Thus, even if the bridge element is exposed to a ship impact with a ship larger than the design ship for that specific structural component, then there is a certain probability that the bridge element does not collapse. This may be due to two effects: (1) The actual impact speed and thus the kinetic energy of the colliding ship at the moment of impact are reduced due to last-minute ship manoeuvres or because the impact, at, for example, a bridge pier, is not a central impact and will result in sliding or partial sliding along the pier surface, and (2) the bridge elements are designed based on a characteristic capacity and will have a reserve ultimate capacity.

Table 1.5 An illustrative example of calculated annual collision frequencies F_{total} for different bridge components.

Frequency (annual)	Ship size class									Total
	1	**2**	**3**	**4**	**5**	**6**	**7**	**8**	**9**	
Pylon 1	3.64E−03	1.27E−04	5.55E−05	5.45E−05	1.28E−05	5.78E−06	3.90E−06	6.29E−07	3.58E−07	3.90E−03
Pylon 2	8.14E−03	4.04E−04	1.60E−04	1.14E−04	3.56E−05	1.60E−05	8.96E−06	2.43E−06	7.66E−07	8.89E−03
Pier 4	1.56E−03	8.11E−04	2.72E−04	1.91E−04	7.30E−05	5.04E−05	2.63E−05	5.75E−06	1.98E−06	2.99E−03
Pier 5	1.32E−03	7.45E−04	4.57E−04	3.32E−04	1.19E−04	9.75E−05	3.95E−05	1.50E−05	2.58E−06	3.13E−03
Bridge deck	8.03E−04	4.19E−05	4.26E−05	4.95E−05	2.25E−05	2.38E−05	1.61E−05	4.02E−06	1.94E−06	1.01E−03

Due to these modifying effects, the probability for a critical ship collision can be expressed as

$$F_{Critical\ failure} = F_{Critical\ failure|Collision} \cdot F_{Collision}$$

where $F_{collision}$ is the probability for a collision with a ship larger than the design ship and $F_{Critical\ failure|\ collision}$ is the Heinrich factor that describes the probability that a given collision from a ship larger than the design ship causes critical failure of the bridge element. This probability will increase with the size of these ships. An accurate calculation of the Heinrich factor is impossible because of the many uncertainties and varying conditions that would have to be considered in practice.

The application of a Heinrich factor can be executed either by a reduction of the calculated collision frequencies or by increasing the global accept criterion.

As mentioned above, then, the momentum and the kinetic energy transferred to the bridge element by the impacting design ship have a certain probability to be less than the momentum and energy estimated simply from an assumed service speed and central collision scenario.

Head-on collisions against bridge elements are most often due to human errors such as distraction, influence of alcohol, sickness, and accident. These mistakes are assumed to cause the ship to continue with unchanged course and velocity towards the bridge. When the ship approaches the bridge, there is likelihood that the navigator detects the error and tries to avoid the collision by mitigation measures such as the following:

- To change speed so that the maximum energy released for crushing of the ship bow or the bridge element is reduced
- To change course so that the likelihood of easy touch and more favourable collision scenario increases

As mentioned in Section 1.2.4, Lützen (2001) made a survey of ship–ship collisions and found that the reported speed of the striking vessel at the moment of impact is reduced and often can be assumed to be uniformly distributed between a very low speed and the service speed. See Fig. 1.14. A reduction in collision speed will have a relatively large impact on the resulting collision energy.

The impact location and the impact angle may also be such that not all the kinetic energy of the striking vessel will be available for crushing of the bridge or the ship. The most critical collisions will be head-on collisions perpendicular to the surface of the bridge pier. For smaller impact angles, sliding can occur between the ship and the bridge pylon or pier. In this case, the energy released for crushing depends on the collision angle and the coefficient of friction. See Section 2.4 where this reduction of energy released for crushing is quantified. For deckhouse collisions against the bridge girder, there is not the same chance for a sliding collision. So, the primary effect for reducing the likelihood of critical failure of the bridge deck will result from a possible speed reduction before the collision.

Besides the contribution to the Heinrich factor from favourable conditions related to the ship impact, the reserve strength capacity of the bridge can also give a

Table 1.6 An example of design ships and energies for different bridge components.

	Design ship—normal speed			Design ship—drifting		
Bridge component	**Length (m)**	**Speed (m/s)**	**Energy (MJ)**	**Length (m)**	**Speed (m/s)**	**Energy (MJ)**
Pylon 1	120	6	170	235	1	35
Pylon 2	130	6	180	240	1	37
Pier 4	90	6	70	200	1	17
Pier 5	85	6	60	200	1	17
Bridge deck	90	6	65	125	1	10

contribution. Provided that the design of the bridge structural elements are based on characteristic loads, then these elements will have an additional reserve capacity as a result of the choice of material parameters and the possibility for a redistribution of load paths in the bridge structure. The AASHTO guideline (2009) gives Heinrich ratios as function of the ratio between the ultimate lateral resistance of a pier and the vessel impact force. In general, the reserve load contribution to the Heinrich factor depends strongly on how wide-ranging and accurate the applied nonlinear collapse load calculation procedure is carried out for the bridge structure.

Even if comprehensive and accurate collapse load estimates of the bridge structure are used to define the design ships, then it is still relevant to take into account that due to the randomness of the ship impact, not all potential ship collisions against the pylons and the bridge piers will lead to the most detrimental impact and cause critical failure of the bridge structure. An exact calculation of the Heinrich factor originating from ship impact speeds and locations is not possible. However, a conservative factor can be estimated when only the ship speed and impact locations can be relied upon. Table 1.6 presents a typical example of design ships for a bridge where the global annual acceptance frequency for disruption is 10^{-4} and the Heinrich factor is simply taken as 50%, irrespective of ship sizes, for collisions against pylons and piers, and 100% for deckhouse collisions against the bridge girder.

1.4.5 Calculation example

Example 1.5: Critical meeting frequencies

The ship traffic histogram distribution in a shipping lane under a bridge is given as follows:

Ship length (m)	Eastbound no. of ship per year	Westbound no. of ship per year
0–100	1800	2500
100–150	1600	2000
150–200	1000	1100
>200	15	30

The average ship speed in the area is 4.0 m/s.

The exceedance frequency for the 10 min mean wind velocity in the vicinity of the bridge opening is given as follows:

Mean wind velocity (m/s)	Exceedance hours per year
10	1500
15	100
20	2

Critical meeting situations can arise when ships with a length larger than 150 m meet within a distance of 1000 m close to the bridge during wind velocities exceeding 15 m/s. The tasks are the following:

(i) Calculate the number of critical meeting situations where one ship length exceeds 150 m.
(ii) Calculate the number of critical meeting situations where both ships have a length exceeding 150 m.

Solution to Example 1.5

(i) Based on the given histogram for the ship traffic, the distribution function for the ship traffic is

Ship length (m)	Eastbound no. of ship per year	Westbound no. of ship per year
>0	4415	5630
>100	2615	3130
>150	1015	1130
>200	15	30

The number of meeting situation per second in all weather conditions and where both ships have a length larger than 150 m can be determined directly from Eq. (1.36):

$$\alpha_{EW} = N_{W\ AC} \cdot Q_E$$

$$\alpha_{EW} = L_w Q_W Q_E \left(\frac{1}{V_W} + \frac{1}{V_E} \right)$$

$$\alpha_{EW} = 1000 \frac{1015 \cdot 1130}{(365 \cdot 24 \cdot 60 \cdot 60)^2} \left(\frac{1}{4} + \frac{1}{4} \right)$$

$$\alpha_{EW} = 0.577 \cdot 10^{-6} \, \text{meetings/s}$$

Assuming that the number of meeting situations and the wind velocity are independent variables, the annual numbers of meeting situation where the wind velocity exceeds 15 m/s and both ships have a length larger than 150 m are found as

$$N_{150,150} = 0.577 \cdot 10^{-6} \cdot \frac{100}{(365 \cdot 24)} \cdot (365 \cdot 24 \cdot 60 \cdot 60)$$

$$N_{150,150} = 0.21 \, \text{meetings/year}$$

The number of meetings per second where one ship has a length larger than 150 m can be determined from

$$\alpha_{EW} = \frac{1000}{2 \cdot (365 \cdot 24 \cdot 60 \cdot 60)^2} \cdot (1015 \cdot 5630 + 1130 \cdot 4415 - 1015 \cdot 1130)$$

or

$$\alpha_{EW} = 4.87 \cdot 10^{-6}\,\text{meetings/s}$$

Thus, the annual number of meetings where one ship has a length above 150 m when the wind velocity exceeds 15 m/s is

$$N_{150,0} = 4.87 \cdot 10^{-6} \cdot \frac{100}{(365 \cdot 24)} \cdot (365 \cdot 24 \cdot 60 \cdot 60)$$

or

$$N_{150,0} = 1.76\,\text{meetings/year}$$

1.5 Probability of ship grounding

Groundings and strandings are the most common types of marine casualties. See Zhu et al. (2002) and Lützen and Simonsen (2003). It is called grounding when a ship strikes the sea bed, while a stranding is when the ship is aground and remains there for some length of time. Both events can damage a vessel depending on the ocean bottom type, that is, whether it is rocky, sandy, or muddy, and on the sea and tide conditions.

Whereas risk from collision accidents has been subjected to extensive research during the years, then there is much less research work available on procedures to determine the probability of grounding events. Procedures for calculating the probability of grounding should include the following:

- Powered grounding where the ship is running on a ground with forward speed
- Drifting grounding for disabled ships due to engine or steering gear failure

The subsequent ship damage for both types of grounding may be followed by a stranding condition due to a receding tide or wave action.

Models for the calculation of the powered grounding probability P_g have been developed in a way similar to the method described above for ship–ship and ship–offshore collisions; see Otto et al. (2002). The difference is that an obstacle, for example, a rock or a shallow area, on which the ship grounds, is fixed in its position and that it is in most cases below the water surface.

Based on the draught for loaded and ballasted situations and the effect of sinkage and trim due to shallow water effects, the actual draught can be calculated and compared with the actual water depth. In principle, the actual draught of a ship is transmitted via its AIS transmitter. When the ship draught distribution and the water depth are known, then the ship classes (see Table 1.1) that can run aground in a certain area are known.

Ship grounding analyses are also needed when the probability for impacts against offshore pipelines or power and communication cables due to ship ploughing into the sea bed must to be analysed. Similarly, the risk associated with ship collisions on submerged tunnel roofs near abutments can be an important hazard leading to tunnel flooding.

1.5.1 Qualitative assessment of the probability of ship grounding events

Again, ship traffic data need to be collected for the number of ships and their spatial distribution in the vicinity of the shoals and most important coast lines. Similarly, a procedure must be developed for the characterization of coast lines and shallow grounds, that is, distribution of rocks, bottom profile data, and tide variations.

For the calculation of the probability of powered grounding again, two categories of human errors are considered. The first is based on an assumed simple spatial distribution of the vessels in combination with a causation factor for ships on a track that can lead to grounding, and the second category of accidents is related to the probability that the vessels not always change course at bends on the shipping route.

Finally, further categories related to drifting vessels due to engine blackout or steering machine failure must be considered for a complete estimation of the probability of grounding in a specific geographical area.

For powered groundings, a study of databases for ship groundings has shown that many vessels are proceeding at the service speed at the moment of impact (Lützen and Simonsen, 2003). The result is that the probability density function can be described by an increasing function. This is in contrast to the situation for ship–ship impacts where the striking vessels have nearly a uniform impact velocity between zero and service speed. See Fig. 1.14B.

1.5.2 Model for estimation of ship grounding frequency

As for the estimation of ship collisions against offshore installations, the grounding frequencies $F_{grounding}$ are estimated as the number of grounding candidates $N_a^{grounding}$ multiplied by causation factors P_c; see Eq. (1.4):

$$F_{grounding} = P_c N_a^{grounding}$$

The model for estimation of the number of groundings in the case of blind navigations, $N_a^{grounding}$, is again based on a division of the ship grounding problem into a number of different phenomena and subsequent application of mathematical models for the quantification of the risk from each category.

Category I: This category includes ships following the ordinary, straight route at a normal speed and with a normal ship track distribution; see Fig. 1.32. For this

Fig. 1.32 Powered grounding candidates with too deep draught from Categories 1 and 2.

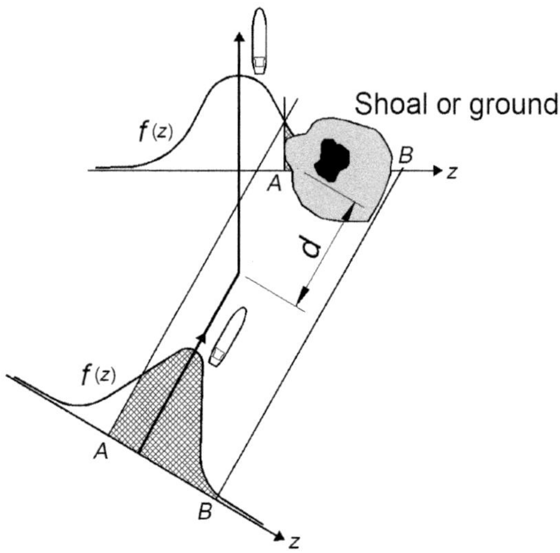

category, the expected number $N_a^{Cat,1}$ of blind groundings per unit time Δt can be determined from the following expression:

$$N_a^{Cat,1} = \frac{1}{\Delta t} \sum_{\substack{Ship \\ class}}^{n} \int_{z=A}^{B} \int_{t=0}^{\Delta t} Q_i(t) f_i(z) dt dz \tag{1.43}$$

where Q_i is the traffic density function of class i ships with a draught equal to or larger than the water depth at the shoal that can lead to grounding, $f_i(z)$ is the spatial distribution (see Eq. 1.5), and $z = A$ and $z = B$ are the transverse coordinates of the shoal boundaries. See Fig. 1.32.

Category 2: This category includes ships that miss a turn at a turning point in the vicinity of the ground or shoal. To determine the added probability for ship grounding in the vicinity of bends of the seaway (see Fig. 1.32) where the vessels may not make the turn, a procedure similar to the one used for Category 1 can be used provided that the procedure is augmented with the probability, P_0, that the ships fail to act on the changing route direction:

$$N_a^{Cat,2} = \frac{1}{\Delta t} \sum_{\substack{Ship \\ class}}^{n} P_0 \int_{z=A}^{B} \int_{t=0}^{\Delta t} Q_i(t) f_i(z) dt dz \tag{1.44}$$

The probability P_0 that a ship fails to change course at a turning point can, for example, be modelled by Eq. (1.22) or (1.23) from Section 1.2.2.

Category 3: This category includes ships drifting due to loss of propulsion or control. For drifting ships or ships with control failure, it is normally assumed that the ships at the time of loss of propulsion are located on the idealized sailing route. Drifting or loss of control is most critical in narrow navigational channels where a ship can run aground after a relatively short time after the technical error.

The number of grounding events due to drifting $F_{drift}^{grounding}$ from a leg of length L_w depends on the wind speed and the current direction and velocity. As for the case of collision against an offshore installation, the frequency of grounding events due to drifting can be estimated as

$$F_{drift}^{grounding} = \frac{1}{\Delta t} \int_0^{\Delta t} \sum_{shipclass\ i} Q_i^{grounding}(t)(1 - P_{anchor}) \int_{\theta=0}^{360} P_{wind,cur}(\theta)$$

$$\sum_{segments} P_{Technicalerror}(L_w) \int_{z=0}^{L_w} \int_{v_{drift}} P_{drift}(t_{col}|X) f(v_{drift}) dv_{drift} dz d\theta dt \tag{1.45}$$

where P_{anchor} is the probability of successful anchoring in a propulsion failure system; $Q_i^{grounding}$ is the number of ships per time unit that have a draught large enough to ground; $P_{wind,cur}$ represents the probability of having the drifting direction θ; $P_{Technical\ error}$ is the probability of a technical error given by Eq. (1.26); P_{drift} is probability for drifting, Eq. (1.27); the parameter X defines the parameters that give the time to collision; and the drifting speed is given by $f(v_{drift})$.

To Category 3 must also be added the contribution from ships suffering a risk of loss of control, $F_{control}^{grounding}$. For a steering gear failure, there is a probability that the rudder may be locked in an extreme position. See Fig. 1.30.

Category 4 can represent groundings that are caused by meeting situations where the ships may feel they are forced to give way and as a result ground. No mathematical models exist at present for this probability.

The total number of ship groundings at a certain geographical is found by adding these contributions:

$$F_{grounding} = P_c \cdot N^{Cat,1} + P_c \cdot N^{Cat,2} + F_{drift}^{grounding} + F_{control}^{grounding} \tag{1.46}$$

where proposed numerical values for the causation factor P_c for powered grounding can be found in Section 1.2.3.

1.5.3 An application of the grounding frequency analysis

The probabilistic procedure represented by expressions like Eqs (1.43)–(1.45) has been established for several geographical regions, and in general, they seem to predict the number of ship accidents with reasonable accuracy. Such a comparison was made in connection with the ship collision study of the East Bridge of the Great Belt Link, Denmark. Here, grounding events and lighthouse collisions registered over a period of 15 years in the Great Belt were used to verify the estimate of the causation probability P_c. Fig. 1.33 shows recorded grounding events in an area just north of the East Bridge

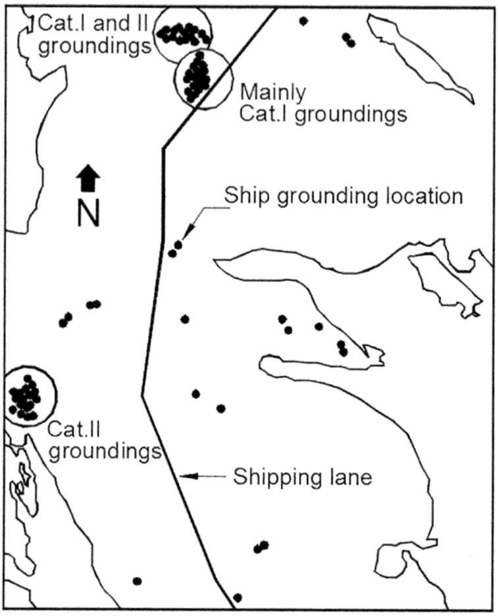

Fig. 1.33 Grounding events during a 15-year period near Hatterbarn, Denmark.
cf. Frandsen, A.G., Olsen, D., Lund, H.T., Bach, P.E., Gerwick, B.C., 1991. Evaluation of minimum bridge span openings applying ship domain theory. Transp. Res. Rec. 1313, 83–90.

alignment and illustrates clearly grounding events that can be categorized as either Category 1 or Category 2. The calculated number of groundings in the study period was found to be approximately equal to the observed number of grounding events.

1.6 Concluding remarks

Risk analysis is a tool that is increasingly applied in the marine and offshore industries to manage safety, health, and environmental protection. Ship collision and grounding risk may be estimated in a given navigational area by the use of mathematical models for evaluation of frequencies and consequences. The consequences can be measured in terms of structural damage, the number of fatalities and injuries, the amount of material released to sea, the immediate impact on environmental resources, and the subsequent costs of restoration.

Risk minimizing measures include a combination of actions that can reduce the frequency and can reduce consequences of accidents. Here, the reduction of the frequency of ship collision and grounding events normally is the most cost-effective risk control option.

The purpose of this present chapter has been to present procedures for the estimation of the frequency of ship–ship collisions, ship collision against offshore installations, ship collisions against bridge structures, and predictions for the frequency of grounding events. Even if ship collisions and groundings are low-probability events (but high-consequence), then statistics give valuable data for identifying typical and critical incident cases. Accident statistics show that human factors contribute to about 80% of all ship collisions and grounding events, and technical errors due to loss of propulsion and loss of steerage capability contribute to the remaining 20%. However, statistics of historical data are not free of errors. The error sources can be due to underreporting, information misinterpretation, and incorrect incident categorization. Furthermore, historical data cannot be used to predict the effect of new navigational conditions or new generations of ship structures and equipment. Thus, mathematical tools are needed for the predictions of future collision and grounding frequencies.

The calculation procedure presented in this chapter to determine the probability of ship collisions and groundings due to human errors is a two-step procedure that first requires determining the potential number of collision candidates or groundings as if no aversive manoeuvres are made. For this part of the analysis, a distribution of the ship traffic must be known. In the second step, a so-called causation factor is determined that models the effect of crew and equipment related actions to avoid the collision or grounding. This causation factor depends strongly on the available navigational aids in the shipping area, the weather conditions, the visibility, the equipment on the ship bridge, and the quality and training of the navigators.

Two categories of accidents due to human errors are considered. The first category is accidents where the ships follow their ordinary route but due to navigational errors collide with other ships or obstacles or ground on shoals. The second category is ships that miss a turn at a bend of the shipping lane.

The model presented for the calculation of the frequency for technical errors leading to collisions or to grounding includes drifting vessels due to loss of propulsion and vessels that lose control due to steering gear failure.

The probabilistic mathematical procedures presented in this chapter have been applied to a number of specific geographical areas. Karlson et al. (1998) compared actual number and locations of casualties registered in Øresund with calculated accident frequencies. Gluver and Olsen (1998) described the application for analysis of ship collisions to bridges. Cucinotta et al. (2017) analysed the frequency of ship collisions in the Strait of Messina using a collision and grounding risk analysis tool, Risk Management Tool for Aids to Navigation and VTS Authorities, IALA Waterways Risk Assessment Program (IWRAP), developed by the International Association of Marine Aids to Navigation and Lighthouse Authorities (IALA, 2009).

The IALA procedure was developed as a procedure for cost-effective design of navigational aids in ports and waterways. The mathematical models behind risk analysis program IWRAP are very similar to the models presented in this chapter. The IWRAP program is a further development of the grounding and collision analysis toolbox (GRACAT) procedure by Friis-Hansen and Simonsen (2002).

References

AASHTO, 2009. Guide Specifications and Commentary for Vessel Collision Design of Highway Bridges, second ed. .

Cucinotta, F., Guglielmino, E., Sfravara, F., 2017. Frequency of ship collisions in the Strait of Messina through regulatory and environmental constraints assessment. J. Navig. The Royal Institute of Navigation 70, 1002–1022.

Eurocode, 2006. 1—Actions on Structures—Part 1-7: General Actions—Accidental Actions. (Incorporated corrigendum February 2010). July 2006 CEN.

Frandsen, A.G., Olsen, D., Lund, H.T., Bach, P.E., Gerwick, B.C., 1991. Evaluation of minimum bridge span openings applying ship domain theory. Transp. Res. Rec. 1313, 83–90.

Friis-Hansen, P., Pedersen, P.T., 1998. Risk Analysis of Conventional and Solo Watch Keeping. Submitted to International Maritime Organization (IMO) Maritime Safety Committee by Denmark at 69th Session.

Friis-Hansen, P., Simonsen, B.C., 2002. GRACAT, software for grounding and collision analysis. Mar. Struct. 15 (4–5), 383–402.

Fujii, Y., Mizuki, N., 1998. Design of VTS systems for water with bridges. In: Proc. Int. Symp on Advances in Ship Collision Analysis, Copenhagen Denmark. A.A. Balkema, pp. 177–193.

Fujii, Y., Shiobara, R., 1971. The analysis of traffic accidents. J. Navig. 24, 534.

Fujii, Y., Yamanouchi, H., Mizuki, N., 1974. Some factors affecting the frequency of accidents in marine traffic, II—the probability of stranding, and III—the effect of darkness on the probability of collision and stranding. J. Navig. 27, 235–247.

Gluver, H., Olsen, D., 1998. Survey of ship tracks in Fehmarn Belt. In: Proc. 2nd International Conference on Collision and Grounding of Ships, Maritime Engineering, Dept. Mechanical Engineering, Technical University of Denmark, pp. 13–22.

Hänninen, M., Kujala, P., 2012. Influences of variables on ship collision probability in a Bayesian belief network model. Reliab. Eng. Syst. Saf. 102, 27–40.

Hansen, M.G., Jensen, T.K., Lehn-Schiøler, T., Melchild, K., Rasmussen, F.M., Ennemark, F., 2013. Empirical ship domain based on AIS data. J. Navig. 66, 931–940.

Hassel, M., Utne, I.B., Vinnem, 2017. Allision risk analysis of offshore petroleum installations on the Norwegian Continental Shelf-an empirical study of vessel traffic patterns. WMU J. Marit. Aff. 16, 175. https://doi.org/10.1007/s13437-016-0123-7.

Heinrich, H.W., 1959. Industrial Accident Prevention, a Scientific Approach, fourth ed. McGraw Hill.

International Association of Lighthouse Authorities (IALA), 2009. Risk Management Tool for Ports and Restricted Waterways (IWRAP). France.

International Association of Oil and Gas Producers, 2010. Risk assessment data directory. Report No. 434/16 March.

International Maritime Organization (IMO), 2013. Revised Guidelines for Formal Safety Assessment (FSA) for Use in the IMO Rule-Making Process. MSC-MEPC.2/Circ.12 8.

Jensen, T.K., Hansen, M.G., Lehn-Schiøler, T., Melchild, K., Rasmussen, F.M., Ennemark, F., 2013. Free flow efficiency of a one-way traffic lane between two pylons. J. Navig. 66, 941–951.

Kaneko, F., Hara, D., 2007. Estimation of dangerous encounters' number from observed ship trajectories. In: 4th International Conference on Collision and Grounding of Ships. Hamburg University of Technology, Schiffbautechnische Gesellschaft, pp. 187–193.

Karlson, M., Rasmussen, F., Frisk, L., 1998. Verification of ship collision frequency model. In: Gluver, Olsen, (Eds.), Proceedings of the International Symposium on Advances in Ship Collision Analysis. Copenhagen Denmark. Balkema, pp. 117–121.

Larsen, O.D., 1993. Ship Collision With Bridges. Structural Engineering Document No. 4IABSE International Association for Bridge and Structural Engineering.

Lehn-Schiøler, T., Hansen, M.G., Melchild, K., Jensen, J.K., Randrup-Thomsen, S., Glibbery, K.A.K., Rasmussen, F.M., Ennemark, F., 2013. VTS a risk reducer: a quantitative study of the effect of VTS in the Great Belt. In: Amdahl, E., Leira, T. (Eds.), Proceeding Collision and Grounding of Ships and Offshore Structures. Taylor and Francis Group, pp. 19–26.

Li, S., Meng, Q., Qu, X., 2012. An overview of maritime waterway quantitative risk assessment models. Risk Anal. 32 (3), 496–512.

Liu, J., Zhou, F., Li, Z., Wang, M., Liu, R.W., 2016. Dynamic ship domain models for capacity analysis of restricted water channels. J. Navig. 69, 481–503.

Lützen, M., 2001. Ship Collision Damage. PhD thesis, Department of Mechanical Engineering, Technical University of Denmark.

Lützen, M., Friis-Hansen, P., 2003. Risk reducing effect of AIS implementation on collision. In: Proceedings of Society of Naval Architects and Marine Engineers: The World Maritime Technology Conference (SNAME). 13 pp.

Lützen, M., Simonsen, B.C., 2003. Grounding damage to conventional vessels. In: Proceedings of Society of Naval Architects and Marine Engineers: The World Maritime Technology Conference (SNAME) (Rosenblatt, B. (red.). 13 s).

Moan, T., Amdahl, J., Ersdal, G., 2017. Assessment of ship impact risk to offshore structures—new NORSOK N-003 guidelines. Mar. Struct. https://doi.org/10.1016/j.marstruc.2017.05.003.

NORSOK Standard N-004, 2004. Design of Steel Structures. Appendix A: Design against accidental actions.

Otto, S., Pedersen, P.T., Samuelidis, M., Sames, P., 2002. Elements of risk analysis for collision and grounding of a RoRo passenger ferry. Mar. Struct. 15 (4–5), 461–475.

Pedersen, P.T., 1995. Collision and grounding mechanics. In: Proc. WEMT'95, Danish Society of Naval Architecture and Marine Engineering, Copenhagen, pp. 125–157.

Pedersen, P.T., 1998. In: Gimsing, N.J. (Ed.), Ship Crushing Load Studies. The Storebælt Publications, East Bridge, pp. 49–57 (Chapter 3.6).

Pedersen, P.T., 2002. Collision risk for fixed offshore structures close to high-density shipping lanes. Proc. Inst. Mech. Eng. M J. Eng. Mar. Environ. 216 (Part M1), 29–44.

Pedersen, P.T., 2014. Risk assessment for ship collisions against offshore structures. In: Guedes Soares, Santos (Eds.), Invited Lecture, Maritime Technology and Engineering. Taylor & Francis Group, London. ISBN: 978-1-138-02727-5.

Pedersen, P.T., 2010. Review and application of ship collision and grounding analysis procedures. Mar. Struct. 23, 241–262.

Rasmussen, F.M., Glibbery, K.A.K., Hansen, M.G., Jensen, T.K., Lehn-Schiøler, T., Randrup-Thomsen, S., 2012. Quantitative assessment of risk to ship traffic in the Fehmarnbelt fixed link project. J. Pol. Saf. Reliab. Assoc., Summer Safety and Reliability Seminars 3(1).

Pedersen, P.T., Valsgaard, S., Olsen, D., Spangenberg, S., 1993. Ship impacts: bow collisions. Int. J. Impact Eng. 13 (2), 163–187.

Sand, S.E., Petersen, J.B., 1998. Use of ship simulations in bridge design. In: Gluver, Olsen, (Eds.), Proc. Ship Collision Analysis. Balkema, Rotterdam, pp. 159–168.

Silveira, P.A.M., Teixeira, A.P., Guedes Soares, C., 2013. Use of AIS data to characterise marine traffic patterns and ship collision risk off the coast of Portugal. J. Navig. 66, 879–898.

Sotiralis, P., Ventikos, N.P., Hamann, R., Golyshev, P., Teixeira, A.P., 2016. Incorporation of human factors into ship collision risk model focusing on human centered design aspects. Reliab. Eng. Syst. Saf. 156, 210–227.

Szlapczynski, R., Szlapczynski, J., 2017. Review of ship domains: models and applications. Ocean Eng. 145, 277–289.

Zhu, L., James, P., Zhang, S., 2002. Statistics and damage assessment of ship grounding. Mar. Struct. 15 (4–5), 515–530.

External dynamics of ship collisions and grounding

2

2.1 Introduction

In the classical ship collision theory, the collision consequence estimation is most often separated into two independent problems, namely, the problem of

external collision dynamics,
inner collision dynamics

where the external collision dynamics deals with the motion of the colliding ships and their interaction with the surrounding water and the inner dynamics involves local crushing of material in the involved ship structures. The crushing damage analysis is discussed in detail in Chapters 3 and 4.

This separation into two independent problems is from a theoretical point of view only valid when the hydrodynamic pressure on the ship hull from the surrounding water can be modelled by constant added masses, and the duration of the impact is so short that the motions of the involved ships during the collision event can be neglected.

The external dynamic analysis can be performed by the following:

Coupled analysis using time simulation procedures
Uncoupled analysis using a classical mechanics approach

In the following, the coupled time simulation procedures will be discussed first briefly, that is, procedures suited for analyses where the coupling between the ship motions and the hydrodynamic forces as well as the collision forces during the impact process need to be considered. However, the main focus of this chapter will be on uncoupled classical mechanics procedures based on momentum and energy considerations for ship–ship collisions, for ship–platform collisions, and for grounding analyses.

2.1.1 *Coupled analysis based on time simulations*

Provided that the energy transmitted to elastic vibrations in the colliding ship hulls can be neglected (see Pedersen and Li, 2009), then the general motion of the rigid ship hulls is described by Newton's law in six degrees of freedom. In a right-hand coordinate system fixed in the ship, the general equations of motion for each of the ships can be expressed as the following (Petersen, 1982):

$$X = M\left[\ddot{u} + \dot{q}\dot{w} - \dot{r}\dot{v} - x_g\left(\dot{q}^2 + \dot{r}^2\right) + y_g(\dot{p}\dot{q} - \ddot{r}) + z_g(\dot{p}\dot{r} + \ddot{q})\right]$$

$$Y = M\left[\ddot{v} + \dot{r}\dot{u} - \dot{p}\dot{w} - y_g\left(\dot{r}^2 + \dot{p}^2\right) + z_g(\dot{q}\dot{r} - \ddot{p}) + x_g(\dot{p}\dot{q} + \ddot{r})\right]$$

Probability and Mechanics of Ship Collision and Grounding. https://doi.org/10.1016/B978-0-12-815022-1.00002-5

$$Z = M\left[\dot{w} + p\dot{v} - \dot{q}\dot{u} - z_g\left(\dot{p}^2 + \dot{q}^2\right) + x_g(\dot{p}\dot{r} - \ddot{q}) + y_g(\dot{r}\dot{q} + \ddot{p})\right]$$

$$K = I_x\ddot{p} + \left(I_z - I_y\right)\dot{q}\dot{r} + M\left[y_g(\ddot{w} + \dot{p}\dot{v} - \dot{q}\dot{u}) - z_g(\ddot{v} + \dot{r}\dot{u} - \dot{p}\dot{w})\right]$$

$$L = I_y\ddot{q} + (I_x - I_z)\dot{p}\dot{r} + M\left[z_g(\ddot{u} + \dot{q}\dot{w} - \dot{r}\dot{v}) - x_g(\ddot{w} + \dot{p}\dot{v} - \dot{q}\dot{u})\right]$$

$$N = I_z\ddot{r} + \left(I_y - I_x\right)\dot{p}\dot{q} + M\left[x_g(\ddot{v} + \dot{r}\dot{u} - \dot{p}\dot{w}) - y_g(\ddot{u} + \dot{q}\dot{w} - \dot{r}\dot{v})\right]$$

where

X, Y, and Z are the external forces on the ship along the three axes;
K, L, and N are the external moments experienced at the origin;
u, v, and w are the respective displacement in the x-, y-, and z-directions;
p, q, and r are the respective rotations about x-, y-, and z-directions;
x_g, y_g, and z_g are the coordinates for the centre of gravity (COG);
I_x, I_y, and I_z are the moments of inertia about the axes through the origin;
M is the mass of the considered ship;
$(\dot{\cdot}) = d/dt$ denotes the derivation with respect to time.

The external forces X, Y, and N acting in the water plane on the ships during a collision are the results of propeller forces, rudder forces, wave-induced forces, forces due to current, hydrodynamic forces, and collision forces arising from the deformation of the material. However, the primary contributions to the external forces and moments originate from the frequency-dependent hydrodynamic loads and the collision forces, and these forces will normally dominate to such an extent that other forces can be neglected.

Based on a set of equations, similar to the equations given above for the rigid ship, Petersen (1982) demonstrated a coupled simulation procedure where the collision forces were modelled by four nonlinear springs and a strip theory was applied to model the time variation of the hydrodynamic forces. Later, Brown (2002) and Le Sourne (2007) coupled the internal and the external analysis through an explicit FEA simulation method in order to obtain a more realistic modelling of the structural damage. For supply ship collisions against rigid plane walls, Yu and Amdahl (2016) derived a coupled simulation procedure where the hydrodynamic loads were calculated through a ship manoeuvring model and the internal collision forces were derived by the use of an explicit FEA procedure. Coupled six DOF calculations with consistent, updated hydrodynamic forces and detailed calculations of the crushing behaviour of the impacted ships are highly demanding in time for modelling and for computer time.

During a collision, the most significant motions of the involved ship hulls will be in the waterplane. If there is no coupling between the motion in the waterplane and the vertical motions and, as indicated in Fig. 2.1, the coordinate systems have their origins in the centreline, that is, X-axes oriented longitudinally positively forwards, Y-axes transversely positively to the port side, and Z-axes positively upwards from the ships, then the equations for horizontal motions take the simple form:

$$X = M\left[\ddot{u} - \dot{\omega}\dot{v} - x_g\dot{\omega}^2\right]$$

$$Y = M\left[\ddot{v} + \dot{\omega}\dot{u} + x_g\ddot{\omega}\right]$$

$$N = I_z\ddot{\omega} + M\left[x_g\left(\ddot{v} + \dot{\omega}\dot{u}\right)\right]$$

where the generalized forces X, Y, and N are composed of hydrodynamic forces and crushing forces

$$X = X^H(t) + X^C(t)$$

$$Y = Y^H(t) + Y^C(t)$$

$$N = N^H(t) + N^C(t)$$

For a ship that is symmetrical about a plane through the longitudinal axis, the time variation of the hydrodynamic pressure forces acting on the hull can, for example, be expressed as

$$X^H(t) = -M_x\ddot{u} + M_y\dot{v}\dot{\omega} + J_{y\omega}\dot{\omega}^2 - \int_0^\infty h_x(\tau)\dot{\omega}(t-\tau)d\tau$$

$$Y^H(t) = -M_y\ddot{v} - M_x\dot{u}\dot{\omega} + J_{y\omega}\ddot{\omega} - \int_0^\infty h_y(\tau)\dot{v}(t-\tau)d\tau - \int_0^\infty h_{y\omega}(\tau)\dot{\omega}(t-\tau)d\tau$$

$$N^H(t) = -J_{y\omega}\ddot{v} - J_{\omega}\ddot{\omega} - J_{y\omega}\dot{u}\dot{\omega} - \left(M_y - M_x\right)\dot{u}\dot{v} + - \int_0^\infty h_{y\omega}(\tau)\dot{v}(t-\tau)d\tau$$
$$- \int_0^\infty h_{\omega\omega}(\tau)\dot{\omega}(t-\tau)d\tau$$

where the terms with constant coefficients express the instantaneous hydrodynamic pressure forces when the free surface effects are absent. Due to the free surface, the ship will set up waves when it accelerates. These waves will cause variations in the hydrodynamic pressure forces that are functions of the earlier motion of the ship. These memory effects are taken into account by the convolution integrals in the above equations.

The constant, added mass coefficients M_α are equal to the frequency-dependent added masses, for infinite frequency:

$$M_\alpha = \lim_{\omega \to \infty} a_{\alpha\alpha}(\omega)$$

and the unit response functions $h_{\alpha\beta}(t)$ can be found by cosine transformation of the frequency-dependent damping coefficients $b_{\alpha\beta}(\omega)$:

$$h_{\alpha\beta} = \frac{2}{\pi} \int_0^\infty b_{\alpha\beta}(\omega) \cos(\omega t) d\omega$$

It will be shown later that when the hydrodynamic forces are approximated by suitable constant added mass terms, reasonable accurate response results can be obtained without going through the computationally demanding inclusion of the exact time-varying procedures. That is, uncoupled procedures will most often be sufficient.

2.1.2 Uncoupled analytical procedures

Analytical methods based on classical mechanics procedures to determine the energy released to provoke structural damage in any given ship collision scenario are preferred for probabilistic models of ship collisions as they are relatively simple and fast.

The analytical method to be presented in Section 2.2 includes the ship motions in the plane of the water surface, and it is valid for arbitrary ship–ship collisions, including different impact locations and collision angles. The analysis procedure is based on rigid body mechanics, where it is assumed that there is negligible strain energy for deformation outside the contact region and that the contact region is local. This implies that the collision can be considered as instantaneous and that each body is assumed to exert an impulsive force on the other at the point of contact. The model can include friction between the impacting surfaces to identify situations with glancing blows. At the start of the calculation, the ships involved in the collision can have forward and sway speeds, and the influence of the hydrodynamic forces due to the sudden acceleration of the ships is approximated by simple constant added mass coefficients.

One important feature of this analytical model is that it is so fast that it can be used together with a Monte Carlo-based ship collision probability model to calculate collision consequence probabilities on various shipping routes.

Since the analytical method to be presented only considers the ship motions in the waterplane, the effect of the roll motions of the struck ship is estimated separately. However, the results show the effect of the roll motion of the struck ship is small, and therefore, it can most often be neglected.

To provide confidence when using the present calculation procedures in engineering applications, validation with numerical and experimental data is presented, including (1) time-domain simulation results reported by Petersen (1982), (2) experimental results summarized by Tabri et al. (2008, 2009, 2010), and (3) two full-scale ship collision experiments reported by Wevers and Vredeveldt (1998) and Wolf (2003). The good agreement with those results indicates that the applied rigid body mechanics procedure is sufficiently accurate for conducting analysis of ship–ship collisions. These experimental results are summarized not only to validate the analytical method but also to provide information and data for better understanding of dynamic ship collision.

The analytical methods presented here are mainly based on the work carried out by Zhang (1999) and Pedersen and Zhang (1998). A similar six DOF analytical uncoupled procedure has been presented by Liu and Amdahl (2010).

2.2 Analytical method for ship–ship collisions

2.2.1 Mathematical modelling and analytical formulation

The collision scenario considers that two ships collide each other where **Ship A** sails with a forward speed V_{ax} and a sway speed V_{ay} and **Ship B** sails with a forward speed V_{b1} and a sway speed V_{b2}.

An *XYZ*-coordinate system is fixed to the sea bottom. The *Z*-axis points in the direction out of the water surface; the X-axis lies in the symmetry plane of Ship A pointing towards the bow; and the origin of the *XYZ* system is placed so that the midship section is in the *YZ* plane at the moment of contact, $t=0$. The origin of a $\xi\eta$-system is located at the impact point C, where the ξ-direction is normal to the impact surface; the angle between the *X*-axis and the η-axis is α; and the collision angle between the two ships is β, that is, between the *X*-axis and the 1-axis, as shown in Fig. 2.1.

2.2.1.1 Motions of Ship A

The equations of motion of Ship A due to the impact force components F_ξ in the ξ-direction and F_η in the η-direction can be expressed as

$$M_a(1+m_{ax})\dot{v}_{ax} = -F_\xi \sin\alpha - F_\eta \cos\alpha \tag{2.1}$$

$$M_a\left(1+m_{ay}\right)\dot{v}_{ay} = -F_\xi \cos\alpha + F_\eta \sin\alpha \tag{2.2}$$

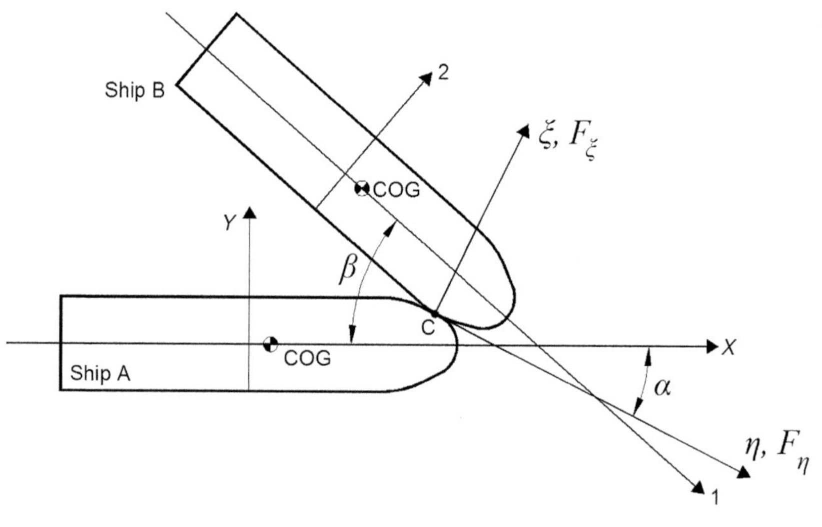

Fig. 2.1 The coordinate system used for analysis of ship–ship collisions (COG is the centre of gravity of the ships).

$$M_aR_a^2(1+j_a)\dot{\omega}_a = F_\xi[y_c\sin\alpha - (x_c - x_a)\cos\alpha] + F_\eta[y_c\cos\alpha + (x_c - x_a)\sin\alpha] \tag{2.3}$$

where M_a is the mass of Ship A, $(\dot{v}_{ax}\dot{v}_{ay})$ denote the accelerations during the collisions of Ship A in the X- and Y-directions, and $\dot{\omega}_a$ is the rotation around the COG. The radius of the ship mass inertia around the COG is R_a, the coordinate of the COG of Ship A is $(x_a, 0)$, the coordinate of the impact point is (x_c, y_c), the added mass coefficient for the surge motion is m_{ax}, the added mass coefficient for the sway motion is m_{ay}, and the added mass coefficient of the moment for the rotation around the centre of the gravity is j_a.

From Eqs (2.1)–(2.3), the accelerations at the COG of Ship A are expressed as

$$\dot{v}_{ax} = -\frac{\sin\alpha}{M_a(1+m_{ax})}F_\xi - \frac{\cos\alpha}{M_a(1+m_{ax})}F_\eta$$

$$\dot{v}_{ay} = -\frac{\cos\alpha}{M_a(1+m_{ay})}F_\xi + \frac{\sin\alpha}{M_a(1+m_{ay})}F_\eta$$

$$\dot{\omega}_a = \frac{y_c\sin\alpha - (x_c - x_a)\cos\alpha}{M_aR_a^2(1+j_a)}F_\xi + \frac{y_c\cos\alpha + (x_c - x_a)\sin\alpha}{M_aR_a^2(1+j_a)}F_\eta$$

The acceleration at the impact point C of Ship A in the ξ-direction can be expressed as

$$\begin{aligned}\ddot{\xi}_a &= (\dot{v}_{ax} - \dot{\omega}_a y_c)\sin\alpha + [\dot{v}_{ay} + \dot{\omega}_a(x_c - x_a)]\cos\alpha \\ &= -\frac{F_\xi}{M_a}\left\{\frac{1}{1+m_{ax}}\sin^2\alpha + \frac{1}{1+m_{ay}}\cos^2\alpha + \frac{1}{1+j_a}\cdot\frac{[y_c\sin\alpha - (x_c - x_a)\cos\alpha]^2}{R_a^2}\right\} \\ &\quad -\frac{F_\eta}{M_a}\left\{\left[\frac{1}{1+m_{ax}} - \frac{1}{1+m_{ay}}\right]\sin\alpha\cos\alpha + \frac{1}{1+j_a}\cdot\frac{[y_c\sin\alpha - (x_c - x_a)\cos\alpha][y_c\cos\alpha + (x_c - x_a)\sin\alpha]}{R_a^2}\right\}\end{aligned}$$

Similarly, the acceleration of Ship A at the impact point C in the η-direction can be written as

$$\begin{aligned}\ddot{\eta}_a &= [\dot{v}_{ax} - \dot{\omega}_a y_c]\cos\alpha - [\dot{v}_{ay} + \dot{\omega}_a(x_c - x_a)]\sin\alpha \\ &= -\frac{F_\xi}{M_a}\left\{\left[\frac{1}{1+m_{ax}} - \frac{1}{1+m_{ay}}\right]\sin\alpha\cos\alpha + \frac{1}{1+j_a}\cdot\frac{[y_c\sin\alpha - (x_c - x_a)\cos\alpha][y_c\cos\alpha + (x_c - x_a)\sin\alpha]}{R_a^2}\right\} \\ &\quad -\frac{F_\eta}{M_a}\left\{\frac{1}{1+m_{ax}}\cos^2\alpha + \frac{1}{1+m_{ay}}\sin^2\alpha + \frac{1}{1+j_a}\cdot\frac{[y_c\cos\alpha + (x_c - x_a)\sin\alpha]^2}{R_a^2}\right\}\end{aligned}$$

2.2.1.2 Motions of Ship B

The motions of Ship B can be expressed as

$$M_b(1+m_{b1})\dot{v}_{b1} = -F_\xi\sin(\beta - \alpha) + F_\eta\cos(\beta - \alpha) \tag{2.4}$$

$$M_b(1+m_{b2})\dot{v}_{b2} = F_\xi\cos(\beta - \alpha) + F_\eta\sin(\beta - \alpha) \tag{2.5}$$

$$M_b R_b^2(1+j_b)\dot{\omega}_b = -F_\xi[(y_c-y_b)\sin\alpha-(x_c-x_b)\cos\alpha] \\ -F_\eta[(y_c-y_b)\cos\alpha+(x_c-x_b)\sin\alpha] \quad (2.6)$$

where the mass of Ship B is M_b, the radius of the ship mass inertia around the COG is R_b, the coordinate of the COG of Ship B is (x_b, y_b), the added mass coefficient for the surge motion is m_{b1}, the added mass coefficient for the sway motion is m_{b2}, and the added mass coefficient of the moment for the rotation around the COG is j_b.

From Eqs (2.4)–(2.6), the accelerations of Ship B at the impact point C and in the ξ-direction and in the η-direction are expressed as

$$\ddot{\xi}_b = \dot{v}_{b1}\sin(\beta-\alpha)+\dot{v}_{b2}\cos(\beta-\alpha)-\dot{\omega}_b[(y_c-y_b)\sin\alpha-(x_c-x_b)\cos\alpha]$$

$$=\frac{F_\xi}{M_b}\left\{\frac{1}{1+m_{b1}}\sin^2(\beta-\alpha)+\frac{1}{1+m_{b2}}\cos^2(\beta-\alpha)+\frac{1}{1+j_b}\cdot\frac{[(y_c-y_b)\sin\alpha-(x_c-x_b)\cos\alpha]^2}{R_b^2}\right\}$$

$$+\frac{F_\eta}{M_b}-\frac{1}{1+m_{b1}}\sin(\beta-\alpha)\cos(\beta-\alpha)+\frac{1}{1+m_{b2}}\sin(\beta-\alpha)\cos(\beta-\alpha)$$

$$+\frac{1}{1+j_b}\cdot\frac{[(y_c-y_b)\sin\alpha-(x_c-x_b)\cos\alpha][(y_c-y_b)\cos\alpha+(x_c-x_b)\sin\alpha]}{R_b^2}\Bigg\}$$

$$\ddot{\eta}_b = -[\dot{v}_{bx}-\dot{\omega}_b(y_c-y_b)]\cos\alpha+[\dot{v}_{by}+\dot{\omega}_b(x_c-x_b)]\sin\alpha$$

$$=\frac{F_\xi}{M_b}\left\{-\frac{1}{1+m_{b1}}\sin(\beta-\alpha)\cos(\beta-\alpha)+\frac{1}{1+m_{b2}}\sin(\beta-\alpha)\cos(\beta-\alpha)\right.$$

$$\left.+\frac{1}{1+j_b}\cdot\frac{[(y_c-y_b)\sin\alpha-(x_c-x_b)\cos\alpha][(y_c-y_b)\cos\alpha+(x_c-x_b)\sin\alpha]}{R_b^2}\right\}$$

$$+\frac{F_\eta}{M_b}\left\{\frac{1}{1+m_{b1}}\cos^2(\beta-\alpha)+\frac{1}{1+m_{b2}}\sin^2(\beta-\alpha)+\frac{1}{1+j_b}\cdot\frac{((y_c-y_b)\cos\alpha+(x_c-x_b)\sin\alpha)^2}{R_b^2}\right\}$$

2.2.1.3 Relative motions of the Ship A and the Ship B

The relative accelerations of Ship A and Ship B at the impact point C are found to be

$$\ddot{\xi}=\ddot{\xi}_a-\ddot{\xi}_b=-\left[\frac{D_{a\xi}}{M_a}+\frac{D_{b\xi}}{M_b}\right]\cdot F_\xi-\left[\frac{D_{a\eta}}{M_a}+\frac{D_{b\eta}}{M_b}\right]\cdot F_\eta \quad (2.7)$$

$$\ddot{\eta}=\ddot{\eta}_a-\ddot{\eta}_b=-\left[\frac{K_{a\xi}}{M_a}+\frac{K_{b\xi}}{M_b}\right]\cdot F_\xi-\left[\frac{K_{a\eta}}{M_a}+\frac{K_{b\eta}}{M_b}\right]\cdot F_\eta \quad (2.8)$$

where

$$D_{a\xi}=\frac{1}{1+m_{ax}}\sin^2\alpha+\frac{1}{1+m_{ay}}\cos^2\alpha+\frac{1}{1+j_a}\cdot\frac{[y_c\sin\alpha-(x_c-x_a)\cos\alpha]^2}{R_a^2}$$

$$D_{a\eta} = \frac{1}{1+m_{ax}} \sin\alpha\cos\alpha - \frac{1}{1+m_{ay}} \sin\alpha\cos\alpha + \frac{1}{1+j_a} \cdot \frac{[y_c \sin\alpha - (x_c - x_a)\cos\alpha][y_c \cos\alpha + (x_c - x_a)\sin\alpha]}{R_a^2}$$

$$D_{b\xi} = \frac{1}{1+m_{b1}} \sin^2(\beta-\alpha) + \frac{1}{1+m_{b2}} \cos^2(\beta-\alpha) + \frac{1}{1+j_b} \cdot \frac{[(y_c - y_b)\sin\alpha - (x_c - x_b)\cos\alpha]^2}{R_b^2}$$

$$D_{b\eta} = -\frac{1}{1+m_{b1}} \sin(\beta-\alpha)\cos(\beta-\alpha) + \frac{1}{1+m_{b2}} \sin(\beta-\alpha)\cos(\beta-\alpha) + \frac{1}{1+j_b} \cdot \frac{[(y_c - y_b)\sin\alpha - (x_c - x_b)\cos\alpha][(y_c - y_b)\cos\alpha + (x_c - x_b)\sin\alpha]}{R_b^2}$$

$$K_{a\xi}3 = \frac{1}{1+m_{ax}} \sin\alpha\cos\alpha - \frac{1}{1+m_{ay}} \sin\alpha\cos\alpha + \frac{1}{1+j_a} \cdot \frac{[y_c \sin\alpha - (x_c - x_a)\cos\alpha][y_c \cos\alpha + (x_c - x_a)\sin\alpha]}{R_a^2}$$

$$K_{a\eta} = \frac{1}{1+m_{ax}} \cos^2\alpha + \frac{1}{1+m_{ay}} \sin^2\alpha + \frac{1}{1+j_a} \cdot \frac{[y_c \cos\alpha + (x_c - x_a)\sin\alpha]^2}{R_a^2}$$

$$K_{b\xi} = -\frac{1}{1+m_{b1}} \sin(\beta-\alpha)\cos(\beta-\alpha) + \frac{1}{1+m_{b2}} \sin(\beta-\alpha)\cos(\beta-\alpha) + \frac{1}{1+j_b} \cdot \frac{[(y_c - y_b)\sin\alpha - (x_c - x_b)\cos\alpha][(y_c - y_b)\cos\alpha + (x_c - x_b)\sin\alpha]}{R_b^2}$$

$$K_{b\eta} = \frac{1}{1+m_{b1}} \cos^2(\beta-\alpha) + \frac{1}{1+m_{b2}} \sin^2(\beta-\alpha) + \frac{1}{1+j_b} \cdot \frac{[(y_c - y_b)\cos\alpha + (x_c - x_b)\sin\alpha]^2}{R_b^2}$$

The simplified forms of Eqs (2.7), (2.8) are as follows:

$$\ddot{\xi} = -D_\xi \cdot F_\xi - D_\eta \cdot F_\eta \tag{2.9}$$

$$\ddot{\eta} = -K_\xi \cdot F_\xi - K_\eta \cdot F_\eta \tag{2.10}$$

where

$D_\xi = \frac{D_{a\xi}}{M_a} + \frac{D_{b\xi}}{M_b}$, $D_\eta = \frac{D_{a\eta}}{M_a} + \frac{D_{b\eta}}{M_b}$, $K_\xi = \frac{K_{a\xi}}{M_a} + \frac{K_{b\xi}}{M_b}$, *and* $K_\eta = \frac{K_{a\eta}}{M_a} + \frac{K_{b\eta}}{M_b}$

2.2.1.4 Relative velocities before and at the end of the collision

To solve Eqs (2.9), (2.10), we need to know the relative velocities of the two ships before the collision and at the end of the collision. At the beginning of the collision, time $t=0$, the relative velocities of Ship A and Ship B at impact point C and in the ξ-direction and the η-direction can be determined from the initial velocities of the two ships:

$$\dot{\xi}(0)=\dot{\xi}_a(0)-\dot{\xi}_b(0)=V_{ax}\sin\alpha+V_{ay}\cos\alpha+V_{b1}\sin(\beta-\alpha)-V_{b2}\cos(\beta-\alpha) \tag{2.11}$$

$$\dot{\eta}(0)=\dot{\eta}_a(0)-\dot{\eta}_b(0)=V_{ax}\cos\alpha-V_{ay}\sin\alpha-V_{b1}\cos(\beta-\alpha)-V_{b2}\sin(\beta-\alpha) \tag{2.12}$$

where V_{ax} is the surge speed of Ship A, V_{ay} is the sway speed of Ship A, V_{b1} is the surge speed of Ship B, and V_{b2} is the sway speed of Ship B.

At the end of the collision, $t=t_\Delta$, we assume that the ships may rebound from each other, in the ξ-direction, that is,

$$\dot{\xi}(t_\Delta)=\dot{\xi}_a(t_\Delta)-\dot{\xi}_b(t_\Delta)=-e\cdot\dot{\xi}(0) \tag{2.13}$$

Here, e $(0\le e\le 1)$ is the coefficient of restitution. For an entirely plastic collision, $e=0$, and for a perfectly elastic collision, $e=1.0$. For the case where the two ships are locked together at the end of a collision, the relative velocities can be written as

$$\begin{aligned}\dot{\xi}(t_\Delta)&=0\\ \dot{\eta}(t_\Delta)&=0\end{aligned} \tag{2.14}$$

In the following sections, we will discuss the case of glancing blows, that is, when the two ships can slide against each other during a collision.

2.2.1.5 Determination of the impact impulse

By integration of equations (2.9) and (2.10) with respect to time and by the use of equations (2.11)–(2.14), the impact impulse in the ξ-direction and in the η-direction can be derived as

$$I_{\xi 0}=\int_0^{\Delta t}F_\xi dt=\frac{K_\eta\dot{\xi}(0)(1+e)-D_\eta\dot{\eta}(0)}{D_\xi\cdot K_\eta-D_\eta\cdot K_\xi}$$

$$I_{\eta 0}=\int_0^{\Delta t}F_\eta dt=\frac{D_\xi\dot{\eta}(0)-K_\xi\dot{\xi}(0)(1+e)}{D_\xi\cdot K_\eta-D_\eta\cdot K_\xi}$$

Thus, the ratio of the impact impulses is expressed as

$$\mu = \frac{I_{\eta 0}}{I_{\xi 0}} = \frac{D_{\xi}\dot{\eta}(0) - K_{\xi}\dot{\xi}(0)(1+e)}{K_{\eta}\dot{\xi}(0)(1+e) - D_{\eta}\dot{\eta}(0)} \tag{2.15}$$

If the collision angle is small or very large, the ships may glance against each other. Therefore, it is necessary to consider an impact with friction. The effective coefficient of friction resistances between the ships is assumed to be μ_0. Then, we have the following:

If $|\mu_0| \geq |\mu|$, the two ships stick together.

If $|\mu_0| < |\mu|$, the two ships slide against each other (however, if the side shell of the struck ship ruptures, the two ships will stick together).

In order to determine the energy to be dissipated by crushing of the ship structures, we further make the assumption that the ratio between the parallel component F_η of the collision force and the perpendicular component F_ξ is constant during the collision, that is, $F_\eta = \mu \cdot F_\xi$. Then, Eqs (2.9), (2.10) can be written as

$$\ddot{\xi} = -\left(D_\xi + \mu D_\eta\right) \cdot F_\xi \tag{2.16}$$

$$\ddot{\eta} = -\left(\frac{1}{\mu}K_\xi + K_\eta\right) \cdot F_\eta \tag{2.17}$$

2.2.1.6 Energy released for crushing of ship structures

Before determining the energy to be dissipated by crushing of the ship structures, we must know the relative velocity $\dot{\eta}(t_\Delta)$ at the end of the collision for the case of sliding. By integrating Eq. (2.17) with respect to time (now $\mu = \mu_0$), we get the glancing speed at the end of the collision:

$$\dot{\eta}(t_\Delta) = \dot{\eta}(0) - \frac{K_\xi + \mu_0 K_\eta}{D_\xi + \mu_0 D_\eta} \cdot \dot{\xi}(0)(1+e) \tag{2.18}$$

By the use of the transformations $\ddot{\xi} = \frac{d\dot{\xi}}{d\xi}\frac{d\xi}{dt} = \dot{\xi}\frac{d\dot{\xi}}{d\xi}$ and $\ddot{\eta} = \frac{d\dot{\eta}}{d\eta}\frac{d\eta}{dt} = \dot{\eta}\frac{d\dot{\eta}}{d\eta}$, Eqs (2.16), (2.17) become

$$\dot{\xi}d\dot{\xi} = -\left(D_\xi + \mu D_\eta\right) \cdot F_\xi d\xi \tag{2.19}$$

$$\dot{\eta}d\dot{\eta} = -\left(\frac{1}{\mu}K_\xi + K_\eta\right) \cdot F_\eta d\eta \tag{2.20}$$

By integrating Eqs (2.19), (2.20) and using the conditions given above, we obtain the results of the energy released for crushing of the ship structures and the impact impulses for the ship sticking case and the sliding case.

Sticking case: the energy released in the ξ-direction E_ξ and the energy released in the η-direction E_η can be expressed as

$$E_\xi = \int_0^{\xi_{\max}} F_\xi d\xi = \frac{1}{2}\frac{1}{D_\xi + \mu \cdot D_\eta}\dot{\xi}(0)^2 \tag{2.21}$$

$$E_\eta = \int_0^{\eta_{\max}} F_\eta d\eta = \frac{1}{2}\frac{1}{\frac{1}{\mu}\cdot K_\xi + K_\eta}\dot{\eta}(0)^2 \tag{2.22}$$

where $\xi_{\max}$ and $\eta_{\max}$ are the respective maximum relative penetrations in the ξ-direction and in the η-direction at the end of the collision. The total released energy is the sum of the energy released in the ξ-direction and in the η-direction: $E_{total} = E_\xi + E_\eta$.

The components of the impact impulse in the ξ-direction and in the η-direction can be calculated from

$$I_\xi = \int_0^{\Delta t} F_\xi dt = \frac{1}{D_\xi + \mu \cdot D_\eta}\dot{\xi}(0) \tag{2.23}$$

$$I_\eta = \int_0^{\Delta t} F_\eta dt = \frac{1}{\frac{1}{\mu}K_\xi + K_\eta}\dot{\eta}(0) \tag{2.24}$$

Sliding case (if $|\mu_0| < |\mu|$ and E_ξ of Eq. (2.21) is less than the rupture energy $E_{\xi CR}$): the energy released in the ξ-direction E_ξ and the energy released in the η-direction E_η can be expressed as

$$E_\xi = \int_0^{\xi_{\max}} F_\xi d\xi = \frac{1}{2}\frac{1}{D_\xi + \mu_0 \cdot D_\eta}\dot{\xi}(0)^2\left(1 - e^2\right) \tag{2.25}$$

$$E_\eta = \int_0^{\eta_{\max}} F_\eta d\eta = \frac{1}{2}\frac{1}{\frac{1}{\mu_0}\cdot K_\xi + K_\eta}\left(\dot{\eta}(0)^2 - \dot{\eta}(t_\Delta)^2\right) \tag{2.26}$$

The components of the impact impulse in the glancing case are

$$I_\xi = \int_0^{\Delta t} F_\xi dt = \frac{1}{D_\xi + \mu_0 \cdot D_\eta}\dot{\xi}(0)(1 + e) \tag{2.27}$$

$$I_\eta = \int_0^{\Delta t} F_\eta dt = \frac{1}{\frac{1}{\mu_0} K_\xi + K_\eta} (\dot{\eta}(0) - \dot{\eta}(t_\Delta)) \tag{2.28}$$

2.2.1.7 Velocities at the end of the collision

At the end of a collision, the two ships still undergo translational and rotational motions. The equations of conservation of momentum for Ship A are expressed as

$$M_a(1+m_{ax})(v_{ax} - V_{ax}) = -I_x$$

$$M_a(1+m_{ay})(v_{ay} - V_{ay}) = -I_y$$

$$M_a R_a^2 (1+j_a) \cdot \omega_a = -I_x y_c + I_y (x_c - x_a)$$

where I_x and I_y are the respective components of the impact impulses in the X-direction and the Y-direction.

Similarly, the equations of conservation of momentum for Ship B can be expressed as

$$M_b(1+m_{b1})(v_{b1} - V_{b1}) = I_1$$

$$M_b(1+m_{b2})(v_{b2} - V_{b2}) = I_2$$

$$M_b R_b^2 (1+j_b) \cdot \omega_b = -I_x (y_c - y_b) + I_y (x_c - x_b)$$

where I_1 and I_2 are the respective components of the impact impulse in the 1-direction and the 2-direction.

By application of the transformations,

$$I_x = I_\xi \sin\alpha + I_\eta \cos\alpha$$

$$I_y = I_\xi \cos\alpha - I_\eta \sin\alpha$$

$$I_1 = I_\xi \sin(\beta - \alpha) - I_\eta \cos(\beta - \alpha)$$

$$I_2 = -I_\xi \cos(\beta - \alpha) - I_\eta \sin(\beta - \alpha)$$

The expressions for the velocities of Ship A at the end of the collision are found to be

$$v_{ax} = V_{ax} - \frac{I_\xi \sin\alpha + I_\eta \cos\alpha}{M_a(1+m_{ax})} \tag{2.29}$$

$$v_{ay} = V_{ay} - \frac{I_\xi \cos\alpha - I_\eta \sin\alpha}{M_a(1+m_{ay})} \tag{2.30}$$

$$\omega_a = -\frac{I_\xi[y_c \cdot \sin\alpha - (x_c - x_a)\cos\alpha] + I_\eta[y_c \cdot \cos\alpha + (x_c - x_a)\sin\alpha]}{M_a R_a^2 (1 + j_a)} \tag{2.31}$$

The velocities of Ship B at the end of the collision are

$$v_{b1} = V_{b1} - \frac{I_\xi \sin(\beta - \alpha) - I_\eta \cos(\beta - \alpha)}{M_b(1 + m_{b1})} \tag{2.32}$$

$$v_{b2} = V_{b2} + \frac{I_\xi \cos(\beta - \alpha) + I_\eta \sin(\beta - \alpha)}{M_b(1 + m_{b2})} \tag{2.33}$$

$$\omega_b = \frac{I_\xi[(y_c - y_b)\sin\alpha - (x_c - x_b)\cos\alpha] + I_\eta[(y_c - y_b)\cos\alpha + (x_c - x_b)\sin\alpha]}{M_b R_b^2 (1 + j_b)}$$

2.2.2 Summary of the formulations for ship–ship collision analysis

This section summarizes the equations required to conduct a ship–ship collision analysis for both the sticking case and the sliding case.

Sticking case:

$$E_\xi = \int_0^{\xi_{max}} F_\xi d\xi = \frac{1}{2} \frac{1}{D_\xi + \mu \cdot D_\eta} \dot{\xi}(0)^2$$

$$E_\eta = \int_0^{\eta_{max}} F_\eta d\eta = \frac{1}{2} \frac{1}{\frac{1}{\mu} \cdot K_\xi + K_\eta} \dot{\eta}(0)^2$$

$$I_\xi = \int_0^{\Delta t} F_\xi dt = \frac{1}{D_\xi + \mu \cdot D_\eta} \dot{\xi}(0)$$

$$I_\eta = \int_0^{\Delta t} F_\eta dt = \frac{1}{\frac{1}{\mu} K_\xi + K_\eta} \dot{\eta}(0)$$

Sliding case (if $|\mu_0| < |\mu|$ and E_ξ of Eq. (2.21) is less than the rupture energy $E_{\xi CR}$):

$$E_\xi = \int_0^{\xi_{max}} F_\xi d\xi = \frac{1}{2} \frac{1}{D_\xi + \mu_0 \cdot D_\eta} \dot{\xi}(0)^2 (1 - e^2)$$

$$E_\eta = \int_0^{\eta_{\max}} F_\eta d\eta = \frac{1}{2}\frac{1}{\frac{1}{\mu_0}\cdot K_\xi + K_\eta}\left(\dot{\eta}(0)^2 - \dot{\eta}(t_\Delta)^2\right)$$

$$I_\xi = \int_0^{\Delta t} F_\xi dt = \frac{1}{D_\xi + \mu_0 \cdot D_\eta}\dot{\xi}(0)(1+e)$$

$$I_\eta = \int_0^{\Delta t} F_\eta dt = \frac{1}{\frac{1}{\mu_0}K_\xi + K_\eta}(\dot{\eta}(0) - \dot{\eta}(t_\Delta))$$

where

$$D_\xi = \frac{D_{a\xi}}{M_a} + \frac{D_{b\xi}}{M_b},\ D_\eta = \frac{D_{a\eta}}{M_a} + \frac{D_{b\eta}}{M_b},\ K_\xi = \frac{K_{a\xi}}{M_a} + \frac{K_{b\xi}}{M_b},\ K_\eta = \frac{K_{a\eta}}{M_a} + \frac{K_{b\eta}}{M_b}$$

$$D_{a\xi} = \frac{1}{1+m_{ax}}\sin^2\alpha + \frac{1}{1+m_{ay}}\cos^2\alpha + \frac{1}{1+j_a}\cdot\frac{[y_c\sin\alpha - (x_c - x_a)\cos\alpha]^2}{R_a{}^2}$$

$$D_{a\eta} = \frac{1}{1+m_{ax}}\sin\alpha\cos\alpha - \frac{1}{1+m_{ay}}\sin\alpha\cos\alpha + \frac{1}{1+j_a}\cdot\frac{[y_c\sin\alpha - (x_c - x_a)\cos\alpha][y_c\cos\alpha + (x_c - x_a)\sin\alpha]}{R_a{}^2}$$

$$D_{b\xi} = \frac{1}{1+m_{b1}}\sin^2(\beta-\alpha) + \frac{1}{1+m_{b2}}\cos^2(\beta-\alpha) + \frac{1}{1+j_b}\cdot\frac{[(y_c - y_b)\sin\alpha - (x_c - x_b)\cos\alpha]^2}{R_b{}^2}$$

$$D_{b\eta} = -\frac{1}{1+m_{b1}}\sin(\beta-\alpha)\cos(\beta-\alpha) + \frac{1}{1+m_{b2}}\sin(\beta-\alpha)\cos(\beta-\alpha) + \frac{1}{1+j_b}\cdot\frac{[(y_c - y_b)\sin\alpha - (x_c - x_b)\cos\alpha][(y_c - y_b)\cos\alpha + (x_c - x_b)\sin\alpha]}{R_b{}^2},$$

$$K_{a\xi} = \frac{1}{1+m_{ax}}\sin\alpha\cos\alpha - \frac{1}{1+m_{ay}}\sin\alpha\cos\alpha + \frac{1}{1+j_a}\cdot\frac{[y_c\sin\alpha - (x_c - x_a)\cos\alpha][y_c\cos\alpha + (x_c - x_a)\sin\alpha]}{R_a{}^2}$$

$$K_{a\eta} = \frac{1}{1+m_{ax}}\cos^2\alpha + \frac{1}{1+m_{ay}}\sin^2\alpha + \frac{1}{1+j_a}\cdot\frac{[y_c\cos\alpha + (x_c - x_a)\sin\alpha]^2}{R_a{}^2}$$

$$K_{b\xi} = -\frac{1}{1+m_{b1}}\sin(\beta-\alpha)\cos(\beta-\alpha) + \frac{1}{1+m_{b2}}\sin(\beta-\alpha)\cos(\beta-\alpha) + \frac{1}{1+j_b}\cdot\frac{[(y_c - y_b)\sin\alpha - (x_c - x_b)\cos\alpha][(y_c - y_b)\cos\alpha + (x_c - x_b)\sin\alpha]}{R_b{}^2}$$

$$K_{b\eta} = \frac{1}{1+m_{b1}}\cos^2(\beta-\alpha) + \frac{1}{1+m_{b2}}\sin^2(\beta-\alpha) + \frac{1}{1+j_b}\cdot\frac{[(y_c - y_b)\cos\alpha + (x_c - x_b)\sin\alpha]^2}{R_b{}^2}$$

2.2.3 Determination of added mass coefficients and effective friction coefficients

2.2.3.1 Added mass coefficients of the surrounding water

The added mass coefficients m_{ax}, m_{ay}, j_a for Ship A and m_{b1}, m_{b2}, j_b for Ship B take into account the interaction effects between the ships and the surrounding water. They depend on the hull form and draught of the ships, as well as the impact duration, among others.

For simplicity, Minorsky (1959) proposed to use a constant value of the added mass coefficient for the sway motion: $m_{ay}=0.4$. Motora et al. (1971) conducted a series of model tests and a hydrodynamic analysis to determine the added mass coefficient for the sway motion. They found that this added mass coefficient varies during the collision and that the value is in the range $m_{ay}=0.4-1.3$. The coefficient depends on the hull shape, and it increases with the impact duration. However, if the collision duration is very short, the value $m_{ay}=0.4$ proposed by Minorsky may be a good approximation. In Petersen and Pedersen (1981), it is shown that the added mass coefficient for the sway motion can be estimated from $m_{ay}=\{m(\infty)+k[m(0)-m(\infty)]\}$, where $m(\infty)$ and $m(0)$ are the threshold values of the added mass coefficient for the sway motion when the frequency of the collision, respectively, approaches infinity or zero. The value of the factor k is a function of the duration of the collision and the ship's draught.

Based on the work conducted by Motora et al. (1971) and Petersen and Pedersen (1981), it is here proposed to determine the equivalent added mass coefficient for sway motions as

$$m_{ay}=\left\{1.1+0.729\left(t_\Delta\sqrt{\frac{g}{B}}\right)-0.088\left(t_\Delta\sqrt{\frac{g}{B}}\right)^2\right\}\cdot\frac{T}{B} \tag{2.34}$$

where t_Δ is the collision duration that can be determined from $t_\Delta=(\pi\xi_{\max})/(2\dot{\xi}(0))$ (see Chapter 3, Section 3.9.1), B is the breadth of the ship, g is the acceleration of gravity ($g=9.81\,\text{m/s}^2$), and T is the draught of the ship.

An example of the equivalent added mass coefficient for sway motion of a ship with breadth $B=19\,\text{m}$ at the full load condition ($T=6.9\,\text{m}$) and the ballast condition ($T=4.5\,\text{m}$) is presented in Fig. 2.2 for reference.

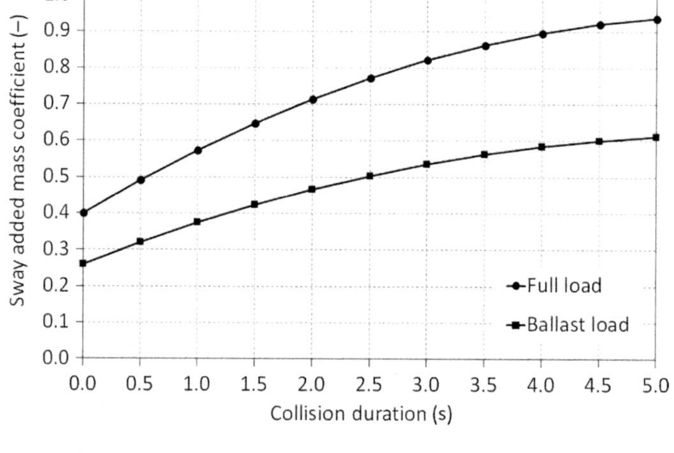

Fig. 2.2 An example of the sway added mass coefficient for a ship of breadth 19 m, in full load and ballast load conditions.

The added mass coefficient m_{ax} related to the forward motion is small compared with the mass of the ship. It is found to be

$$m_{ax}=0.02 \text{ to } 0.07$$

The added mass coefficient for the yaw motion of the ship, j_a, can be taken as

$$j_a=0.21$$

If the lack of detailed information and/or for simplicity, the following data may be used for the added mass coefficients:

$$m_{ax}=m_{b1}=0.05 \quad \text{for the surge motion}$$

$$m_{ay}=m_{b2}=0.85 \quad \text{for the sway motion at full load condition (deep draught)}$$

$$m_{ay}=m_{b2}=0.55 \quad \text{for the sway motion at ballast condition (shallow draught)}$$

$$j_a=j_b=0.21 \quad \text{for the yaw motion}$$

The radius of inertia of the ship can be approximated by the expression suggested by Pianc (1984) $R=(0.11+0.19C_b)L$, where C_b is the block coefficient and L is the ship length. However, for simplicity, the radius of inertia can simply be taken as a quarter of the ship length $R=L/4$.

2.2.3.2 Effective mass coefficients of internal liquid on ship

Tabri et al. (2009) conducted ship collision model tests where the striking ships carried liquid with free surface. The study was performed to investigate the effects of the liquid on the collision energy by comparing with tests of ships carrying the same amount of solid cargo. It was shown that due to the motion of the liquid, the collision energy decreases. This shows that the liquid is not fully effective as a solid weight. Therefore, an effective mass concept was introduced by Zhang et al. (2017) to deal with this situation; they defined an effective mass coefficient m_{in} for internal liquid as the ratio of the effective mass M_{eff} and the total mass M_{total} of the liquid (as solid): $m_{in}=\frac{M_{eff}}{M_{total}}$.

For liquid tank that is fully loaded (100%), the effective mass coefficient is $m_{in}=1.0$.

Zhang and Suzuki (2007) and Lee (2014) studied theoretically and numerically the effect of the liquid cargo by assuming a mass-spring system. Based on these studies and for ship collision analysis purpose, the effective mass coefficient can be determined from

$$m_{in}=1-\frac{\pi}{11}\frac{a}{h}\tanh\left(\frac{h}{a}\pi\right) \tag{2.35}$$

where a is the tank length in the direction of the collision and h is the height of the liquid with free surface. For example, for a typical double-hull Aframax oil tanker of length 234m and an oil cargo tank of length 30m and width 19m with an assumed free surface height of 10m (half of the tank height), the effective mass coefficients are found to be

$$m_{in} = 0.33 \quad \text{for surge motion (note : take tank length } a = 30\text{m in this case)}$$

$$m_{in} = 0.50 \quad \text{for sway motion (note : take tank length } a = 19\text{m in this case)}$$

2.2.3.3 Effective friction coefficient for ship collisions in sliding case

If the collision angle is small or very large, the ships may slide against each other without rupture of the side shell structure. Thus, the effective coefficient of friction resistances μ_0 between the two ships needs to be determined for the collision energy analysis. The general physics on dry friction operates with two values. There is a static coefficient of friction, where the static friction force keeps a body at rest and a dynamic friction that slows down an object sliding on a surface; the static friction is usually higher than the dynamic friction. It should be noted that the coefficients of friction vary significantly for different materials and surfaces. It should also be noted that strictly speaking, then the assumption made in the applied procedure concerning a fixed collision point during the impact is violated for glancing impacts.

For a ship collision analysis, we assume that the static and the dynamic frictions are the same (or similar). Here, the interest is put in determining the ratio between the perpendicular force F_ξ and the parallel force F_η in the sliding case ($F_\eta = \mu_0 F_\xi$). Fig. 2.3 illustrates that for a collision with sliding, the parallel force includes two parts: the friction force and the force to overcome the deformation resistance, where the respective friction coefficients are μ_f and μ_R. Due to the large deformations in the contact area, the friction will also be large and may cause scratching on the surfaces. These coefficients can be taken as

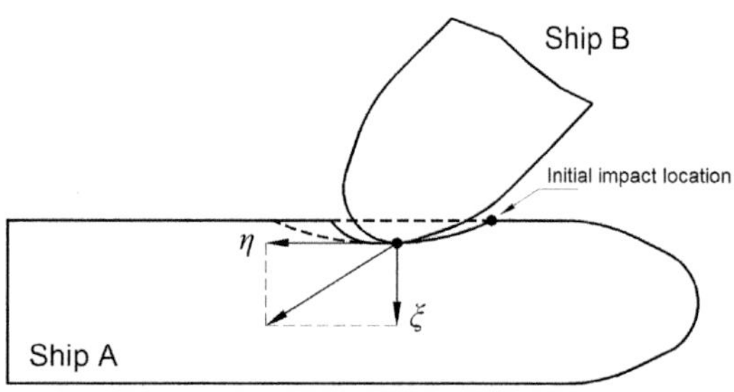

Fig. 2.3 Illustration of the effective friction in a ship collision with sliding case.

$$\mu_f = 0.35 \ to \ 0.45 \text{ and } \mu_R = 0.15 \ to \ 0.2$$

Eventually, the effective coefficient of friction resistance for ship collision analysis in sliding case (steel to steel) can be taken as

$$\mu_0 = \mu_f + \mu_R = 0.50 \ to \ 0.65 \tag{2.36}$$

This approach is similar, in physics, to the grounding experiments on soft sandy beaches performed by Pedersen (1994) and Sterndorff and Pedersen (1996). In those grounding experiments, the average effective friction coefficient (steel to sands) was about 0.6, while the typical friction coefficient between steel and sand is about 0.35. This implies that the determined effective coefficient of friction of 0.6 includes the additional resistance coefficient of about 0.25 on the ship bow from sand pressure in these experiments.

2.2.4 Comparison between the analytical method and numerical simulations

Petersen (1982) carried out ship–ship collision analyses using time-domain simulations taking into account large structural deformations, large rotations, and the effects of the varying hydrodynamic pressure. Two identical ships were considered in the collision simulations using the procedure described in Section 2.1.1. The length of the ships is 116.0m, the breadth is 19.0m, the draught is 6.9m, and the mass is 10,340 tons; the radius of inertia for yaw motion is 29.0m.

The coefficients of the added mass used in the present analytical calculation are taken as 0.05 for surge motion and 0.21 for yaw motion, while the coefficient for sway motion is determined using Eq. (2.34). The impact location x_c is measured from the centre of gravity of the struck ship. The collision scenarios evaluated here are the same in Petersen (1982), and an entirely plastic collision is assumed, that is, the two ships are locked together during the collision. The results are presented in Table 2.1. It can be seen that good agreement is achieved, except for case No. 4. It is difficult to explain the difference for case No. 4 except that the high value determined by Petersen (1982) for this case does not seem to be reasonable.

Liu et al. (2017) have compared the present theory with fully coupled finite element analysis (FEA) of collisions of two identical container ships (length 248m). For the perpendicular collision (collision angle of 90 degrees) and oblique collision (collision angle of 60 degrees), the agreement of the collision energy loss between the two methods is excellent where the difference is less than 3.5%. For the oblique collision with a small collision angle of 30 degrees, the energy loss calculated by the present theory (31.7MJ) is 22% lower than that of the FEA result (40.5MJ), when the same friction coefficient of 0.3 is used for both the FEA and the present theory. In this case, the friction coefficient will affect the calculation results because the collision corresponds to the sliding case. However, the effective friction coefficient should be used in the present theory as discussed in Section 2.2.3. Therefore, if the effective friction

Table 2.1 Comparison between the present analytical method and the time-domain simulations of Petersen (1982).

	Collision parameter					E_ξ(MJ)		E_η(MJ)		E_{total}(MJ)	
Case ID	V_{ax} **(m/s)**	V_{b1} **(m/s)**	**Collision angle** β **(degrees)**	x_c **(m)**	m_{ay} **(–)**	**Petersen (1982)**	**Analytical**	**Petersen (1982)**	**Analytical**	**Petersen (1982)**	**Analytical**
1	0	4.5	90	0.00	0.92	69.60	71.10	0.00	0.00	69.60	71.10
2	4.5	4.5	90	0.00	0.90	64.10	70.84	24.70	21.49	88.80	92.33
3	4.5	4.5	60	0.00	0.78	29.79	34.82	5.24	0.19	35.03	35.01
4	4.5	4.5	30	0.00	0.92	71.90	7.51	49.30	0.23	121.20	7.74
5	4.5	4.5	120	0.00	0.91	60.46	65.54	93.14	90.82	153.60	156.36
6	4.5	4.5	120	38.67	0.89	49.16	43.02	90.72	85.53	139.88	128.55
7	4.5	4.5	120	19.33	0.92	64.86	60.53	91.58	92.79	156.44	153.32
8	4.5	4.5	120	-38.67	0.87	26.30	30.86	86.71	68.07	113.01	98.93
9	0	4.5	120	0.00	0.88	54.01	50.37	9.78	15.03	63.79	65.40
10	2.25	4.5	120	0.00	0.90	60.32	57.98	40.45	45.31	100.77	103.29
11	9.5	4.5	120	0.00	0.95	50.67	82.54	258.00	246.81	308.67	329.35

coefficient $\mu_0 = \mu_f + \mu_R = 0.3 + 0.15 = 0.45$ is used, the agreement between the present theoretical calculation (41.7 MJ) and FEA result (40.5 MJ) is very good, and the difference is 3.0%.

2.2.5 Collision experiments and the analytical results

While the analytical closed-form solution has been validated with the results of time-domain numerical simulations in Section 2.2.4, it is also essential to verify the analytical method with experimental results. Here, the results of 60 experiments are presented and compared with the analytical results. These experiments were introduced here with a view that they also provide valuable data and information to understand the ship collision phenomenon.

The comparison includes the results of the model-scale experiments published by Tabri et al. (2008, 2009, 2010) and the full-scale ship collision experiments reported by Wevers and Vredeveldt (1998) and Wolf (2003). The model-scale tests include perpendicular ($\beta = 90$ degrees) and oblique collision experiments and a set of model tests that evaluated the effect of liquids with free surface carried on board the ships during the collision. The two full-scale experiments represent a perpendicular collision with and without liquids on board.

2.2.5.1 Model-scale perpendicular collision experiment

The results of the perpendicular collision experiments were reported by Tabri et al. (2008). The main particulars of the ship models are given in Table 2.2. The general arrangement of the model tests is provided in Fig. 2.4 where the collision angle for the perpendicular collision experiments is $\beta = 90$ degrees.

In the experiments, a rigid bulbous bow equipped with a force sensor to measure the contact force was mounted on the bow of the striking ship. The vertical position of the bulb was adjustable to control the height of the contact point (Z_c in Fig. 2.4). Part of the port side of the struck ship was made of polyurethane foam, acting as the contact area to record the bulbous penetration history. The impact velocity was imposed to the striking ship by using the impulsive loading from a pneumatic cylinder. The contact force was only measured in the longitudinal direction since the vertical and transverse components were assumed to be small for perpendicular collisions. The experimentally obtained force-penetration curve provided the deformation energy at the end of the collision process.

Table 2.2 Main particulars of the ship models by Tabri et al. (2008).

Ship model	Length (m)	Breadth (m)	Depth (m)
Ship A, struck ship	2.290	0.271	0.120
Ship B, striking ship	2.290	0.234	0.120

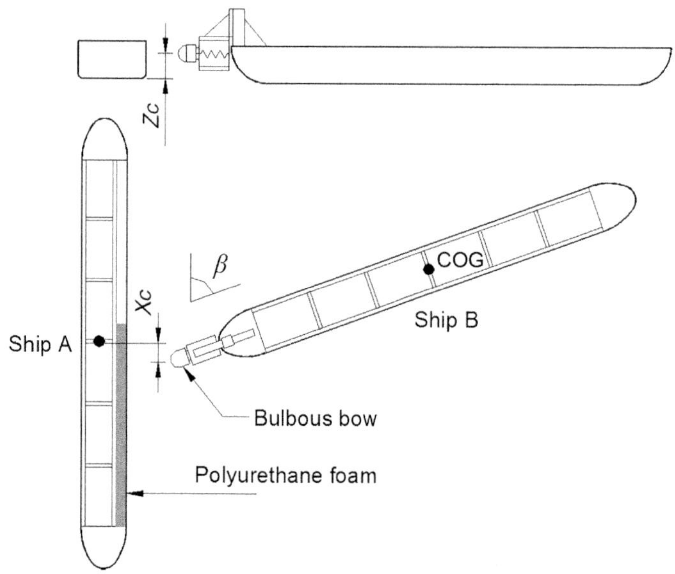

Fig. 2.4 General arrangement of the collision model tests.
Based on Tabri, K., Määttänen, J., Ranta, J., 2008. Model-scale experiments of symmetric ship collisions. J. Mar. Sci. Technol. 13, 71–84. Tabri, K., Varsta, P., Matusiak, J., 2010. Numerical and experimental motion simulations of non-symmetric ship collisions. J. Mar. Sci. Technol. 15, 87–101.

Three loading conditions for the striking and struck ship were evaluated. The respective draught, mass, vertical COG (Z_a), radius of inertia around the COG, and added mass coefficients for sway and yaw motions are presented in Table 2.3. The longitudinal COG (x_a) for both ships is located at the midship. The added mass coefficients were estimated by Tabri et al. (2008) using strip theory.

The characteristics of each model test experiment and the comparison of the experimental and analytical total energy absorbed during the impact are summarized in

Table 2.3 Physical parameters of the ship models: Perpendicular collisions by Tabri et al. (2008).

Ship model	Draught (m)	Mass (kg)	Z_a (m)	R_a (m)	m_{ay}	j_a
Ship A, struck ship	0.04 0.06 0.08	20.5 30.5 44.5	0.074 0.073 0.051	0.93 0.83 0.70	0.16 0.21 0.27	0.10 0.17 0.25
Ship model	**Draught (m)**	**Mass (kg)**	Z_a **(m)**	R_b **(m)**	m_{b2}	j_b
Ship B, striking ship	0.04 0.06 0.08	20.5 28.5 40.5	0.074 0.064 0.051	0.77 0.72 0.70	0.17 0.23 0.28	0.14 0.20 0.27

Table 2.4 Analytical and experimental results for the 13 model-scale perpendicular collision tests.

						E_{total} (J)		
ID	Case	M_a (kg)	M_b (kg)	V_{b1} (m/s)	x_c (m)	Model test	Analytical	Ratio
1	101	30.5	28.5	0.39	0.110	1.50	1.25	0.83
2	102	30.5	28.5	0.86	−0.015	6.20	6.11	0.99
3	103	30.5	28.5	0.91	0.003	7.10	6.84	0.96
4	104	30.5	28.5	0.45	0.064	1.70	1.67	0.98
5	105	30.5	28.5	0.66	0.023	3.60	3.60	1.00
6	106	30.5	20.5	0.90	0.033	7.70	5.50	0.71
7	107	30.5	40.5	0.83	0.020	7.30	6.80	0.93
8	108	20.5	40.5	0.83	0.010	7.20	5.25	0.73
9	109	20.5	40.5	0.45	0.060	1.80	1.54	0.86
10	110	20.5	28.5	0.92	0.050	7.00	5.60	0.80
11	111	44.5	28.5	0.93	0.020	9.20	8.46	0.92
12	112	44.5	20.5	1.01	0.025	7.80	7.95	1.02
13	113	44.5	20.5	0.58	0.048	2.60	2.62	1.01

Table 2.4 and Fig. 2.5. The ratio between the analytical and model test results is also given for reference (Table 2.4). It is seen from the comparison that the difference between the analytical and the model test results is good and generally less than 5% (6 of 13 cases), except for few test cases in which the differences reach more than 20% (cases 106, 108, and 110).

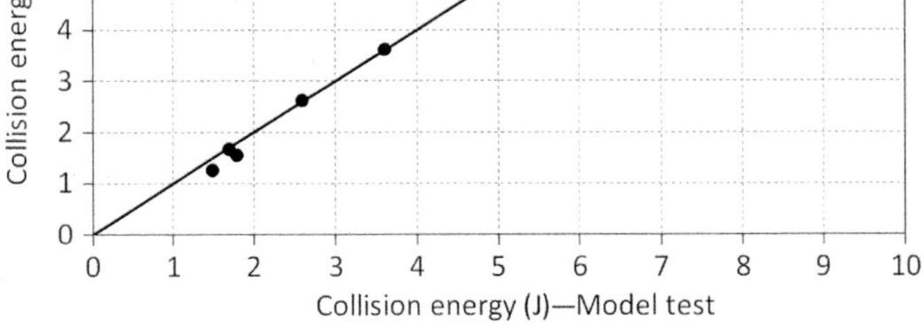

Fig. 2.5 Comparison of the analytical and experimental results for the 13 model-scale perpendicular collision tests.

Taking test case 106 as an example, the energy is similar to that of test case 103, in which the mass of the striking ship is larger, the impact location is closer to the COG, and the collision speed is slightly higher. Thus, the collision energy of case 106 is to be less than that of case 103, but the model test results show the opposite. The largest difference obtained for case 106 cannot be explained completely here.

2.2.5.2 Model-scale oblique collision experiments

The experimental details and results of the oblique collision experiments were presented by Tabri et al. (2010). The ship models and test set-up were the same as described in the previous section, but for the oblique collision cases, the collision angle between the two ships varied from 60 to 145 degrees (cases for $\beta = 90$ degrees were also included) as shown in Fig. 2.4. During the experiments, the longitudinal and the transverse contact force components were measured in the striking ship. Although the vertical component was not measured (due to the measuring instruments' limitations), only small magnitudes of this force were expected. The physical parameters of the ship models used for these tests are summarized in Table 2.5.

The results of the model experiments and the comparison with the analytical method are presented in Table 2.6. For the eight perpendicular collision cases (cases 201–208), the comparison shows a good agreement where the differences are generally less than 10%, except for test cases 205 and 207.

For the oblique collisions with angles 60 and 120 degrees, the deviations are also around 10%, except for test cases 304 and 316. This agreement was anticipated since the collision angle deviations are relatively small, and sliding between the colliding ships was not observed during the experiments. On the other hand, the largest difference between the analytical formulation and the model test results is observed for the collision angle $\beta = 145$ degrees. This deviation was expected due to the additional uncertainty of the sliding phenomenon.

Table 2.5 Physical parameters of the ship models: Oblique collisions by Tabri et al. (2010).

Ship model	Draught (m)	Mass (kg)	Z_a (m)	R_a (m)	m_{ay}	j_a
Ship A, struck ship	0.04	20.5	0.074	0.77	0.16	0.10
	0.06	30.5	0.073	0.69	0.21	0.17
	0.08	44.5	0.051	0.65	0.27	0.25
Ship model	**Draught (m)**	**Mass (kg)**	Z_a **(m)**	R_b **(m)**	m_{b2}	j_b
Ship B, striking ship	0.06	28.5	0.064	0.67	0.23	0.20

Table 2.6 Analytical and experimental results for the model-scale oblique collision tests.

							E_{total} (J)		
ID	**Case**	M_a **(kg)**	M_b **(kg)**	β **(degrees)**	V_{b1} **(m/s)**	x_c **(m)**	**Model test**	**Analytical**	**Ratio**
14	201	30.5	28.5	90	0.87	0.820	3.91	3.79	0.97
15	202	30.5	28.5	90	0.71	0.830	2.36	2.50	1.06
16	203	30.5	28.5	90	0.38	0.830	0.75	0.72	0.96
17	204	30.5	28.5	90	0.91	0.450	6.30	5.72	0.91
18	205	30.5	28.5	90	0.91	0.480	6.30	5.60	0.89
19	206	30.5	28.5	90	0.38	0.380	0.95	1.05	1.10
20	207	20.5	28.5	90	0.90	0.800	4.20	3.29	0.78
21	208	20.5	28.5	90	0.89	0.410	4.92	4.50	0.92
22	301	20.5	28.5	120	0.87	0.370	4.20	4.66	1.11
23	302	20.5	28.5	120	0.30	0.320	0.51	0.56	1.11
24	303	44.5	28.5	120	0.84	0.300	6.50	6.58	1.01
25	304	44.5	28.5	120	0.37	0.380	1.01	1.25	1.23
26	305[a]	20.5	28.5	145	0.34	0.320	0.54	0.46	0.86
27	306[a]	20.5	28.5	145	0.87	0.440	3.91	2.89	0.74
28	307[a]	44.5	28.5	145	0.84	0.380	5.47	3.75	0.69
29	308	44.5	28.5	145	0.28	0.340	0.52	0.42	0.81
30	309[a]	20.5	28.5	145	0.87	0.460	3.19	2.86	0.90
31	310[a]	20.5	28.5	145	0.88	0.440	3.19	2.95	0.93
32	311	20.5	28.5	120	0.88	0.420	4.25	4.67	1.10
33	312	20.5	28.5	120	0.86	0.410	4.64	4.48	0.96
34	313	20.5	28.5	60	0.76	0.290	3.14	3.43	1.09
35	314	20.5	28.5	60	0.36	0.320	0.81	0.76	0.94
36	315	44.5	28.5	60	0.75	0.380	4.35	4.79	1.10
37	316	44.5	28.5	60	0.43	0.400	1.17	1.56	1.33

[a]Sliding in the model tests; the coefficient of friction between the two ships is taken as $\mu_0 = 0.65$.

2.2.5.3 Model-scale collision experiments carrying liquid with free surface

Tabri et al. (2009) conducted ship collision model tests where the ships carried liquid with free surface. The study was performed to investigate the effects of the liquid on the collision energy by comparing with tests in which the ships carried the same amount of solid cargo.

The same ship models presented in Section 2.2.5.1 were used for these experiments. However, the striking ship was equipped with partially filled tanks and collided with an initially motionless struck ship without any liquid on board (hereafter, these experiments are named 'wet' tests). To better understand the liquid effects, most wet tests were repeated with a loading condition in which the water in the tanks was replaced with rigid masses of the same weight (these experiments are referred as 'dry' tests).

Two tanks were installed on board the striking ship model as illustrated in Fig. 2.6. The length of each liquid tank is 0.6 m, the width is 0.18 m, and the height is 0.25 m. For the wet

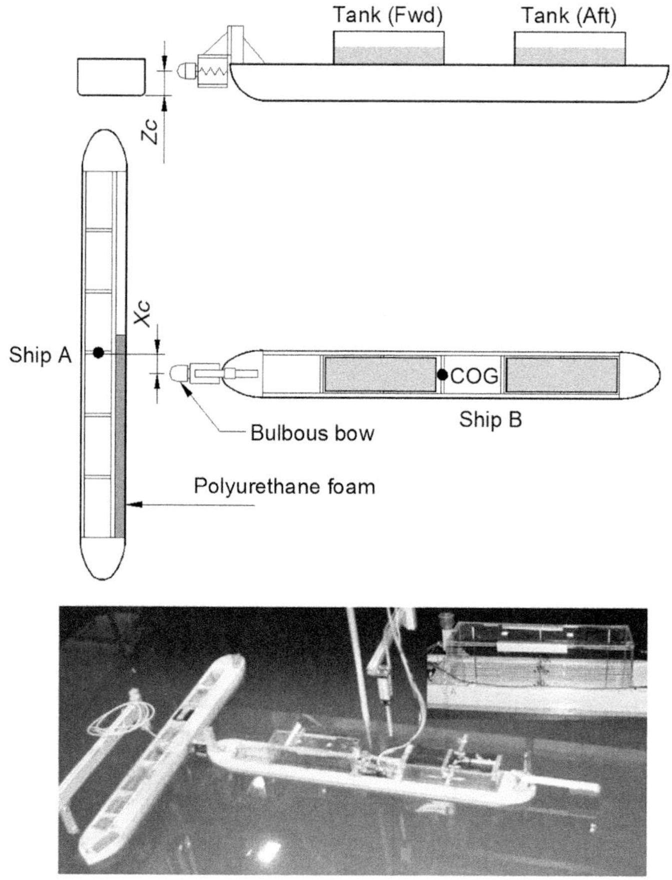

Fig. 2.6 General arrangement of the model tests with water tanks.
Based on Tabri, K., Matusiak, J., Varsta, P., 2009. Sloshing interaction in ship collisions—an experimental and numerical study. Ocean Eng. 36, 1366–1376.

Table 2.7 Physical parameters of the ship models: dry and wet tests by Tabri et al. (2009).

Ship model	**Draught (m)**	**Mass (kg)**	Z_a **(m)**	R_a **(m)**	m_{ay}	j_a
Ship A, struck ship	0.06	30.5	0.073	0.69	0.21	0.17
Ship model	**Draught (m)**	**Mass (kg)**	Z_a **(m)**	R_b **(m)**	m_{b2}	j_b
Ship B, striking ship (dry)	0.065	33.5	0.087	0.94	0.27	0.11
	0.0675	35.1	0.068	0.82	0.28	0.15
	0.075	41.1	0.068	0.78	0.31	0.18
Ship B, striking ship (wet)	0.055	28.1	0.081	0.90	0.23	0.10
	0.065	33.1	0.089	0.94	0.27	0.11
	0.0675	35.0	0.075	0.82	0.28	0.15
	0.075	40.7	0.070	0.78	0.31	0.18

tests, four loading conditions were evaluated with different amounts of water in the tanks. For the dry test, the rigid masses were positioned so that the position of the ship's COG remains unchanged with respect to the corresponding wet test. The physical parameters of the ship models used for these tests are presented in Table 2.7. The striking ship was connected to a carriage close to the COG, and it was accelerated smoothly until the desired impact velocity, to prevent sloshing before the collision and to avoid initial pitch motion. The contact force was measured only in the longitudinal direction since the other two components were expected to be negligible for such perpendicular collisions.

The results of the dry model test experiments and the comparison with the analytical method are presented in Table 2.8. The dry model test should be comparable with those validated in the previous sections where both exhibit similar agreement; for most cases, the differences range between 5% and 10%. It should be mentioned that some of the experimental results are not clearly understood; for example, cases with larger initial kinetic energy result in the same collision energy (compare N1-V4 with N4-V4 and N1-V10 with N2-V10). This could explain the largest deviation observed for case N2-V10.

Here, Eq. (2.35) is used to determine the effective mass coefficient of the internal liquid. The effective mass coefficients for the model test cases where the striking ship carried water in an individual tank are 0.12 for 5.0 kg of liquid cargo, 0.13 for 6.0 kg, 0.13 for 6.8 kg, 0.15 for 8.4 kg, and 0.17 for 10.5 kg.

Using the effective mass coefficients, the wet model test experiments were calculated using the analytical method, and the comparison is presented in Table 2.9. It is seen that the agreement for the wet tests is very good and similar to that for the dry tests. The mass of the striking ship given in Table 2.9 differs to that in Table 2.8 (solid mass), since the liquid mass has been reduced taking into account its effectiveness based on Eq. (2.35).

Table 2.8 Analytical and experimental results for model-scale collision experiments (dry tests) by Tabri et al. (2009).

					E_{total} (J)		
ID	**Case**	**M_a (kg)**	**M_b (kg)**	**V_{b1} (m/s)**	**Model test**	**Analytical**	**Ratio**
38	N1-V4	30.5	35.1	0.40	1.30	1.48	1.13
39	N1-V7	30.5	35.1	0.70	4.10	4.52	1.10
40	N1-V10	30.5	35.1	1.00	8.60	9.22	1.07
41	N2-V4	30.5	41.1	0.40	1.50	1.59	1.06
42	N2-V7	30.5	41.1	0.70	4.50	4.87	1.08
43	N2-V10	30.5	41.1	1.00	8.60	9.95	1.16
44	N4-V4	30.5	33.5	0.40	1.30	1.44	1.11
45	N4-V7	30.5	33.5	0.70	4.10	4.41	1.08
46	N4-V10	30.5	33.5	1.00	8.20	9.00	1.10

Table 2.9 Analytical and experimental results for model-scale collision experiments (wet tests) by Tabri et al. (2009).

					E_{total} (J)		
ID	**Case**	**M_a (kg)**	**M_b (kg) including effective liquid**	**V_{b1} (m/s)**	**Model test**	**Analytical**	**Ratio**
47	S1-V4	30.5	23.9	0.40	1.10	1.19	1.09
48	S1-V7	30.5	23.9	0.70	3.30	3.66	1.11
49	S1-V10	30.5	23.9	1.00	6.70	7.47	1.11
50	S2-V4	30.5	25.2	0.40	1.30	1.23	0.95
51	S2-V7	30.5	25.2	0.70	3.50	3.77	1.08
52	S2-V10	30.5	25.2	1.00	6.50	7.70	1.18
53	S4-V4	30.5	22.9	0.40	1.20	1.16	0.97
54	S4-V7	30.5	22.9	0.70	3.60	3.56	0.99
55	S4-V10	30.5	22.9	1.00	7.30	7.27	1.00
56	S3-V4	30.5	23.0	0.40	1.10	1.17	1.06
57	S3-V7	30.5	23.0	0.70	3.20	3.58	1.12
58	S3-V10	30.5	23.0	1.00	6.40	7.30	1.14

Table 2.10 Main particulars of the striking and struck ship of the full-scale collision tests.

	X-core experiment		Y-core experiment	
Particulars	**Striking ship**	**Struck ship**	**Striking ship**	**Struck ship**
Length (m)	80.00	76.40	80.00	80.00
Breadth (m)	8.20	11.40	8.20	9.50
Depth (m)	2.62	4.67	2.62	2.80
Draught (m)	1.30	3.32	1.45	2.15
Displacement (t)	721.0	2465.0	774.0	1365.0
Ballast water with free surface (t)	44.6	0.0	303.5	545.0

2.2.5.4 Full-scale experiments

Two full-scale collision experiments using inland vessels by Wevers and Vredeveldt (1998) and Wolf (2003) are analysed here: one collision experiment used a struck ship with X-core side structure and the other with a Y-core side structure. For the X-core experiment, the striking ship had a small amount of ballast water with free surface, while for the Y-core experiment, the striking ship carried a large amount of ballast water with free surface.

The striking ship was the same for both experiments, though different displacements (loading conditions) were used. On the other hand, the struck ships were different. The main dimensions of the striking and struck ships are given in Table 2.10. In both experiments, the striking ship was equipped with a rigid bulbous bow and impacted the struck ship at the midship section resulting in very small yaw motions.

For the analytical calculations, the radius of inertia is estimated as one-quarter of the ship length, the respective added mass coefficients for surge and yaw motion are considered as 0.05 and 0.21, and the coefficient for sway motion is calculated with Eq. (2.34). The effective mass coefficient of the internal liquid with free surface was taken as 0.15 (Zhang et al., 2017). The results of the full-scale experiments and the analytical calculations are presented in Table 2.11. Very good agreement is achieved.

Table 2.11 Analytical and experimental results for the two full-scale experiments.

							E_{total} (MJ)		
ID	**Case**	M_a **(ton) (struck ship)**	m_{ay} **(–)**	M_b **(ton) (striking ship)**	m_{b2} **(–)**	V_{b1} **(m/s)**	**Model test**	**Analytical**	**Ratio**
59	X-core	2465.0	0.40	683.1	0.22	3.33	3.19	3.29	1.03
60	Y-core	901.8	0.31	516.0	0.24	3.51	2.15	2.29	1.06

2.2.6 Calculation examples

This section provides detailed calculation examples using the present analytical method for readers to verify their calculations. We consider two identical supply vessels colliding at different locations and impact angles. The length of the vessels is 82.5 m, the breadth is 18.8 m, the draught is 5.7 m, and the displacement is 7000 tons. The radius of inertia for yaw motion is 20.6 m. The coefficients of the added mass are taken as 0.05 for surge motion, 0.50 for sway motion, and 0.21 for yaw motion. The effective coefficient of friction is taken as 0.60, and the critical energy ($E_{\xi CR}$) in the ξ-direction of rupturing the side shell is assumed to be 12 MJ.

Two collision speeds are considered: (a) Both ships have a forward speed of V = 1.5 m/s, and (b) both ships have a high forward speed of V = 4.0 m/s.

The calculated results for the total energy loss are presented in Table 2.12 for V = 1.5 m/s and in Table 2.13 for V = 4.0 m/s. Fig. 2.7 shows the collision scenarios and the ratios between the energy loss and the total initial kinetic energy of the two ships. It should be noted that the coordinate of the centre of gravity of Ship B is determined from $X_b = X_c - 0.5L_b\cos\beta$ and $Y_b = Y_c + 0.5L_b\sin\beta$. It can be seen that the energy loss is larger when the collision occurs at the fore area of the struck ship and that the collision angle has a strong effect on the energy loss. It is noticed that this collision model explains the observed distribution of severe damages in the forward part of the ship hull (IMO A265A, 1974).

2.2.7 Effect of the roll motion of the struck ship

The analytical formulations in Section 2.2.1 only consider the ship motions in the water surface plane. The roll motions of Ship A (treated here as the struck ship) may have effects on the results if the impact location is not at the vertical COG of the ship. The following expressions are further derived to take into account this effect; other parameters in Section 2.2.1 are unchanged:

$$D_{a\xi} = \frac{1}{1+m_{ax}}\sin^2\alpha + \frac{1}{1+m_{ay}}\cos^2\alpha + \frac{1}{1+j_a}\frac{[y_c\sin\alpha - (x_c - x_a)\cos\alpha]^2}{R_a^2} + \frac{1}{1+j_{ax}}\frac{[(z_c - z_a)\cos\alpha]^2}{R_{ax}^2}$$

$$D_{a\eta} = \frac{1}{1+m_{ax}}\sin\alpha\cos\alpha - \frac{1}{1+m_{ay}}\sin\alpha\cos\alpha + \frac{1}{1+j_a}\frac{[y_c\sin\alpha - (x_c - x_a)\cos\alpha][y_c\cos\alpha + (x_c - x_a)\sin\alpha]}{R_a^2} - \frac{1}{1+j_{ax}}\frac{(z_c - z_a)^2\sin\alpha\cos\alpha}{R_{ax}^2}$$

Table 2.12 Collision energy loss between two identical supply vessels colliding at a speed $V = 1.5\,m/s$ (see Fig. 2.7 for impact locations).

Geometric parameter					Total energy loss (MJ), collision speed $V = 1.5\,m/s$				
Impact location	X_c **(m)**	Y_c **(m)**	**Angle α (degrees)**	X_c/L	$\beta = 150$	$\beta = 120$	$\beta = 90$	$\beta = 60$	$\beta = 30$
1 (bow)	40.00	2.60	45.00	0.48	13.36	7.40	3.40	1.29	0.29
2	38.50	3.00	42.00	0.47	13.53	7.59	3.51	1.33	0.30
3	36.60	4.10	37.50	0.44	13.60	7.94	3.69	1.39	0.31
4	34.60	5.60	32.50	0.42	13.02	8.37	3.91	1.47	0.33
5	30.80	7.50	21.70	0.37	9.72	9.09	4.29	1.60	0.35
6	27.00	9.00	14.40	0.33	7.34	9.62	4.69	1.74	0.38
7	23.10	9.40	7.30	0.28	5.04	9.26	5.07	1.88	0.41
8	19.30	9.40	0.00	0.23	2.98	7.89	5.43	2.01	0.43
9	15.40	9.40	0.00	0.19	3.17	8.65	5.79	2.15	0.45
10	11.60	9.40	0.00	0.14	3.32	9.25	6.11	2.28	0.47
11	7.70	9.40	0.00	0.09	3.42	9.66	6.35	2.39	0.50
12	3.90	9.40	0.00	0.05	3.46	9.79	6.46	2.46	0.53
13 (midship)	0.00	9.40	0.00	0.00	3.43	9.62	6.41	2.49	0.54
14	−3.90	9.40	0.00	−0.05	3.35	9.19	6.19	2.46	0.55
15	−7.70	9.40	0.00	−0.09	3.21	8.57	5.85	2.38	0.54
16	−11.60	9.40	0.00	−0.14	3.03	7.83	5.41	2.24	0.53
17	−15.40	9.40	0.00	−0.19	2.84	7.06	4.95	2.08	0.51
18	−19.30	9.40	0.00	−0.23	2.62	6.28	4.49	1.91	0.48
19	−23.10	9.40	0.00	−0.28	2.41	5.58	4.07	1.74	0.45
20	−27.00	9.40	0.00	−0.33	2.20	4.93	3.70	1.58	0.41
21	−30.80	9.40	0.00	−0.37	2.00	4.36	3.37	1.43	0.37
22	−34.60	9.40	0.00	−0.42	1.82	3.86	3.07	1.30	0.34
23	−38.50	9.40	0.00	−0.47	1.65	3.42	2.79	1.19	0.31
24 (stern)	−40.00	9.40	0.00	−0.48	1.59	3.27	2.68	1.15	0.29

Table 2.13 Collision energy loss between two identical supply vessels colliding at a speed $V = 4.0\,m/s$ (see Fig. 2.7 for impact locations).

Geometric parameter					Total energy loss (MJ), collision speed $V = 4.0\,m/s$				
Impact location	X_c **(m)**	Y_c **(m)**	**Angle** α **(degrees)**	X_c/L	$\beta = 150$	$\beta = 120$	$\beta = 90$	$\beta = 60$	$\beta = 30$
1 (bow)	40.00	2.60	45.00	0.48	94.98	52.59	24.20	9.16	2.07
2	38.50	3.00	42.00	0.47	96.22	53.98	24.97	9.45	2.13
3	36.60	4.10	37.50	0.44	98.59	56.45	26.23	9.90	2.23
4	34.60	5.60	32.50	0.42	101.36	59.53	27.77	10.43	2.33
5	30.80	7.50	21.70	0.37	104.83	64.61	30.54	11.39	2.52
6	27.00	9.00	14.40	0.33	107.02	69.39	33.38	12.37	2.71
7	23.10	9.40	7.30	0.28	107.84	73.10	36.04	13.34	2.89
8	19.30	9.40	0.00	0.23	108.02	76.15	38.60	14.31	3.06
9	15.40	9.40	0.00	0.19	107.69	78.78	41.19	15.31	3.23
10	11.60	9.40	0.00	0.14	106.78	80.53	43.45	16.22	3.38
11	7.70	9.40	0.00	0.09	105.23	81.14	45.16	17.00	3.58
12	3.90	9.40	0.00	0.05	103.15	80.37	45.93	17.52	3.74
13 (midship)	0.00	9.40	0.00	0.00	24.40	78.14	45.58	17.70	3.85
14	−3.90	9.40	0.00	−0.05	23.80	74.68	44.03	17.47	3.89
15	−7.70	9.40	0.00	−0.09	22.84	70.49	41.59	16.94	3.86
16	−11.60	9.40	0.00	−0.14	21.57	65.80	38.47	15.94	3.76
17	−15.40	9.40	0.00	−0.19	20.16	61.21	35.20	14.82	3.61
18	−19.30	9.40	0.00	−0.23	18.63	56.77	31.92	13.58	3.40
19	−23.10	9.40	0.00	−0.28	17.12	52.84	28.96	12.37	3.17
20	−27.00	9.40	0.00	−0.33	15.62	49.29	26.28	11.21	2.92
21	−30.80	9.40	0.00	−0.37	14.24	46.31	24.02	10.17	2.66
22	−34.60	9.40	0.00	−0.42	12.96	27.47	22.10	9.26	2.42
23	−38.50	9.40	0.00	−0.47	11.76	24.33	19.82	8.44	2.17
24 (stern)	−40.00	9.40	0.00	−0.48	11.33	23.24	19.08	8.16	2.08

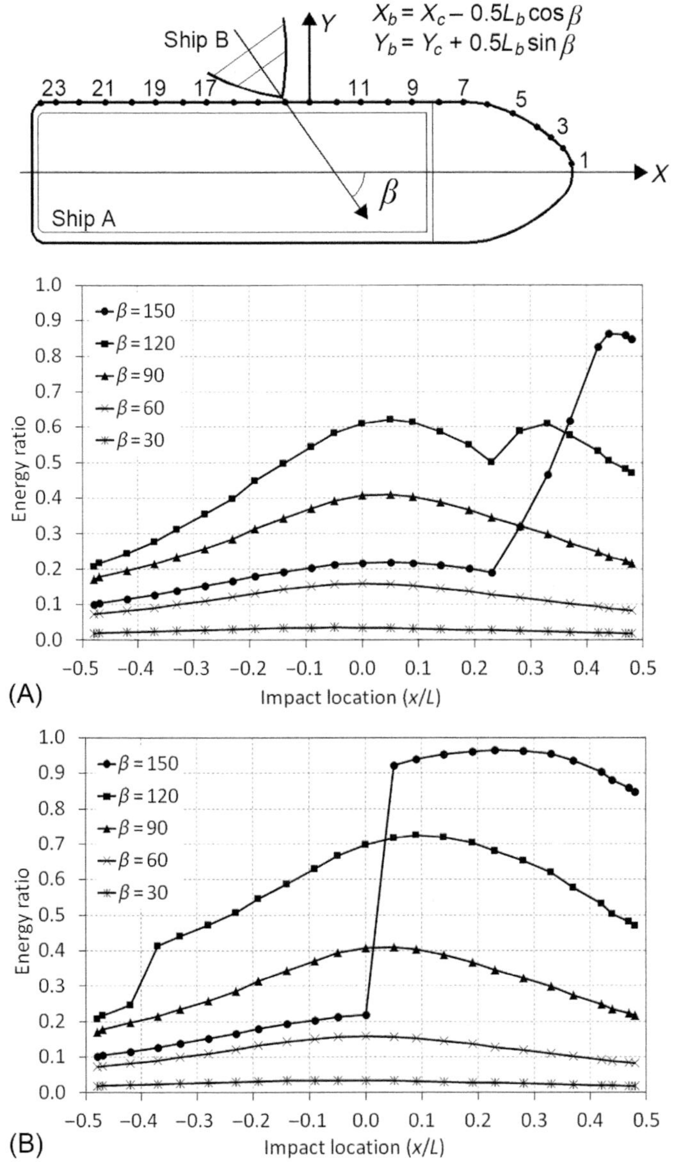

Fig. 2.7 Collision scenarios of the two identical supply vessels. The energy ratio is defined as the ratio between the energy released for crushing and the total kinetic energy of the two ships before collision (collision angles $\beta = 30, 60, 90, 120$, and 150 degrees), (A) for $V = 1.5\,\text{m/s}$ and (B) for $V = 4.0\,\text{m/s}$.

$$K_{a\xi} = \frac{1}{1+m_{ax}}\sin\alpha\cos\alpha - \frac{1}{1+m_{ay}}\sin\alpha\cos\alpha + \frac{1}{1+j_a}\frac{[y_c\sin\alpha-(x_c-x_a)\cos\alpha][y_c\cos\alpha+(x_c-x_a)\sin\alpha]}{R_a^2} - \frac{1}{1+j_{ax}}\frac{(z_c-z_a)^2\cos\alpha\sin\alpha}{R_{ax}^2}$$

$$K_{a\eta} = \frac{1}{1+m_{ax}}\cos^2\alpha + \frac{1}{1+m_{ay}}\sin^2\alpha + \frac{1}{1+j_a}\cdot\frac{[y_c\cos\alpha+(x_c-x_a)\sin\alpha]^2}{R_a^2} + \frac{1}{1+j_{ax}}\frac{[(z_c-z_a)\sin\alpha]^2}{R_{ax}^2}$$

where z_a is the vertical COG of the struck ship from the baseline and z_c is the vertical coordinate of impact location measured from the baseline of the struck ship as illustrated in Fig. 2.8. R_{ax} is the roll radius of gyration around the longitudinal axis, and j_{ax} is the associated added mass coefficient.

This new formulation is used to estimate the collision energy of the perpendicular collision cases summarized in Table 2.4 (cases 101–113). Table 2.14 compares these results considering roll motion and without considering the roll motion. The coordinate of the impact location is displaced from the vertical COG, that is, $(z_c - z_a) \neq 0$. The ratio $(z_c - z_a)/D$ is also given for indicating the distance from the centre (D is the depth of the struck ship). In the calculations, it is assumed that $R_{ax} = 0.45B$ (B is the breadth of the struck ship) and $j_{ax} = 0.21$. It can be concluded that despite some large displacements of the impact point, $(z_c - z_a)/D > 0.3$, the effect of the roll motion on the energy released is negligible. It should be mentioned that a similar conclusion was obtained in the study conducted by Liu and Amdahl (2010). However, it is noted that the excitation of the roll motion can lead to a second impact.

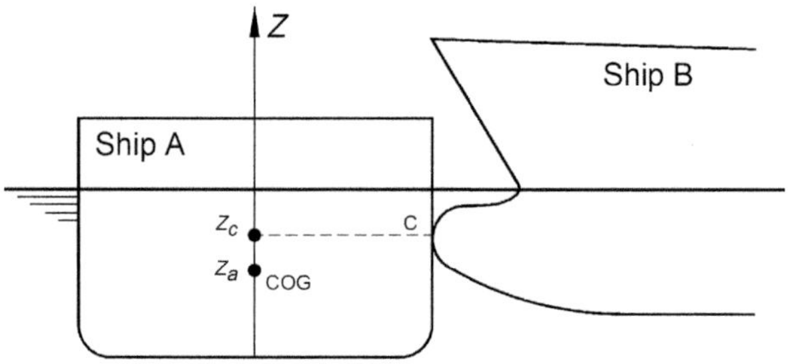

Fig. 2.8 Representation of the impact location displaced from the ship's vertical centre of gravity.

Table 2.14 Effect of the roll motion of the struck ship on collision energy.

					E_{total} (J)		
ID	**Case**	z_a **(m)**	$z_c - z_a$ **(m) (distance away from VCG)**	$(z_c - z_a)/D$	**Analytical (not consider roll)**	**Analytical (consider roll)**	**Ratio**
1	101	0.073	−0.002	−0.02	1.25	1.25	1.00
2	102	0.073	−0.005	−0.04	6.11	6.11	1.00
3	103	0.073	−0.005	−0.04	6.84	6.84	1.00
4	104	0.073	−0.001	−0.01	1.67	1.67	1.00
5	105	0.073	−0.003	−0.03	3.60	3.60	1.00
6	106	0.073	−0.012	−0.10	5.50	5.48	1.00
7	107	0.073	−0.008	−0.07	6.80	6.79	1.00
8	108	0.074	−0.026	−0.22	5.25	5.11	0.97
9	109	0.074	−0.027	−0.23	1.54	1.49	0.97
10	110	0.074	−0.026	−0.22	5.60	5.47	0.98
11	111	0.051	0.044	0.37	8.46	8.08	0.95
12	112	0.051	0.034	0.28	7.95	7.77	0.98
13	113	0.051	0.036	0.30	2.62	2.55	0.98

2.2.8 Hand calculation exercise: Simplified outer collision mechanics for ship–ship collisions

In a central perpendicular collision, the yaw motions can be neglected, and the three DOF equations for horizontal motion, Eqs (2.1)–(2.3), can be reduced to two uncoupled one-dimensional equations of motion for surge and sway. For such a central and perpendicular striking ship impact against a resting struck vessel, it is seen from Fig. 2.1 that conservation of momentum in the direction perpendicular to the struck ship can be expressed as

$$M_a(1+m_{ax})v_{ax} = [M_a(1+m_{ax}) + M_b(1+m_{b1})]U$$

Here, U is the common velocity of the ships at the end of the collision.

The loss of kinetic energy in the direction perpendicular to the struck ship is then

$$\Delta E_k = \frac{1}{2}M_a(1+m_{ax})v_{ax}{}^2 - \frac{1}{2}[M_a(1+m_{ax}) + M_b(1+m_{b1})]U^2$$

Elimination of the common velocity U yields

$$\Delta E_k = \frac{1}{2}\frac{M_a(1+m_{ax})M_b(1+m_{b1})}{M_a(1+m_{ax}) + M_b(1+m_{b1})} \cdot v_{ax}{}^2$$

Fig. 2.9 Representation of a central, right angle collision between two similar ships.

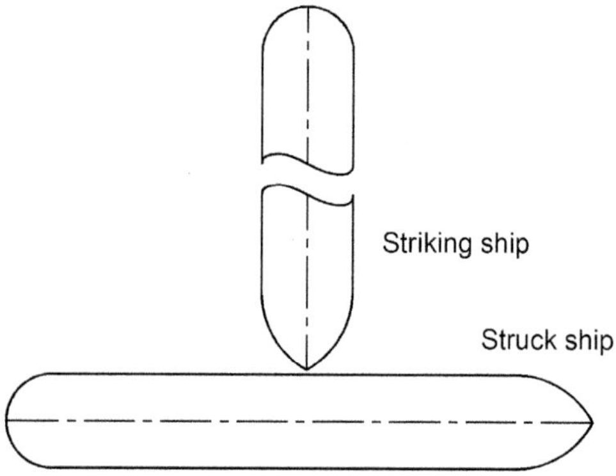

which is the loss in kinetic energy to be dissipated by crushing of the structural materials of the colliding ships. This expression was derived by Minorsky (1959).

Questions:
(a) Derive the same expression for the loss of kinetic energy for central perpendicular impacts using the equations derived in Section 2.2.

(b) Consider a central, right angle collision between two similar ships (see Fig. 2.9), with the following main particulars:

Length, 116 m; breadth, 19 m; draught, 6.9 m; displacement, $10.34 \cdot 10^6$ kg; and radius of inertia for yaw, 29.0 m

The hydrodynamic mass in the longitudinal direction of the rigid striking ship is 5.0% of the ship mass, and the crushing strength of the struck vessel is constant, and it is approximated to be equal to 7.022×10^6 N. At the moment of impact, the speed of the striking vessel is 4.5 m/s, and the struck vessel is at standstill.

Assume that the added mass of the struck ship is constant and equal to 40% of the ship mass and calculate the relative acceleration, the relative velocity, and the relative displacement for the two ships as function of time.

Solution to question (b)
The accelerations of the striking and the struck ship, $\dot{v}_{ax}$ and $\dot{v}_{by}$, respectively, can be determined as

$$-M_a(1+m_{ax})\dot{v}_{ax} = F$$

$$M_b(1+m_{by})\dot{v}_{by} = F$$

where F is the constant collision force. The constant relative acceleration is

$$\ddot{s} = \dot{v}_r = \dot{v}_{ax} - \dot{v}_{by}$$

and it can be determined from

$$\ddot{s} = \dot{v}_r = \dot{v}_{ax} - \dot{v}_{by} = F\left\{\frac{1}{M_a(1+m_{ax})} + \frac{1}{M_b(1+m_{by})}\right\}$$
$$= -7.022 \cdot 10^6 \frac{1}{10.34 \cdot 10^6}\left\{\frac{1}{1.05} + \frac{1}{1.4}\right\} = -1.13\,\text{m/s}^2$$

that is, a constant relative acceleration caused by the constant masses and force.

The relative velocity is

$$\dot{s} = \ddot{s}t + \dot{v}_{ax}(0) = -1.13t + 4.5$$

It is seen that the relative velocity equals zero at t=4.5/1.13=3.98 s.

The penetration as function of time of the rigid bow into the struck vessel is found as

$$s = -\frac{1}{2} \cdot 1.13t^2 + 4.5t$$

The maximum penetration at $t=3.98$ s is 8.96 m.

Alternatively, the loss in kinetic energy is

$$\Delta E_k = \frac{1}{2} \cdot \frac{\left(10.34 \cdot 10^6\right)^2 \cdot 1.05 \cdot 1.40}{10.34 \cdot 10^6 \cdot (1.05 + 1.40)} \cdot 4.5^2 = 62.82 \times 10^6\,\text{Nm}$$

With a constant crushing force $F = 7.022 \times 10^6$ N, the crushing distance is determined as $s_{\max} = \frac{62.82 \cdot 10^6}{7.022 \cdot 10^6}$ or $s_{\max} = 8.95$ m

Results from a time-domain numerical hydrodynamic calculation procedure (see Petersen, 1982), which takes into account the memory forces due to the set-up of surface waves described in Section 2.1, are compared with these hand-calculated results based on the constant added masses in Fig. 2.10. This figure shows (A) the collision force, (B) the relative acceleration, (C) the relative velocity, and (D) the penetration into the struck vessel. It is noticed that due to the variation of the hydrodynamic forces with time, the relative acceleration is not constant in the time simulation. It is also seen from the hand calculations that the relative velocity equals zero at $t=4.5/1.13=3.98$ s. The comparison with the results determined by the numerical simulation of the hydrodynamic forces shows that in this case, the collision time based on varying hydrodynamic mass is $(4.52-3.98) \cdot 100/4.52 = 12\%$ higher than when a constant added mass is used.

2.3 Ship collision with offshore platforms or wind turbines

Offshore platforms, drilling rigs, or wind farms are constantly serviced by supply vessels during their operations, and thus, ship collisions are hazards to be considered. The key concern during the design and operation of offshore installations is to make sure that they have sufficient safety in case of a collision incident.

The current design practice is described, for example, in the *Guidance Notes for Collision Analysis* (Lloyd's Register, 2014) where an analysis procedure is described and a number of recognized standards and practices for collision analyses are listed.

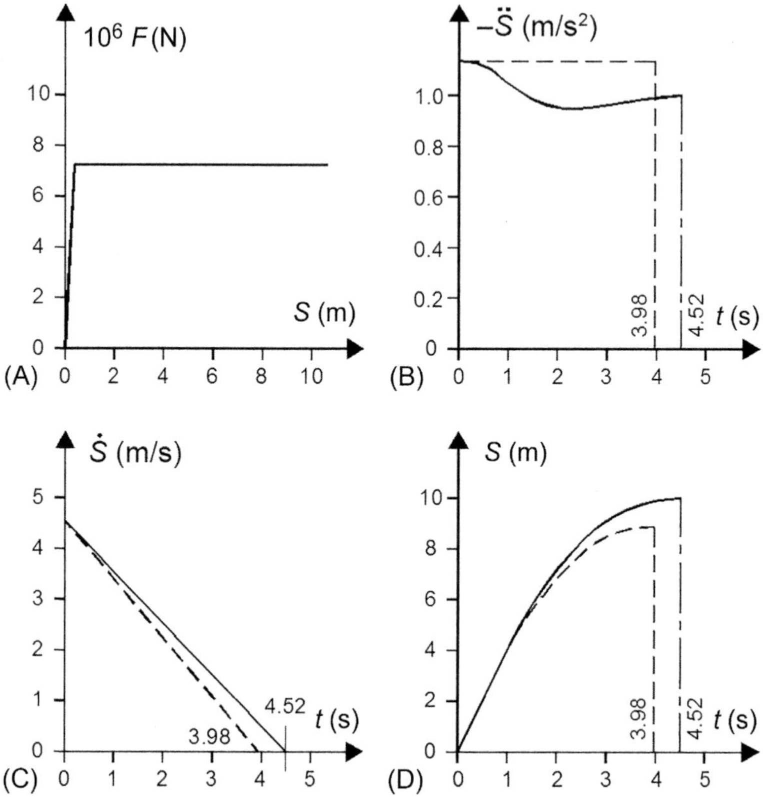

Fig. 2.10 (A) Collision force, (B) relative acceleration, (C) relative velocity, and (D) penetration of the struck vessel calculated taking into account hydrodynamic memory forces *(continuous line)* and compared with results using a constant added mass *(dashed line)*.

The basic impact design philosophy is that the offshore structure should be able to withstand accidental conditions from supply ships drifting out of control with a speed of 2 m/s. Such an impact may require a major repair after the incident, but it should not lead to the total collapse even in a storm with a return period of 1 year. While the displacement of traditional typical supply vessels operating in the North Sea is about 5000 tons, vessels servicing the installations in the Norwegian Shelf have increased significantly up to 7500–10,000 tons displacement (Storheim and Amdahl, 2014). With the increase in vessel's size, the collision energy will also increase, and thus, the risk of damages in case of collisions with supply vessels could be significant. Therefore, such risks should be carefully assessed and reviewed (Pedersen, 2014).

This section presents an analysis of the importance of taking into account the flexibility of the offshore installation in order to estimate accurately the impact energy to be absorbed for structural damage. An application example where a vessel collides with a jack-up rig is analysed and evaluated.

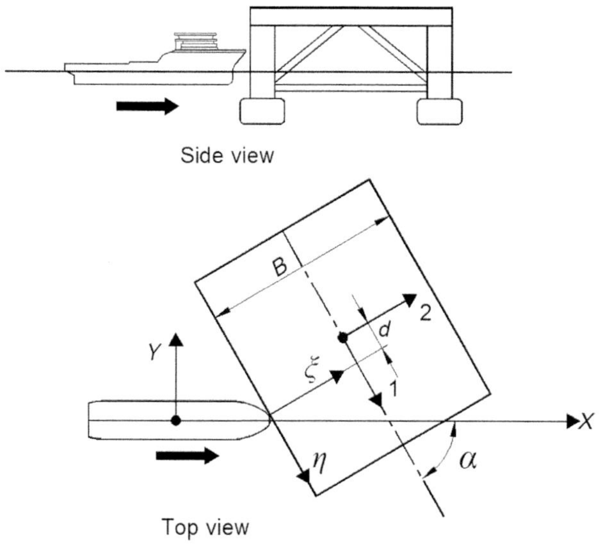

Fig. 2.11 Ship collision with a semisubmersible.

2.3.1 Analytical formulation for ship collision with floating offshore structures

In the concept design stage of an offshore structure, one of the important failure types to be considered is a collision between a ship and the offshore platform. For example, it is an industry practice and requirement to consider collisions with supply vessels and visiting oil tankers. Collision analysis can be performed using advanced numerical simulations, but these analyses are quite time consuming, and the results depend heavily on the chosen inputs, detailed design, and analyst's experience. Thus, simple analytical models are still very useful design tools (Jonge and Laukeland, 2013).

This section describes how the theory developed in Section 2.2 for ship–ship collisions can be used in analysis of ship collisions with floating offshore structures such as semisubmersibles and spar platforms.

2.3.1.1 Ship collision with semisubmersibles

Fig. 2.11 illustrates a ship colliding with a semisubmersible platform. In this collision situation, the theory in Section 2.2.1 or the formulations for ship–ship collision analysis summarized in Section 2.2.2 can be used directly.

If the bow shape of the ship is ignored, the ship can be simplified as a slender beam. For such collision case, we have the simplified relations:

$$\beta = \alpha$$

$$x_c - x_a = \frac{L_a}{2}, y_c = 0$$

$$x_c - x_b = d \cdot \cos\alpha - \frac{B}{2} \cdot \sin\alpha$$

$$y_c - y_b = -d \cdot \sin\alpha - \frac{B}{2} \cdot \cos\alpha$$

$$R_a = \frac{L_a}{4}, R_b = \frac{L_b}{4}$$

Therefore, the coefficients $D_{a\xi}$, $D_{a\eta}$, $D_{b\xi}$, $D_{b\eta}$, $K_{a\xi}$, $K_{a\eta}$, $K_{b\xi}$, and $K_{b\eta}$ in Section 2.2.1 become

$$D_{a\xi} = \frac{1}{1+m_{ax}} \sin^2\alpha + \frac{1}{1+m_{ay}} \cos^2\alpha + \frac{4}{1+j_a} \cos^2\alpha$$

$$D_{a\eta} = \left(\frac{1}{1+m_{ax}} - \frac{1}{1+m_{ay}} - \frac{4}{1+j_a} \right) \sin\alpha\cos\alpha$$

$$D_{b\xi} = \frac{1}{1+m_{b2}} + \frac{16}{1+j_b} \cdot \left(\frac{d}{L_b} \right)^2$$

$$D_{b\eta} = \frac{8}{1+j_b} \cdot \left(\frac{B \cdot d}{L_b{}^2} \right)$$

$$K_{a\xi} = \left(\frac{1}{1+m_{ax}} - \frac{1}{1+m_{ay}} - \frac{4}{1+j_a} \right) \sin\alpha\cos\alpha$$

$$K_{a\eta} = \frac{1}{1+m_{ax}} \cos^2\alpha + \frac{1}{1+m_{ay}} \sin^2\alpha + \frac{4}{1+j_a} \sin^2\alpha$$

$$K_{b\xi} = \frac{8}{1+j_b} \cdot \left(\frac{B \cdot d}{L_b{}^2} \right)$$

$$K_{b\eta} = \frac{1}{1+m_{b1}} + \frac{4}{1+j_b} \cdot \left(\frac{B}{L_b} \right)^2$$

Here, L_b is the length of the semisubmersible, B is the breadth of the semisubmersible, m_{b1} is the added mass coefficient for the surge motion, m_{b2} is the added mass coefficient for the sway motion, j_b is the added mass coefficient of the momentum for the rotation around the COG, and d is the distance from the impact point to the COG of the semisubmersible along the 1-axis. For example, the energy losses can be determined by Eqs (2.21), (2.22) or (2.25), (2.26). For perpendicular central collision to the semisubmersible, where $\alpha=90$ degrees and $d=0$, the formula for energy loss can be simplified as

$$E_{\xi}=\frac{1}{2}\left(\frac{M_a(1+m_{ax})}{1+\dfrac{M_a(1+m_{ax})}{M_b(1+m_{b2})}}\right)\cdot\dot{\xi}(0)^2 \tag{2.37}$$

$$E_{\eta}=0 \tag{2.38}$$

This expression (Minorsky, 1959) is the same as the one given in design standards, such as DNV (2010).

2.3.1.2 Ship collision with spar platforms

Fig. 2.12 illustrates a ship colliding with a spar platform. Jonge and Laukeland (2013) extended the theory in Section 2.2.1 for ship–ship collision analysis to include a ship and spar collision analysis.

In this collision situation, the ship will move transversely to one side in the waterplane, and the spar will move in the opposite direction. The formulae given in Section 2.2.1 or in the above section can be used with slight modifications.

For explanations on the physics, as shown in Fig. 2.12, two force components act at the impact location on the spar: One is the perpendicular component F_{ξ}, and the other is the parallel (or tangent) component F_{η}. The equations of motions of the spar due to

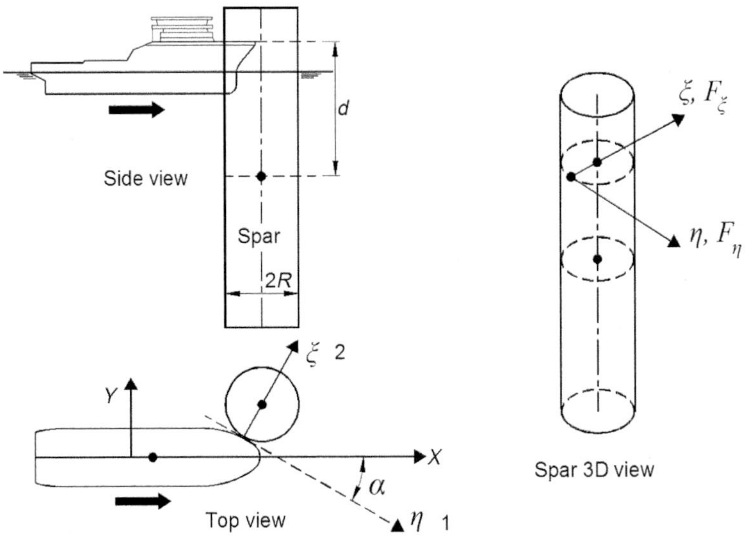

Fig. 2.12 Illustration of a ship colliding with a spar platform.

the impact force components F_ξ in the ξ-direction and F_η in the η-direction can be expressed as

$$M_b(1+m_b)\dot{v}_{b\xi}=F_\xi \tag{2.39}$$

$$M_b(1+m_b)\dot{v}_{b\eta}=F_\eta \tag{2.40}$$

$$M_b R_b^2(1+j_b)\dot{\omega}_{b\eta}=F_\xi d \tag{2.41}$$

$$M_b R_b^2(1+j_b)\dot{\omega}_{b\xi}=F_\eta d \tag{2.42}$$

$$M_b R_R^2(1+j_R)\dot{\omega}_R=F_\eta R \tag{2.43}$$

where d is the distance from the impact point to the centre of the gravity of the spar, R_R is the radius of gyration of the spar around its vertical axis, R_b is the radius of the inertia of the spar around the COG, j_R is the spar added mass coefficient of the momentum for the rotation around its vertical axis, and j_b is the spar added mass coefficient of the momentum for the rotation around the COG. The accelerations of the spar at the impact location in the ξ-direction and in the η-direction are then expressed as

$$\ddot{\xi}_b=\dot{v}_{b\xi}+\dot{\omega}_{b\eta}d=\frac{F_\xi}{M_b}\left(\frac{1}{1+m_b}+\frac{1}{1+j_b}\cdot\left(\frac{d}{R_b}\right)^2\right)$$

$$\ddot{\eta}_b=\dot{v}_{b\eta}+\dot{\omega}_{b\xi}d+\dot{\omega}_R R=\frac{F_\eta}{M_b}\left(\frac{1}{1+m_b}+\frac{1}{1+j_b}\cdot\left(\frac{d}{R_b}\right)^2+\frac{1}{1+j_R}\cdot\left(\frac{R}{R_R}\right)^2\right)$$

Following the same procedure as in Section 2.2.1, the coefficients $D_{a\xi}$, $D_{a\eta}$, $K_{a\xi}$, $K_{a\eta}$, $D_{b\xi}$, $D_{b\eta}$, $K_{b\xi}$, and $K_{b\eta}$ can be determined from

$$D_{a\xi}=\frac{1}{1+m_{ax}}\sin^2\alpha+\frac{1}{1+m_{ay}}\cos^2\alpha+\frac{1}{1+j_a}\cdot\frac{[y_c\sin\alpha-(x_c-x_a)\cos\alpha]^2}{R_a^2}$$

$$D_{a\eta}=\frac{1}{1+m_{ax}}\sin\alpha\cos\alpha-\frac{1}{1+m_{ay}}\sin\alpha\cos\alpha+\frac{1}{1+j_a}\cdot\frac{[y_c\sin\alpha-(x_c-x_a)\cos\alpha][y_c\cos\alpha+(x_c-x_a)\sin\alpha]}{R_a^2}$$

$$K_{a\xi}=\frac{1}{1+m_{ax}}\sin\alpha\cos\alpha-\frac{1}{1+m_{ay}}\sin\alpha\cos\alpha+\frac{1}{1+j_a}\cdot\frac{[y_c\sin\alpha-(x_c-x_a)\cos\alpha][y_c\cos\alpha+(x_c-x_a)\sin\alpha]}{R_a^2}$$

$$K_{a\eta} = \frac{1}{1+m_{ax}}\cos^2\alpha + \frac{1}{1+m_{ay}}\sin^2\alpha + \frac{1}{1+j_a}\cdot\frac{[y_c\cos\alpha + (x_c - x_a)\sin\alpha]^2}{R_a^{\ 2}}$$

$$D_{b\xi} = \frac{1}{1+m_b} + \frac{1}{1+j_b}\cdot\left(\frac{d}{R_b}\right)^2$$

$$K_{b\eta} = \frac{1}{1+m_b} + \frac{1}{1+j_b}\cdot\left(\frac{d}{R_b}\right)^2 + \frac{1}{1+j_R}\cdot\left(\frac{R}{R_R}\right)^2$$

$$D_{b\eta} = 0$$

$$K_{b\xi} = 0$$

If it is considered that the ship is a slender beam and the bow shape of the ship does not affect the energy loss calculation (or the effect is small and can be ignored), then only the collision angle comes into effect. The coefficients $D_{a\xi}, D_{a\eta}, K_{a\xi}, K_{a\eta}, D_{b\xi}, D_{b\eta}, K_{b\xi}$, and $K_{b\eta}$ can be further simplified as

$$D_{a\xi} = \frac{1}{1+m_{ax}}\sin^2\alpha + \frac{1}{1+m_{ay}}\cos^2\alpha + \frac{4}{1+j_a}\cos^2\alpha$$

$$D_{a\eta} = \left(\frac{1}{1+m_{ax}} - \frac{1}{1+m_{ay}} - \frac{4}{1+j_a}\right)\sin\alpha\cos\alpha$$

$$K_{a\xi} = \left(\frac{1}{1+m_{ax}} - \frac{1}{1+m_{ay}} - \frac{4}{1+j_a}\right)\sin\alpha\cos\alpha$$

$$K_{a\eta} = \frac{1}{1+m_{ax}}\cos^2\alpha + \frac{1}{1+m_{ay}}\sin^2\alpha + \frac{4}{1+j_a}\sin^2\alpha$$

$$D_{b\xi} = \frac{1}{1+m_b} + \frac{1}{1+j_b}\cdot\left(\frac{d}{R_b}\right)^2$$

$$K_{b\eta} = \frac{1}{1+m_b} + \frac{1}{1+j_b}\cdot\left(\frac{d}{R_b}\right)^2 + \frac{1}{1+j_R}\cdot\left(\frac{R}{R_R}\right)^2$$

$$D_{b\eta} = 0$$

$$K_{b\xi} = 0$$

Eventually, the energy losses can be determined by Eqs (2.21), (2.22) or (2.25), (2.26) in Section 2.2.1 as before.

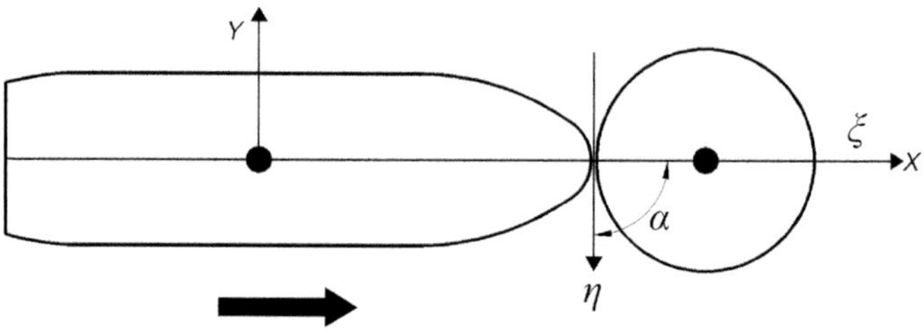

Fig. 2.13 Illustration of a ship colliding with a spar platform at 90 degrees.

For a particular case where the collision angle is $\alpha=90$ degrees, as shown in Fig. 2.13, the coefficients $D_{a\xi}$, $D_{a\eta}$, $K_{a\xi}$, $K_{a\eta}$, $D_{b\xi}$, $D_{b\eta}$, $K_{b\xi}$, and $K_{b\eta}$ are further simplified as

$$D_{a\xi}=\frac{1}{1+m_{ax}},\ D_{a\eta}=0,\ K_{a\xi}=0,\ \textit{and}\ K_{a\eta}=\frac{1}{1+m_{ay}}+\frac{4}{1+j_a}$$

$$D_{b\xi}=\frac{1}{1+m_b}+\frac{1}{1+j_b}\cdot\left(\frac{d}{R_b}\right)^2,\ K_{b\eta}=\frac{1}{1+m_b}+\frac{1}{1+j_b}\cdot\left(\frac{d}{R_b}\right)^2+\frac{1}{1+j_R}\cdot\left(\frac{R}{R_R}\right)^2$$

$$D_{b\eta}=0,\ \textit{and}\ K_{b\xi}=0$$

In this case, the formula for energy losses can be simplified as

$$E_\xi=\frac{1}{2}\left(\frac{M_a(1+m_{ax})}{1+\frac{M_a(1+m_{ax})}{M_b(1+m_b)}+\frac{M_a(1+m_{ax})}{M_b(1+j_b)}\cdot\left(\frac{d}{R_b}\right)^2}\right)\cdot\dot{\xi}(0)^2 \tag{2.44}$$

$$E_\eta=0 \tag{2.45}$$

This formula is similar to the one derived by Jonge and Laukeland (2013) using conservation of momentums in the collision for the collision angle $\alpha=90$ degrees. If it is assumed that $d=0$, representing the impact at the centre of the gravity of the spar, the equation reduces to that given before for central collision between a ship and a semi-submersible (Eqs 2.37, 2.38).

2.3.2 Calculation examples for ship collision with flexible floating offshore structures

2.3.2.1 Calculation examples for a ship collision with a Tension Leg Platform

For a Tension Leg Platform (TLP) installation, a surplus of buoyancy in the hull provides the forces needed to carry the vertical load of the installation and to establish axial forces in the tethers (see Fig. 2.14). The tensioned tethers are secured to the bottom with suction anchors, gravity anchors, or piles.

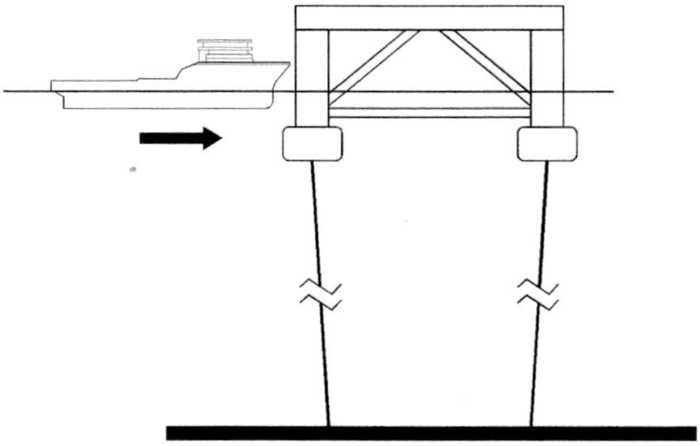

Fig. 2.14 A schematic TLP configuration.

Transverse restoring forces are provided by the tension in the tethers, as increasing tether tension reduces transverse displacements. Due to the large vertical stiffness of the tethers, the system rotates very little around a horizontal axis when the hull is exposed to a transverse force, that is, the installation is upright even for large transverse displacements.

Consider a TLP installation where the mass of the installation including hydrodynamic mass for transverse dynamic motions can be taken as 160,000 tons. The combined tether tension forces are 300 MN, and the length of the tethers is 600 m.

This installation is struck in a central collision by a drifting tanker with a mass plus added mass equal to 100,000 tons and a speed of 2 m/s.

Questions:

(1) Calculate the energy released for crushing of the tanker and/or the TLP structure.

(2) Calculate the maximum amplitude of the sway motion of the struck TLP installation.

Solution to question (a)

First, the restoring forces from the tethers shall be calculated for transverse displacements. From Fig. 2.15, it is seen that for a tension force P in the tether system and a transverse displacement d, which is small compared with the water depth H, the stiffness for the horizontal displacement is

$$k_{tether} = F_H/d = P/H$$

$$k_{tether} = 300\,\text{MN}/600\text{m} = 0.5\,\text{MN/m}$$

Approximating the installation as a one degree of freedom system for oscillations in the transverse plane, the period can be determined as

$$T = 2\pi/\sqrt{\frac{k_{tether}}{M_{TLP}}} = 2\pi/\sqrt{\frac{500}{160{,}000}} = 112.4\,\text{s}$$

This period is so long compared with the impact duration for normal ships and installation structures that the motions and thus the restoring forces during the impact can be neglected. Then, the impact energy released for crushing can be determined from Eq. (2.37):

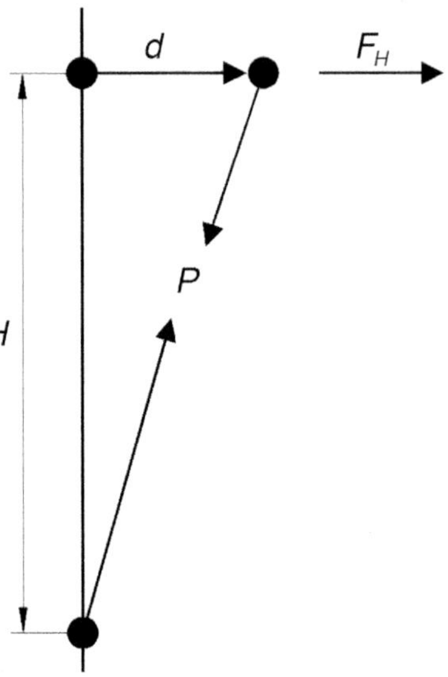

Fig. 2.15 Free body diagram for the collision with the TLP installation.

$$E_\xi = \frac{1}{2}\left(\frac{M_a(1+m_{ax})}{1+\dfrac{M_a(1+m_{ax})}{M_b(1+m_{b2})}}\right)\cdot\dot{\xi}(0)^2 = \frac{1}{2}\left(\frac{100}{1+\dfrac{100}{160}}\right)\cdot 2^2 = 123.1\,\mathrm{MJ}$$

This is the approximation to the energy that must be absorbed by crushing of structures.

Solution to question (b)
For an impact with a ship of displacement 100,000 tons and impact speed $V_{ship}=2.0\,\mathrm{m/s}$, we find that the momentum equilibrium requirement gives the TLP a velocity at the end of the impact, which is

$$V_{TLP} = \frac{M_{ship}\cdot V_{ship}}{M_{Ship}+M_{TLP}} = 0.77\,\mathrm{m/s}$$

Now, neglecting damping and assuming that the impact duration is so short that the TLP has not moved during the ship bow crushing period, then the maximum transverse deflection of the struck TLP is found as

$$\delta_{\max} = \frac{V_{TLP}}{\omega} = V_{TLP}\Big/\sqrt{\frac{k_{tether}}{M_{TLP}}} = 13.77\,\mathrm{m}$$

This is the approximate value for the maximum amplitude for the struck TLP.

2.3.2.2 Calculation examples for a ship collision with a spar platform

The same examples as presented by Jonge and Laukeland (2013) are used here. Consider a tanker with a length of 280m, a total mass of 130,000 tons, and an added mass coefficient of 0.1 that collides with a spar platform at 90 degrees. The spar is a cylindrical body with a radius of 20m and a draught of 160m. The radius of gyration of the spar around the horizontal axis is 70m (it includes the added mass effects). The COG is positioned 80m below the waterline, while the point of impact is 10m above the waterline. The spar mass is 200,000 tons, and the added mass coefficient is taken as 1.0. The topside is part of the spar, and it is included in its mass and inertia. The initial velocity of the tanker is 2.0m/s. For the perpendicular impact, the collision energy to be dissipated is calculated from (2.44):

$$E_\xi = \frac{1}{2}\left(\frac{M_a(1+m_{ax})\cdot\dot{\xi}(0)^2}{1+\dfrac{M_a(1+m_{ax})}{M_b(1+m_b)}+\dfrac{M_a(1+m_{ax})}{M_b(1+j_b)}\left(\dfrac{d}{R_b}\right)^2}\right)$$

$$= \frac{1}{2}\left(\frac{130(1+0.1)\cdot 2^2}{1+\dfrac{130(1+0.1)}{200(1+1)}+\dfrac{130(1+0.1)}{200}\left(\dfrac{90}{70}\right)^2}\right) = 112.6(\text{MJ})$$

Jonge and Laukeland (2013) also performed explicit nonlinear FEA calculations for a ship collision with a spar. A simple model with only three nodes, one beam element, a rigid contact plane, and two mass and inertia elements was used, and the same parameters as in the above example were selected. The added mass was not taken into account for the ship due to the difficulty in implementing added directional mass in the FE analysis. Thus, the added mass of the spar was included by increasing the mass from 200,000 to 400,000 tons; the radius of gyration of the spar around the horizontal axis was 49.5m, while around the vertical axis, it was 9.35m. The friction coefficient was taken as 0.3. The FEA calculations were performed for different collision angles between 15 and 90 degrees. The FEA results with the present analytical results are presented in Fig. 2.16; very good agreement is achieved.

2.3.3 Analytical formulation for ship collisions with fixed offshore structures (or wind turbines)

The assumptions behind the following analysis method are that the global platform natural period (typically about 8.0s) is larger than the duration of the initial contact force between the colliding vessel and the platform and that the global displacement of the topside of the jack-up is small after the initial maximum contact forces between the colliding bodies is achieved. The procedure is also based on the assumption of ship

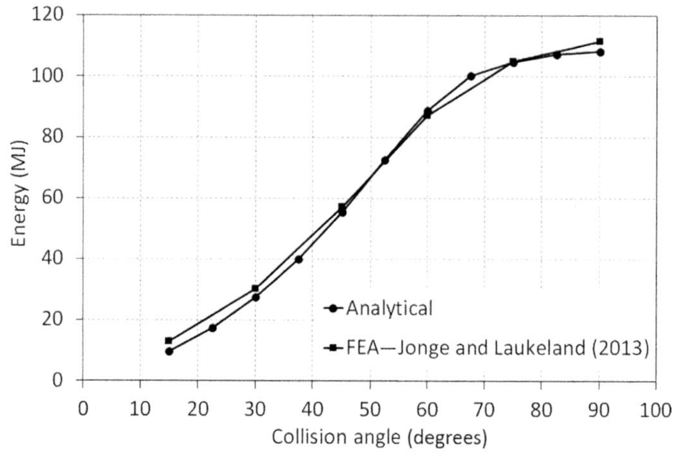

Fig. 2.16 Comparison of collision energy loss between analytical results and FEA results for a tanker collision with a spar at various collision angles.

translations in the horizontal plane only. The crushing load-deflection behaviour is approximately linear, and the structural response of the platform has a linear structural response. With these assumptions fulfilled, the colliding system can be approximated as a two-mass system where one generalized mass represents the supply vessel and the other represents the platform, as shown in Fig. 2.17. If these assumptions are not fulfilled, a time simulation procedure must be applied.

Similar to Eqs (2.1)–(2.3), the equations of motion of the ship due to the impact force F_ξ in the ξ-direction can be expressed as

$$M_a(1+m_{ax})\dot{v}_{ax} = -F_\xi \sin\alpha$$

$$M_a\left(1+m_{ay}\right)\dot{v}_{ay} = -F_\xi \cos\alpha$$

$$M_a R_a^2(1+j_a)\dot{\omega}_a = F_\xi[y_c \sin\alpha - (x_c - x_a)\cos\alpha]$$

where M_a is the mass of the supply vessel. The radius of the ship mass inertia around the COG is R_a, the coordinate of the COG of the striking ship is $(x_a, 0)$, the coordinate of the impact point is (x_c, y_c), and the added mass coefficient for the surge motion is m_{ax}. The added mass coefficient for the sway motion is m_{ay}, and the added mass coefficient of moment for the rotation around the centre of the gravity is j_a.

For the platform response, the following force-stiffness relations are found to be acting:

$$F_\xi = k_{11}\xi_b + k_{12}\xi_p$$

$$F_p = k_{21}\xi_b + k_{22}\xi_p = -M_p\ddot{\xi}_p$$

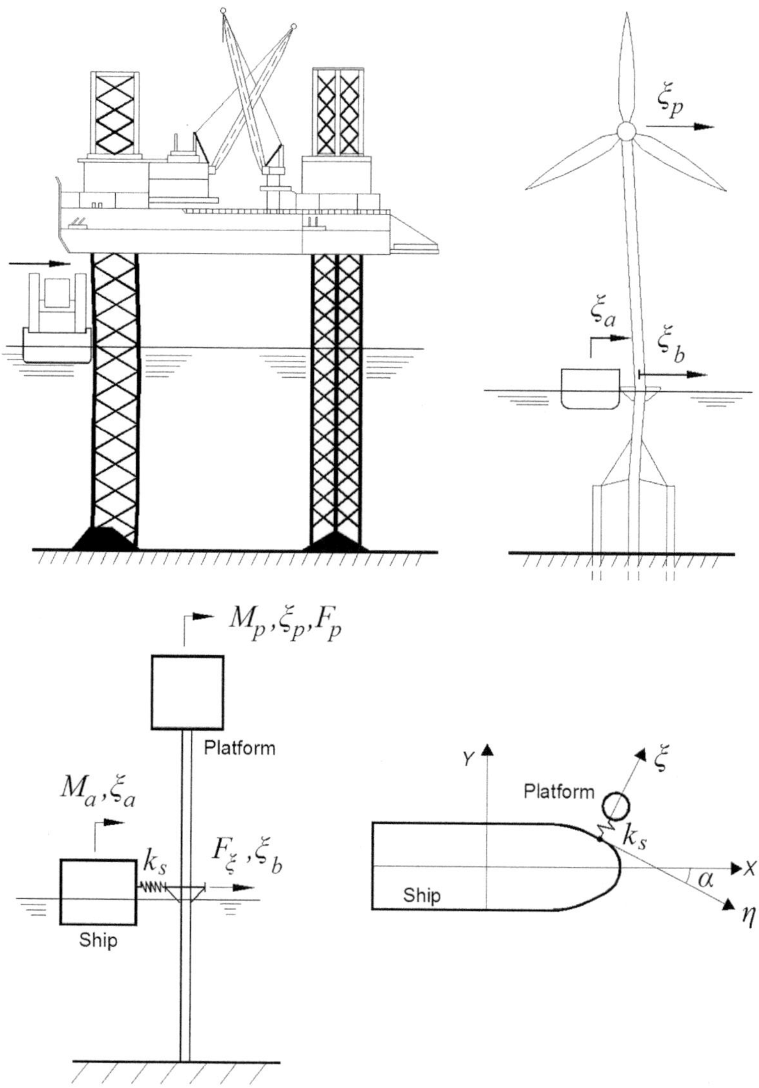

Fig. 2.17 Simplified model of a supply vessel impacting an offshore platform (or wind turbine).

where F_ξ is the collision force between the supply vessel and the platform, F_p is the transmitted force acting on the generalized topside mass M_p of the jack-up, ξ_b is the displacement of the collision point on a leg, and ξ_p is the displacement of the topside mass M_p.

The interaction between the ship and the jack-up is simplified as

$$F_\xi = \begin{cases} k_s(\xi_a - \xi_b), & \text{for } \dot{\xi}_a - \dot{\xi}_b \geq 0 \\ 0, & \text{for } \dot{\xi}_a - \dot{\xi}_b < 0 \end{cases}$$

where k_{11}, k_{22}, k_{12}, k_{21}, k_s are stiffness coefficients and ξ_a is the displacement at the collision point of the supply vessel.

In Pedersen and Jensen (1991), it is assumed that at the end of the collision, where the velocity $\dot{\xi}_a = \dot{\xi}_b$, the displacement ($\xi_p$) of the platform topside can be taken to be small. Then, we get the generalized force at this moment by setting $\xi_p = 0$:

$$F_p = \frac{k_{21}}{k_{11}} F_\xi$$

The impact impulse of the collision between the supply vessel and the platform can be expressed as

$$I_\xi = \frac{M_a}{D_{a\xi}} \left[\dot{\xi}(0) - \dot{\xi}_a\right]$$

The acceleration at the impact point on the ship in the ξ-direction can be expressed as

$$\ddot{\xi}_a = -\frac{F_\xi}{M_a} D_{a\xi}$$

where

$$D_{a\xi} = \frac{1}{1+m_{ax}} \sin^2\alpha + \frac{1}{1+m_{ay}} \cos^2\alpha + \frac{1}{1+j_a} \cdot \frac{\left[y_c \sin\alpha - (x_c - x_a)\cos\alpha\right]^2}{R_a{}^2}$$

The impact impulse on the generalized platform mass can be expressed as

$$I_p = -M_p \dot{\xi}_p = \frac{k_{21}}{k_{11}} I_\xi$$

At the end of the crushing, the velocity of the supply vessel and the velocity of the platform at the collision point are equal; thus, we get

$$\dot{\xi}_a = -\frac{k_{12}}{k_{11}} \dot{\xi}_p$$

The velocity of the topside of the platform is obtained as

$$\dot{\xi}_p = -\frac{\dot{\xi}(0)}{\dfrac{k_{12}}{k_{11}} + \dfrac{k_{11}}{k_{21}} \dfrac{M_p D_{a\xi}}{M_a}}$$

The velocity of the ship's centre of gravity at the end of the collision can be expressed as

$$v_{ax} = \dot{\xi}(0)\sin\alpha - \frac{\sin\alpha}{D_{a\xi}(1+m_{ax})}\left(\dot{\xi}(0) - \dot{\xi}_a\right)$$

$$v_{ay} = \dot{\xi}(0)\cos\alpha - \frac{\cos\alpha}{D_{a\xi}(1+m_{ay})}\left(\dot{\xi}(0) - \dot{\xi}_a\right)$$

$$\omega_a = \frac{y_c\sin\alpha - (x_c - x_a)\cos\alpha}{{R_a}^2(1+j_a)D_{a\xi}}\left(\dot{\xi}(0) - \dot{\xi}_a\right)$$

where

$$\dot{\xi}(0) - \dot{\xi}_a = \frac{\dot{\xi}(0)}{1 + \dfrac{k_{12}k_{21}}{{k_{11}}^2}\dfrac{M_a}{M_p D_{a\xi}}}$$

The energy to be absorbed by crushing of the supply vessel and deformation of the jack-up rig is

$$E_c = E_0 - \left(E_s + E_p\right) \tag{2.46}$$

where

$$E_0 = \frac{1}{2}M_a\left[(1+m_{ax})\sin^2\alpha + \left(1+m_{ay}\right)\cos^2\alpha\right]\dot{\xi}(0)^2$$

is the initial kinetic energy of the supply vessel,

$$E_p = \frac{1}{2}\frac{M_p\dot{\xi}(0)^2}{\left(\dfrac{k_{12}}{k_{11}} + \dfrac{k_{11}}{k_{21}}\dfrac{M_p D_{a\xi}}{M_a}\right)^2}$$

is the kinetic energy of the jack-up topside at the end of the collision, and

$$E_s = E_0 + \frac{1}{2}M_a\dot{\xi}(0)^2\left(\frac{1}{D_{a\xi}\left(1 + \dfrac{k_{12}k_{21}}{{k_{11}}^2}\dfrac{M_a}{M_p D_{a\xi}}\right)^2} - \frac{2}{D_{a\xi}\left(1 + \dfrac{k_{12}k_{21}}{{k_{11}}^2}\dfrac{M_a}{M_p D_{a\xi}}\right)}\right)$$

is the kinetic energy of the supply vessel at the end of the collision.

The energy to be dissipated by the crushing of the ship structure and/or the jack-up rig is

$$E_{crush} = \frac{1}{2}k_s(\xi_a - \xi_b)^2 = \frac{F_\xi^{\,2}}{2k_s} = \frac{k_{11}}{k_{11}+k_s}\left(E_0 - E_s - E_p\right) \tag{2.47}$$

The energy stored in the deformation of the platform is

$$E_{platform} = \frac{k_s}{k_{11}+k_s}\left(E_0 - E_s - E_p\right) \tag{2.48}$$

The maximum collision force can be determined from

$$F_\xi = \left(\frac{2k_{11}k_s}{k_{11}+k_s}E_c\right)^{0.5} \tag{2.49}$$

2.3.4 Analytical and numerical result comparison for fixed platforms

As a first example to evaluate the accuracy of the analytical method, we use the example in Pedersen and Jensen (1991), which is a supply vessel involved in a central collision with the leg of a jack-up rig. The platform stiffness coefficients are

$$k_{11} = 34.5\,\mathrm{MN/m},\ k_{22} = 48.9\,\mathrm{MN/m},\ k_{12} = k_{21} = -27.8\,\mathrm{MN/m},\ \text{and}\ k_s = 18.0\,\mathrm{MN/m}.$$

The mass of the supply vessel, including the added mass, is $M_a = 7.77 \cdot 10^6$ kg. The jack-up rig mass is $M_p = 19.7 \cdot 10^6$ kg. The ship sways into the jack-up rig at a speed of 2.0 m/s. The energy to be dissipated by the crushing is calculated from Eq. (2.46) as 8.13 MJ, which agrees well with the 8.1 MJ obtained by time simulation in Pedersen and Jensen (1991).

2.3.5 Calculation example for fixed platforms

As another illustration, this section provides a calculation example using the present analytical method. Let us consider an ice-strengthened supply vessel colliding with the leg of a jack-up rig at different locations along the ship with a velocity of $V = 2$ m/s in a direction normal to the shipside. The length of the supply vessel is 64.45 m, the breadth is 17.25 m, the draught is 5.5 m, and the displacement is 5000 tons. The added mass coefficients are 0.05 for surge motion, 0.45 for sway motion, and 0.21 for yaw motion. The added mass coefficient for sway motion of 0.45 is estimated using Eq. (2.34) with an impact duration of 0.6 s (Pedersen, 2013).

The jack-up rig is the self-elevating drilling rig described in Section 2.3.4 with a topside overall length of 84 m, width 90 m, and depth 9.4 m, capable of operating in water depths up to 100 m. The generalized mass of the jack-up is 19,700 tons, which was determined from the calculation of the lowest natural frequency of the jack-up

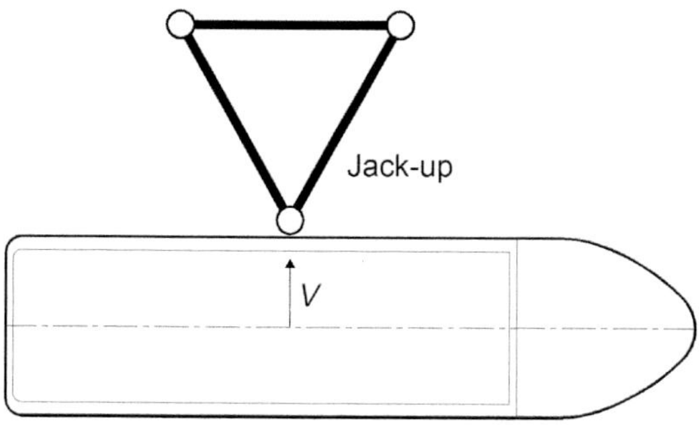

Fig. 2.18 A supply vessel impacting a jack-up rig.

using a beam model. The fixity at the spud cans was estimated, the flexibility at the clamping mechanism was determined from a FEA, and the legs of the rig were modelled as equivalent Timoshenko beams.

The collision scenario is illustrated in Fig. 2.18. Two collision cases were considered to show the effect of the structural flexibility of the jack-up rig on the impact energy: one considers the jack-up rig as rigid and the other as flexible. For the flexible case, the stiffness coefficients are determined from the same beam model used for estimation of the generalized mass for the platform. These stiffness coefficients are

$$k_{11} = 34.5\,\text{MN/m},\ k_{22} = 48.9\,\text{MN/m},\ k_{12} = k_{21} = -27.8\,\text{MN/m, and}\ k_s = 45.0\,\text{MN/m}.$$

For the different impact locations, Table 2.15 and Fig. 2.19 show the total kinetic energy of the supply vessel before the impact (E_k) and the energy released for crushing of the ship and/or platform for both cases (E_c), rigid jack-up and flexible jack-up, and also the elastic energy stored in the global deformations of the jack-up (E_p) for the flexible case.

The results show that the energies released for crushing decrease considerably if the flexibility of the jack-up rig is considered. For the case where the supply vessel collides sideways to the rig (impact amidship), the collision energy to be absorbed by structural crushing is 14.5 MJ for the rigid case, which is equal to the total initial kinetic energy. On the other hand, the crushing energy is only 5.08 MJ for the flexible case (35% of the rigid case). The platform absorbs 6.62 MJ in elastic bending of the collided leg. The remaining energy of 2.8 MJ is stored in the supply vessel and the platform in kinetic form. For the amidship side collision, the collision force between the ship and the leg of the platform is calculated as 36.12 MN in the case of a rigid jack-up where the indentation to the shipside is 803 mm. On the other hand, the collision force between the ship and the leg of the platform is obtained as 21.38 MN where the indentation to the shipside is 475 mm and the deformation of the leg is 620 mm in the flexible case. This shows the importance of considering the position of the collision point and the flexibility of the platform (or wind turbine) in ship–platform collision analysis.

Table 2.15 Results of a 5000 ton supply vessel collision with a jack-up (treated as a rigid rig and a flexible rig).

					Initial kinetic energy of ship	E_{crush} **(rigid rig)**	E_{crush} **(flexible rig)**	$E_{platform}$ **(flexible rig)**
Impact point	x_c **(m)**	y_c **(m)**	**Angle** α **(degrees)**	*x/Lpp*	E_k **(MJ)**	E_{ship} **(MJ)**	E_{ship} **(MJ)**	$E_{platform}$ **(MJ)**
1 (bow)	31.25	2.39	45.00	0.48	12.50	4.66	1.88	2.45
2	30.08	2.75	42.00	0.47	12.71	4.66	1.88	2.45
3	28.59	3.76	37.50	0.44	13.02	4.74	1.91	2.49
4	27.03	5.14	32.50	0.42	13.35	4.89	1.97	2.56
5	24.06	6.88	21.70	0.37	13.95	5.06	2.03	2.64
6	21.09	8.26	14.40	0.33	14.25	5.61	2.23	2.91
7	18.05	8.62	7.30	0.28	14.44	6.28	2.47	3.22
8	15.08	8.62	0.00	0.23	14.50	7.08	2.75	3.59
9	12.03	8.62	0.00	0.19	14.50	8.69	3.30	4.30
10	9.06	8.62	0.00	0.14	14.50	10.51	3.89	5.07
11	6.02	8.62	0.00	0.09	14.50	12.42	4.48	5.84
12	3.05	8.62	0.00	0.05	14.50	13.90	4.91	6.40
13 (mid)	**0.00**	**8.62**	**0.00**	**0.00**	**14.50**	**14.50**	**5.08**	**6.62**
14	−3.05	8.62	0.00	−0.05	14.50	13.90	4.91	6.40
15	−6.02	8.62	0.00	−0.09	14.50	12.42	4.48	5.84
16	−9.06	8.62	0.00	−0.14	14.50	10.51	3.89	5.07
17	−12.03	8.62	0.00	−0.19	14.50	8.69	3.30	4.30
18	−15.08	8.62	0.00	−0.23	14.50	7.08	2.75	3.59
19	−18.05	8.62	0.00	−0.28	14.50	5.79	2.29	2.99
20	−21.09	8.62	0.00	−0.33	14.50	4.75	1.91	2.49
21	−24.06	8.62	0.00	−0.37	14.50	3.95	1.61	2.10
22	−27.03	8.62	0.00	−0.42	14.50	3.32	1.36	1.78
23	−30.08	8.62	0.00	−0.47	14.50	2.80	1.16	1.52
24 (stern)	−31.25	8.62	0.00	−0.48	14.50	2.63	1.09	1.43

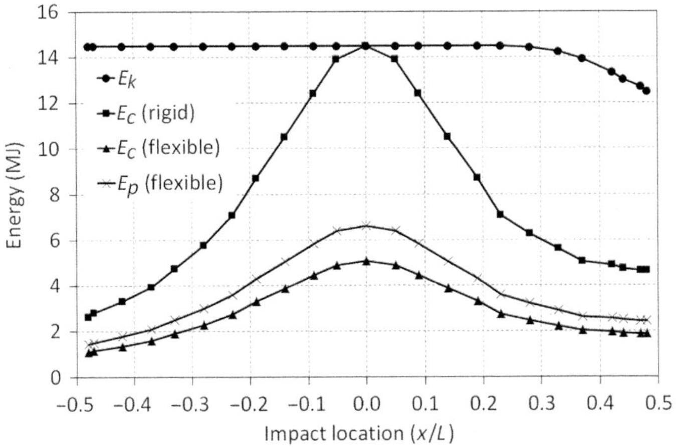

Fig. 2.19 Results of a 5000 ton supply vessel collision with a jack-up leg (treated as a rigid rig and a flexible rig). E_c is the energy released for crushing, E_p is the elastic bending energy in the jack-up leg, and E_k is the initial kinetic energy of the colliding ship.

2.4 Ship collision with a rigid bridge pier or rigid wall

Bridges are exposed to the risk of collision with passing ships that can cause a catastrophic disaster. One of the biggest loss of life occurred in 1983 when a passenger ship collided with a railroad bridge over the Volga River, Russia (176 people lost their life). Besides, there have been numerous vessel collision accidents with bridges that have caused significant damages. On average, some 35 vessel collision incidents are reported every day to the US Coast Guard Headquarters in Washington, DC (Knott and Pruca, 2000). For design assessments, one of the well-known design standards is the US-Guide (1991). The purpose of the present section is to use the analytical theory described in Section 2.2 to analyse the collision energy of ship collision with a bridge pier or a rigid wall.

2.4.1 Analytical formulation

When a ship collides with a rigid wall, the rigid wall can be treated as a fixed object and its mass as infinite (see Fig. 2.20); a bridge pylon can also often be assumed as rigid.

For the case in Fig. 2.20, the analysis procedure in Section 2.2.1 can be simplified as the following:

Sticking case ($|\mu_0| \geq |\mu|$), the energy loss can be expressed as

$$E_\xi = \int_0^{\xi_{max}} F_\xi d\xi = \frac{1}{2} \cdot \frac{M_a}{D_{a\xi} + \mu \cdot D_{a\eta}} \dot{\xi}(0)^2 \tag{2.50}$$

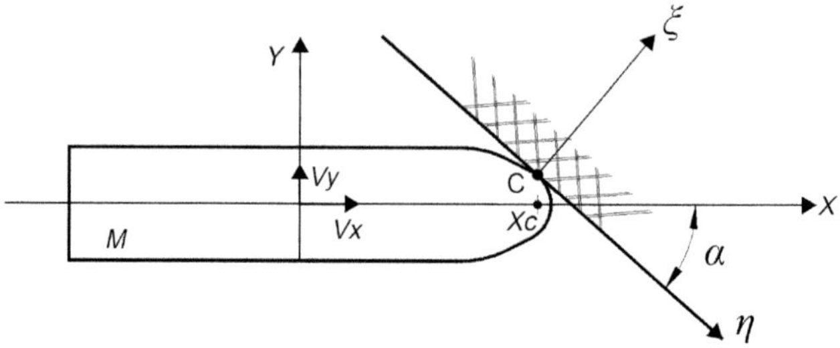

Fig. 2.20 Ship collision with a rigid wall.

$$E_\eta = \int_0^{\eta_{\max}} F_\eta d\eta = \frac{1}{2} \cdot \frac{M_a}{\frac{1}{\mu} K_{a\xi} + K_{a\eta}} \dot{\eta}(0)^2 \tag{2.51}$$

and the impact impulse is

$$I_\xi = \int_0^{\Delta t} F_\xi dt = \frac{M_a}{D_{a\xi} + \mu \cdot D_{a\eta}} \dot{\xi}(0) \tag{2.52}$$

$$I_\eta = \int_0^{\Delta t} F_\eta dt = \frac{M_a}{\frac{1}{\mu} D_{a\xi} + D_{a\eta}} \dot{\xi}(0) \tag{2.53}$$

Sliding case ($|\mu_0| < |\mu|$), the energy loss can be expressed as

$$E_\xi = \int_0^{\xi_{\max}} F_\xi d\xi = \frac{1}{2} \cdot \frac{M_a}{D_{a\xi} + \mu_0 \cdot D_{a\eta}} \cdot \dot{\xi}(0)^2 \left(1 + e^2\right) \tag{2.54}$$

$$E_\eta = \int_0^{\eta_{\max}} F_\eta d\eta = \frac{1}{2} \cdot \frac{M_a}{\frac{1}{\mu_0} K_{a\xi} + K_{a\eta}} \cdot \dot{\eta}(0)^2 \left(1 + e^2\right) \tag{2.55}$$

and the impact impulse is

$$I_\xi = \int_0^{\Delta t} F_\xi dt = \frac{M_a}{D_{a\xi} + \mu_0 \cdot D_{a\eta}} \cdot \dot{\xi}(0)(1 + e) \tag{2.56}$$

$$I_\eta = \int_0^{\Delta t} F_\eta dt = \frac{M_a}{\frac{1}{\mu_0} D_{a\xi} + D_{a\eta}} \cdot \dot{\xi}(0)(1 + e) \tag{2.57}$$

where

$$\dot{\xi}(0) = V_{ax}\sin\alpha + V_{ay}\cos\alpha$$

$$\dot{\eta}(0) = V_{ax}\cos\alpha - V_{ay}\sin\alpha$$

$$\dot{\xi}(\Delta t) = -e \cdot \dot{\xi}(0)$$

$\dot{\eta}(\Delta t) = \dot{\eta}(0) - \frac{K_{a\xi} + \mu_0 \cdot K_{a\eta}}{D_{a\xi} + \mu_0 \cdot D_{a\eta}} \cdot \dot{\xi}(0)(1+e)$ V_{ax} and V_{ay} are the surge velocity and the sway velocity of the ship before collisions, respectively:

$$\mu = \frac{D_{a\xi}\dot{\eta}(0) - K_{a\xi}\dot{\xi}(0)(1+e)}{K_{a\eta}\dot{\xi}(0)(1+e) - D_{a\eta}\dot{\eta}(0)}$$

2.4.1.1 Ship velocities at the end of the collision

The velocities of the colliding ship at the end of the collision can be calculated from

$$v_{ax} = V_{ax} - \frac{I_\xi \sin\alpha + I_\eta \cos\alpha}{M_a(1+m_{ax})}$$

$$v_{ay} = V_{ay} - \frac{I_\xi \cos\alpha - I_\eta \sin\alpha}{M_a(1+m_{ay})}$$

$$\omega_a = -\frac{I_\xi[y_c \cdot \sin\alpha - (x_c - x_a)\cos\alpha] + I_\eta[y_c \cdot \cos\alpha + (x_c - x_a)\sin\alpha]}{M_a R_a^2(1+j_a)}$$

2.4.2 Verification

Here, we consider an example from Brach (1993) who used a slender rod (free in air) impacting a surface with infinite mass, as illustrated in Fig. 2.21. The physical parameters are presented in Table 2.16. A comparison of the present analytical results with the results of Brach is given in Table 2.17, where the initial velocity of the rod in the normal direction of the surface is $\dot{\xi}(0) = 1.0$ m/s. The initial rotational velocity of the

Table 2.16 Physical parameters of the slender rod.

Mass	$M = 1.0$kg
Length	L = 1.0m
Moment of inertia	$I = (1/12)ML^2$
Impact angle α	45 degrees
Initial velocities (m/s)	$\dot{\xi}(0) = 1.0$, $\omega(0) = 0$; $\dot{\eta}(0) = -0.2$, -0.6, 0, 1.0

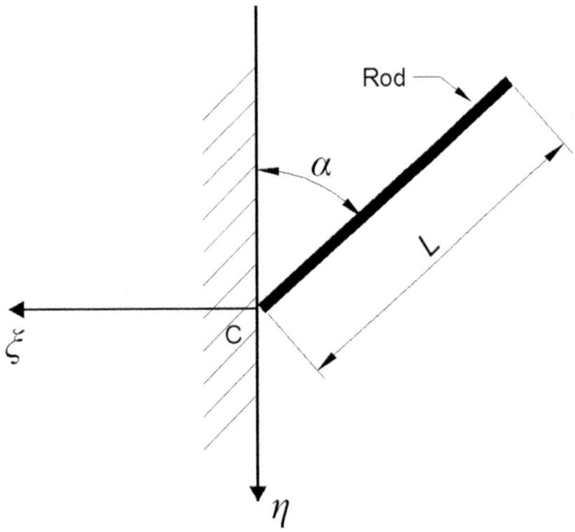

Fig. 2.21 Diagram of a slender rod striking a rigid plane at the point C.

Table 2.17 Comparison of the present results with Brach's results ($\dot{\xi}(0) = 1.0$ m/s).

Initial velocity $\dot{\eta}(0)$	**Coefficient of restitution** $e = -\frac{\dot{\xi}(T)}{\dot{\xi}(0)}$	**Impulse ratio** μ	**Normal impulse** I_ξ**(Nm)**		$\frac{\text{Energy loss}}{\text{Initial energy}} \cdot 100\%$	
			Present	**Brach**	**Present**	**Brach**
0.0	$e=0.5$	0.600[a]	0.938	0.938	46.9	46.9
0.0	$e=0.05$	0.600[a]	0.656	0.656	62.3	62.3
−0.2	$e=0.5$	0.507[a]	0.862	0.862	33.1	33.3
−0.2	$e=0.05$	0.462[a]	0.581	0.580	47.9	47.9
−0.6	$e=0.5$	0.263[a]	0.712	0.732	17.9	17.9
−0.6	$e=0.05$	0.043[a]	0.431	0.569	29.3	29.3
1.0	$e=0.5$	0.905[a]	1.313	1.395	92.2	92.1
1.0	$e=0.05$	0.988[a]	1.031	1.031	99.9	99.6

[a]Critical value that just causes the glancing to stop.

rod is zero, and the initial velocities in the direction parallel to the surface are $\dot{\eta}(0) = 0.0$, -0.2, -0.6, and $1.0\,\text{m/s}$. The coefficients of restitution are $e=0.5$ and 0.05, respectively. The comparison shows that the present analytical results agree quite well with the results of Brach.

2.4.3 Calculation example

Ship damage is often caused by accidents during berthing of the vessel. Furthermore, the design of fender systems, harbour quays, and bridge piers must be based on the statistical distribution of the impact momentum and the energy associated with ship impacts. Therefore, let us consider an incident where a Ro-Ro ship approaches a rigid wall with a forward speed of 2.5 m/s. The main dimensions of the Ro-Ro ship are length 157 m, breadth 24.6 m, and draught 7.25 m. The displacement is 20,000 tons. The coefficients of friction between the ship and the rigid wall are assumed as $\mu_0 = 0.35$ and $\mu_0 = 0.6$, respectively. The added mass coefficients are 0.05 for surge motion, 0.85 for sway motion, and 0.21 for yaw motion.

The calculated results for the energy loss are presented in Table 2.18. Fig. 2.22 presents the ratio between the total energy loss and the total kinetic energy for the ship before the collision. The results show that the larger the friction is, the more energy will be lost due to the crushing of the ship structure. When the collision angle is smaller than 50 degrees ($\mu_0 = 0.6$), the ship will slide against the wall, and when the collision angle is greater than 60 degrees ($\mu_0 = 0.6$), all the kinetic energy of the ship will be lost in crushing.

2.4.4 Ship collision with gravity supported foundations

Gravity-supported offshore wind turbines, bridge piers, break waters, and quays are often held in place without drilling or anchoring. The base consists of a concrete or steel container, which is positioned on the sea bed. The container is then filled with sand or rocks and kept in place by gravity and friction.

In the following, a simplified procedure will be described for a capacity calculation where part of the initial kinetic energy of the striking ship and the surrounding water has to be spent on structural damage to the ship, local crushing damage to the foundation, and sliding of the integrated gravity-supported foundation, such as a wind turbine structure.

The method was described in Pedersen (2013). Assume that the plastic crushing force P associated with the relative displacement between the striking ship and the COG of the foundation is constant and that also the shear sliding force F between the sea bed and the wind turbine foundation to be approximately constant. If $P < F$, the impact will not cause any global displacement of the gravity structure, and the available kinetic energy will be spent on local crushing. Therefore, let us here assume that $P > F$ (see Fig. 2.23).

For the time interval, $0 \le t \le t_0$, where t_0 is the time where the ship velocity $\dot{u}_s$ equals the velocity of the wind turbine foundation structure $\dot{u}_c$, the following simple relations for the ship acceleration, velocity, and displacement can be considered:

$$\frac{M_s \ddot{u}_s}{D_{a\xi}} = -P \Rightarrow \dot{u}_s = \frac{PD_{a\xi}}{M_s} t + V_0$$

$$u_s = -\frac{1}{2} \frac{PD_{a\xi}}{M_s} t^2 + V_0 t$$

As in Section 2.2.1, $D_{a\xi}$ is given by

$$D_{a\xi} = \frac{1}{1+m_{ax}} \sin^2\alpha + \frac{1}{1+m_{ay}} \cos^2\alpha + \frac{1}{1+j_a} \frac{\left[y_c \sin\alpha - \left(x_c - x_g\right)\cos\alpha\right]^2}{R_a^2}$$

Table 2.18 Calculated collision energy of the Ro-Ro ship colliding with a rigid wall at different collision angles ($V = 2.5\,m/s$).

	E_ξ (MJ)		E_η (MJ)		E_{total} (MJ)		$\dot{\eta}(T)$ (m/s)	
Collision angle (degrees)	$\mu = 0.35$	$\mu = 0.6$	$\mu = 0.35$	$\mu = 0.6$	$\mu = 0.35$	$\mu = 0.6$	$\mu = 0.35$	$\mu = 0.6$
10	0.53	0.54	2.09	3.69	2.62	4.24	2.48	2.45
20	2.30	2.48	4.54	8.25	6.84	10.73	2.48	2.39
30	5.82	6.59	7.56	14.11	13.38	20.70	2.48	2.30
40	12.00	14.38	11.33	21.22	23.33	35.60	2.42	2.04
50	22.24	28.36	15.41	26.38	37.64	54.75	2.18	1.36
60	37.89	49.22	17.03	16.41	54.91	65.63	1.53	0.00
70	57.17	57.95	8.40	7.68	65.56	65.63	0.13	0.00
80	63.65	63.65	1.98	1.98	65.63	65.63	0.00	0.00
90	65.63	65.63	0.00	0.00	65.63	65.63	0.00	0.00

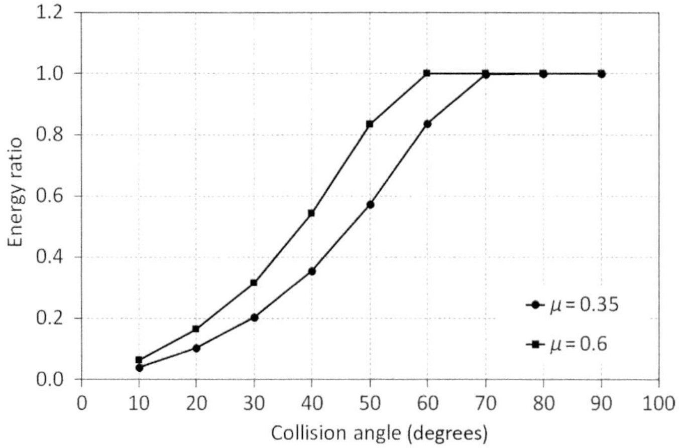

Fig. 2.22 The energy loss ratio of a Ro-Ro vessel colliding with a rigid wall with two different coefficients of friction as function of the collision angle.

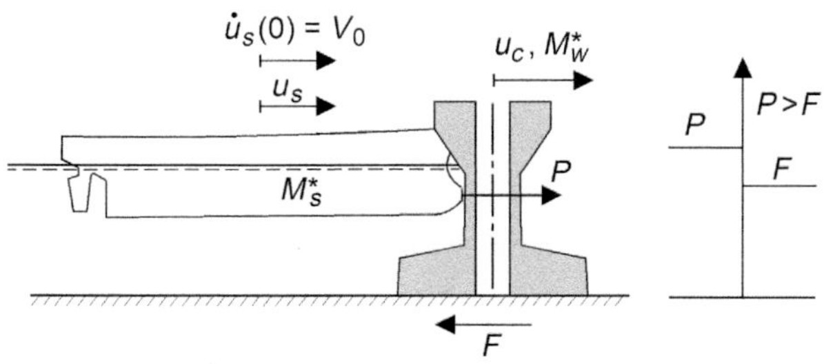

Fig. 2.23 Ship impact with force P against a rigid wind turbine foundation with sliding resistance F.
Based on Pedersen, P.T., 2013. Ship collisions against wind turbines, quays and bridge piers. In: Amdahl, J., Ehlers, S., Leira, B., (Eds.), Collision and Grounding of Ships and Offshore Structures. Taylor & Francis Group, London, ISBN 978-1-138-00059-9.

Similarly, we have for the foundation

$$M_w^* \ddot{u}_c + F = P \Rightarrow \dot{u}_c = \frac{1}{2}\frac{P-F}{M_w^*}t \Rightarrow u_c = \frac{1}{2}\frac{P-F}{M_w^*}t^2$$

where M_w^* denotes the mass of the foundation and wind turbine structure plus the added mass effect of soil and water. The time t_0 is determined from the condition that $\dot{u}_s(t_0) = \dot{u}_c(t_0)$, that is,

$$-\frac{PD_s}{M_s}t_0 + V_0 = \frac{P-F}{M_w^*}t_0$$

or

$$t_0 = V_0 \frac{M_w^* M_s}{P(M_s + M_w^* D_{a\xi}) - F M_s}$$

The maximum crushing distance is given by

$$u_s(t_0) - u_c(t_0) = \frac{1}{2} V_0 t_0$$

Then, the absorbed plastic crushing energy is found as

$$E_{ship} = P[u_s(t_0) - u_c(t_0)] = \frac{V_0^2 M_w^* M_s}{2\left[(M_s + M_w^* D_{a\xi}) - \frac{F}{P} M_s\right]} = E_k \frac{M_w^*}{\left[M_w^* + \frac{M_s}{D_{a\xi}}\left(1 - \frac{F}{P}\right)\right]}$$

where $E_k = \frac{1}{2}\frac{M_s}{D_{a\xi}} V_0^2$ is the initial ship kinetic energy available for structural damage.

The energy balance expresses that the initial kinetic energy, E_k, must be equal to the energy absorbed by crushing of structure, E_a, plus the energy used to move the windmill structure a distance d_{max}:

$$E_k = \frac{1}{2}\frac{M_s}{D_{a\xi}} V_0^2 = F d_{max} + E_{ship}$$

From this expression, the maximum sliding distance is determined as

$$d_{max} = E_k \frac{M_s\left(1 - \frac{F}{P}\right)}{F\left[M_w^* D_{a\xi} + M_s\left(1 - \frac{F}{P}\right)\right]}$$

2.4.5 Numerical example for gravity supported foundations

Let us consider a numerical example where the wind turbine foundation is a concrete gravity caisson foundation consisting of a base plate, chambers with a centrepiece shaft, and an ice cone that is introduced to reduce ice loads (see Fig. 2.23). The foundation has a footprint of 17 × 17 m, a central shaft diameter of 4.25 m, and an empty weight of 1300 tons. The foundations are secured in place by the use of approximately 1400 tons of ballast stones in the bottom cells and approximately 300 tons as scour protection such that the total mass, that is, structural mass plus soil and hydrodynamic mass, equals 3000 tons. The shear resistance of the foundation is taken as 15 MN in the transverse direction.

It is assumed that a drifting vessel or a vessel out of control collides against the wind turbine foundation with a speed of $V_0 = 2.0$ m/s. The impact is assumed

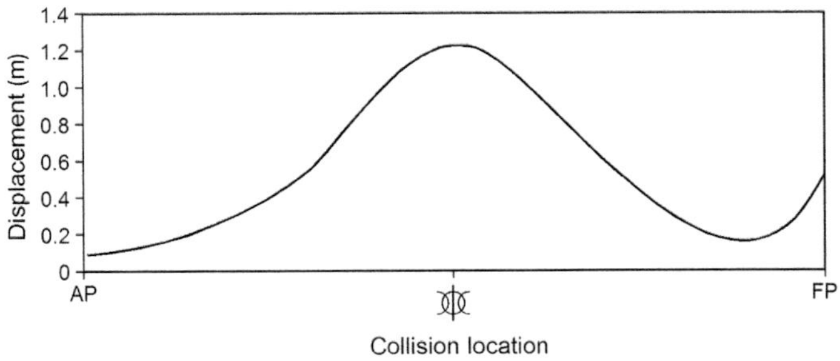

Fig. 2.24 Sliding distance of foundation d_{max} as function of ship impact location.

perpendicular to the side shell of the ship. The total mass of the ship is in this example assumed to be 10,000 tons, and the collision force is taken to be constant and equal to 20 MN.

For this example, Fig. 2.24 shows the foundation displacement as function of the impact location along the hull of the vessel. The outlined calculation procedure shows that for a central midship impact, the crushing of the ship structure together with possible local crushing of the foundation structure will have to absorb 38% of the available kinetic energy for damage and the remaining part of the energy will be spent in moving of the structure. The maximum sliding distance is estimated to be $d_{max} = 1.16$ m for a central collision amidship.

It is seen that provided a limited transverse motion of the foundation is acceptable, the allowable collision loads can be increased significantly above the sliding force for the foundation. Fig. 2.24 also shows that it is important to take into account the actual collision impact location.

2.5 Ship collision with ice

The Arctic is estimated to hold about 20% of the world's remaining recoverable hydrocarbon reserves. Further exploration and transportation may take place off the coasts of Alaska, Canada, Greenland, and Russia. In addition to hydrocarbons, there are also the possibility of shipping large quantities of minerals from various locations in the Arctic to ports in Europe and Asia. The Arctic also sees a growing tourist inflow with the operation of many cruise ships close to the polar circles and numerous ongoing scientific expeditions.

With the greater trade into the Arctic, there is an increasing demand for larger ice-class vessels, particularly oil tankers and liquefied natural gas (LNG) carriers. The global ice-classed fleet has increased significantly in the last 15 years. The harsh environment encountered in cold climates affects the ship safety and operation and requires a focused review and assessment of the ship capability and structural integrity. One important aspect is the strength of the hull structure of ships navigating in ice. The aim of this section is to use the analytical theory described in Section 2.4 to analyse ship collision with ice.

2.5.1 Analytical formulation

Fig. 2.25 illustrates a ship collision with level ice or ice channel edge. It is assumed that the level ice is infinite, and therefore, its mass is also infinitively large. Thus, the formulations in Section 2.4 apply to this purpose and can be used directly for collision analysis.

Ice-classed ships are strengthened in the ice belt of the hull structure to resist the ice loads. For the present analysis, it is assumed that the ship hull is rigid and that the collision energy will be absorbed by crushing the ice. The compressive strength of ice is defined as σ_c for ice with thickness of h_c. The average contact pressure between the ship hull and ice can be determined from $p_\xi = \sigma_c A^\kappa$ (IACS, 2016) where A is the contact area normalized to $1.0\,\text{m}^2$, which can be determined from $A = h_c b_c$ where $b_c = \xi(\tan\theta_1 + \tan\theta_2)$ is the contact length, and κ is the pressure–area coefficient. Therefore, the contact force F_ξ in the ξ-direction (normal to the hull) can be determined from $F_\xi = p_\xi A = \sigma_c A^{\kappa+1}$. Then, the collision energy absorbed by crushing can be further expressed as

$$E_\xi = \int_0^{\xi_{\max}} F_\xi d\xi = \int_0^{\xi_{\max}} \sigma_c A^{\kappa+1} d\xi = \int_0^{\xi_{\max}} \sigma_c (h_c b_c)^{\kappa+1} d\xi$$
$$= \frac{\sigma_c (h_c(\tan\theta_1 + \tan\theta_2))^{\kappa+1}}{\kappa+2} \xi_{\max}^{\kappa+2}$$

Thus, the maximum crushing distance to ice in the ξ-direction is determined from

$$\xi_{\max} = \left(\frac{(\kappa+2)E_\xi}{\sigma_c (h_c(\tan\theta_1 + \tan\theta_2))^{\kappa+1}} \right)^{\frac{1}{\kappa+2}} \tag{2.58}$$

The crushing force F_ξ is determined from

$$F_\xi = \sigma_c A^{\kappa+1} = \sigma_c (h_c \xi_{\max} (\tan\theta_1 + \tan\theta_2))^{\kappa+1}$$
$$= \sigma_c^{\frac{1}{\kappa+2}} ((\kappa+2) h_c E_\xi (\tan\theta_1 + \tan\theta_2))^{\frac{\kappa+1}{\kappa+2}} \tag{2.59}$$

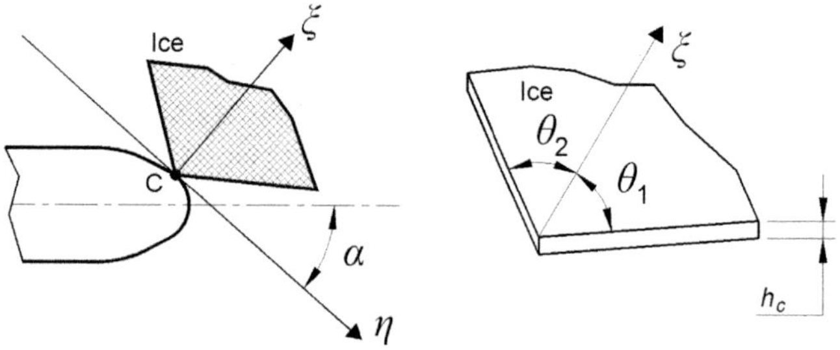

Fig. 2.25 Illustration of a ship collision with level ice.

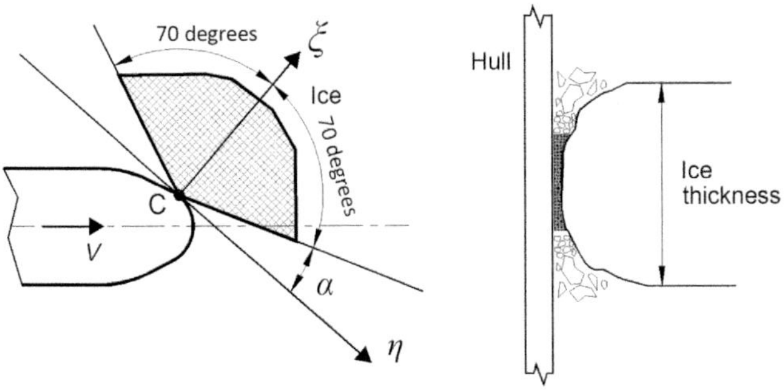

Fig. 2.26 An example of an LNG carrier collision with infinite size of level ice.

The crushing force per contact length is determined from

$$F_{line} = \frac{F_\xi}{b_c} \tag{2.60}$$

2.5.2 Calculation example

It is assumed that an LNG carrier of length 275 m, breadth 44 m, and draught 12 m collides with an edge of level ice of infinite size with forward speed 2.5 m/s (5 knots is a typical sailing speed in ice). The coordinate of the collision location is ($x_c = 84.14$ m, $y_c = 20.09$ m), and angle $\alpha = 12$ degrees, which is at the shoulder of the ship bow as illustrated in Fig. 2.26. The displacement of the ship is 111,600 tons, and the added mass coefficients are taken as 0.05 for surge motion, 0.50 for sway motion, and 0.21 for yaw motion.

The thickness of the level ice is assumed to be 1.0 m with compressive strength of 5.6 MPa. The ice wedge angle is assumed to be $\theta_1 = \theta_2 = 70$ degrees, the pressure–area coefficient κ is taken as -0.50, and the effective friction coefficient between the ship hull and the ice is taken as 0.10.

The results of the calculated collision energy loss are $E_\xi = 8.86$ MJ and $E_\eta = 8.51$ MJ. The total energy loss (17.37 MJ) is only about 4.7% of the initial kinetic energy (366.19 MJ), and the ship slides after the collision.

The crushing distance to the ice in the ξ-direction is calculated as 1.01 m, and the crushing force is $F_\xi = 13.18$ MN. The contact length between the ship hull and the ice is calculated as $b_c = 5.54$ m, the crushing force per contact length is 2.38 MN/m, and the average contact pressure is 2.38 MPa.

Table 2.19 presents the results for the LNG carrier when it collides with the ice at various speeds. The results show that both the crushing distances to the ice and the crushing contact forces increase with the collision speed. However, both the contact force per contact length (line force) and contact pressure decrease when the collision speeds increase (see Fig. 2.27). This is because the larger contact area is, the smaller the contact pressure will be.

Table 2.19 LNG carrier collision with level ice at various collision speeds.

Collision speed (m/s)	Penetration ξ_{max} (m)	Contact length (m)	Force (MN)	Line force (kN/m)	Contact pressure (MPa)
1.0	0.297	1.633	7.16	4382	4.38
1.5	0.510	2.804	9.38	3344	3.34
2.0	0.749	4.115	11.36	2761	2.76
2.5	1.008	5.541	13.18	2379	2.38
3.0	1.286	7.066	14.89	2107	2.11
3.5	1.579	8.678	16.50	1901	1.90
4.0	1.887	10.370	18.03	1739	1.74

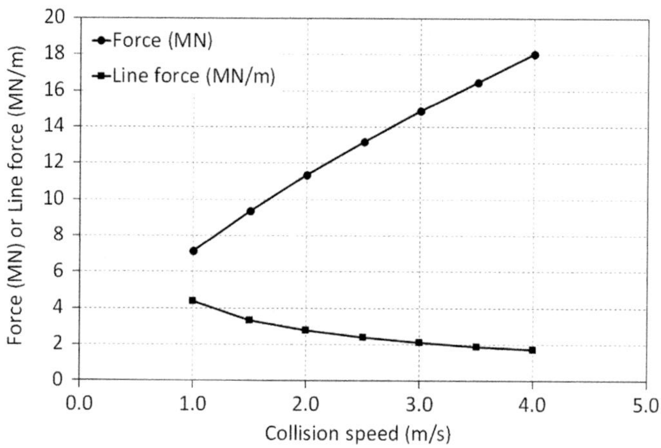

Fig. 2.27 Collision force and line contact force of an LNG carrier collision to level ice (1.0m thick) at different speeds.

2.6 External mechanics of ship grounding events

When ships ground on narrow sharp rocks, the initial kinetic energy is mainly spent by tearing of the bottom plating, and when the ships ground on larger surfaces, the energy will be absorbed by friction and lifting of the vessel.

In the present section, the external mechanics of grounding on relatively plane slopes are first analysed with emphasis on the evaluation of the overall forces on the ship hull girder. The presentation falls into three parts.

Firstly, a mathematical analysis model will be presented for estimation of the contact pressure between the grounded ship and the sea bottom. These ground contact forces can then be compared with the forces that will crush the forward bottom of

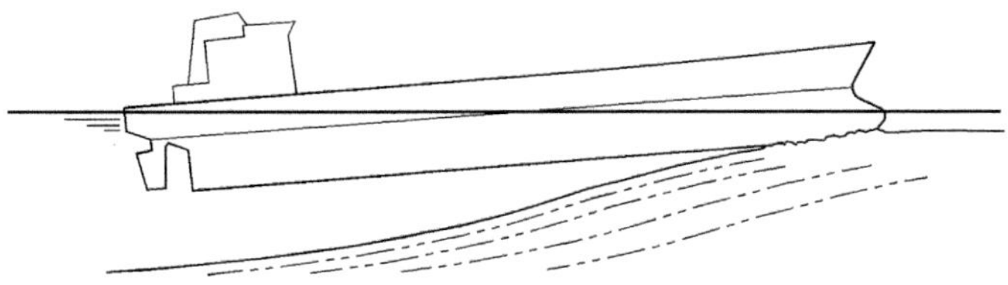

Fig. 2.28 Ship running aground on a plane sandy bottom.

the ship. To get as much insight as possible into the mechanics of ship grounding, simplifications to the problem are sought such that analytical solutions can be derived.

In the second part, the sectional shear forces and bending moments due to grounding of rigid ships will be determined and compared with the ultimate hull girder sectional forces. In the final section, a time simulation procedure is briefly described where a consistent soil model is used for the interaction between the ship and the ground, see Fig. 2.28.

2.6.1 Simplified outer mechanics of ship grounding on hard sea beds

In this section, analysis is presented of the overall response of a ship sailing towards a sloping hard sea bottom in the form of rock, fine-grained sand, or clay. The procedure was developed in Pedersen (1994).

As shown in Fig. 2.29, the ship with mass M is assumed to sail in a direction that is perpendicular to a subsea plane with slope α. The slope ends at a horizontal plane at level h above the still water surface.

Also shown in Fig. 2.29 is an XYZ-coordinate system fixed with respect to the sea bottom. The Z-axis is pointing in a direction opposite to the gravity field; the X-axis lies in the symmetry plane of the ship pointing towards the bow, and at the moment of ground contact, at time $t=0$, the contact point is on the X-axis (x_c, 0, 0). The origin of the XYZ system is placed such that the midship section is in the YZ plane at $t=0$.

The ship centre of gravity has the coordinates (x_g, 0, z_g), and the location of centre of flotation, LCF, is placed a distance x_f forwards of the midship section.

In this simplified procedure, the grounding event will be separated into two phases. In the first phase, the ship is subjected to an impulse caused by the sudden contact with the ground. The soil and the structural response are assumed to be such that the impulse is completely inelastic, and since the ship hull is assumed stiff, this impulse leads to a rapid change of the initial forward speed V such that the ship after the impact has a set of respective surge, heave, and pitch velocity components, v_x, v_z, and $\dot{\theta}_y$, which are compatible with the motion of the contact point along the sloping bottom. In the second phase, the ship is sliding with continuous contact with the ground. Here, the kinetic energy that is available after the end of the first impulse phase is transformed into potential energy and into friction in the contact surface between the ground and the ship as the ship moves up on the sloping bottom.

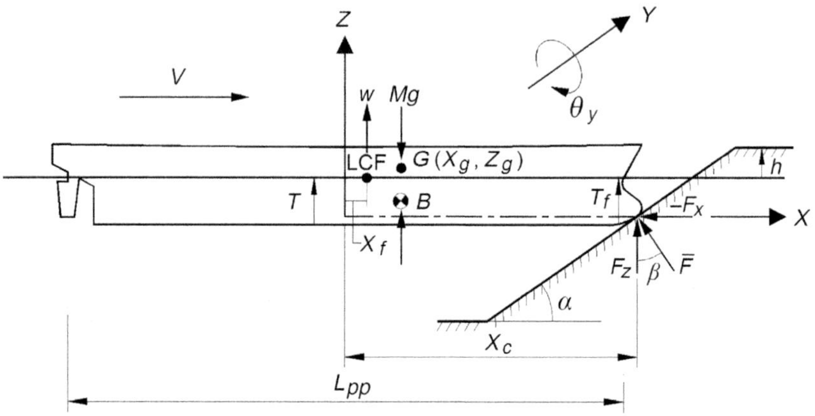

Fig. 2.29 Initial contact with the ground.
Based on Pedersen, P.T., 1994. Ship grounding and hull-girder strength. Mar. Struct. 7, 1–29.

In both phases, the forces exerted on the ship bow are assumed to be governed by a Coulomb's friction law with a constant coefficient of friction μ.

2.6.1.1 *Phase I. Change in momentum at time* t=0

At the first contact between the ship and the slope, the motion of the ship must change instantaneously to be compatible with the imposed kinematic restriction.

Fig. 2.30 shows the forward part of the ship at the moment of impact. In the *XZ*-coordinate system, the components of the impact force $\overline{F}$ are F_x and F_z, and the direction of the impact force is given by the angle defined through

$$|F_x| = \tan\beta F_z$$

The angle β will depend on the unknown detailed interaction between the ship and the slope. As an approximation, a simple Coulomb's friction law is here assumed to control the direction of the impulse.

To determine the angle β, the force components perpendicular to and parallel with the sloping bottom, F_g and F_f, are considered. Assuming that Coulomb's friction law $F_f = \mu F_g$ is valid, then the direction of the impact force and thereby the direction of the impulse *I* are given by

$$\tan\beta = \frac{\mu\cos\alpha + \sin\alpha}{-\mu\sin\alpha + \cos\alpha} \tag{2.61}$$

When the direction of the acting impulse is known, then equations can be established, which express that momentum is conserved in a direction perpendicular to the acting impulse and angular momentum around the contact point, that is,

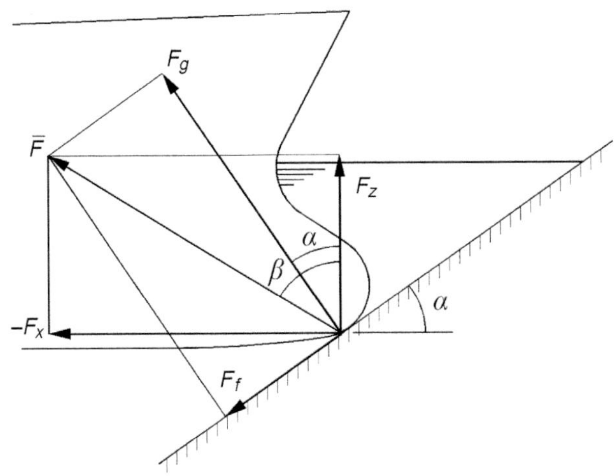

Fig. 2.30 Resulting force acting on the bow at initial impact.
Based on Pedersen, P.T., 1994. Ship grounding and hull-girder strength. Mar. Struct. 7, 1–29.

$$V \cdot M(1+m_{xx})\cos\beta = v_x \cdot M(1+m_{xx})\cos\beta + v_z \cdot M(1+m_{zz})\sin\beta$$

and

$$VM(1+m_{xx})z_g = MR^2(1+j_{yy})\dot{\theta}_y + v_xM(1+m_{xx})z_g + v_zM(1+m_{zz})(x_c - x_g)$$

Here, R denotes the radius of inertia of mass around the centre of gravity, m_{xx} and m_{zz} are dimensionless virtual masses of water for surge and heave motions, and j_{yy} is the dimensionless virtual mass of water moment for the pitch motion. Since the impulse is acting during a very short time compared with $(T/g)^{1/2}$, where T is the mean draught of the ship, these hydrodynamic inertia terms can be evaluated with a high degree of accuracy for infinitely high frequency taking into account the actual water depth.

Finally, the compatibility equation that expresses that the collision point is sliding on the sloping bottom can be expressed as

$$v_z - (x_c - x_g)\dot{\theta}_y = (v_x - z_g\dot{\theta}_y)\tan\alpha$$

From these equations, the velocities v_x, v_z, and $\dot{\theta}_y$ immediately after the moment of impact can be determined:

$$v_x = V\left[1 - \frac{R^2 a_1 a_2 \tan\alpha \tan\beta}{Dx_\alpha}\right]$$

$$v_z = V\left[\frac{R^2 a_2 \tan\alpha}{Dx_\alpha}\right]$$

$$\dot{\theta}_y = -V\left[\frac{a_1 x_\beta \tan\alpha}{D x_\alpha}\right]$$

where a_1 and a_2 are dimensionless added mass ratios:

$$a_1 = \frac{1+m_{zz}}{1+m_{xx}}, a_1 = \frac{1+m_{zz}}{1+m_{xx}}, \text{ and } a_2 = \frac{1+j_{yy}}{1+m_{xx}}$$

The longitudinal x values x_α and x_β are two normally nearly equal effective distances between the centre of gravity and the contact point:

$$x_\alpha = (x_c - x_g) - z_g \tan\alpha$$

$$x_\beta = (x_c - x_g) - z_g \tan\beta$$

and finally,

$$D = a_1 x_\beta + \frac{R^2 a_2 (a_1 \tan\alpha \tan\beta + 1)}{x_\alpha}$$

Since the velocities immediately after the impact have been calculated, the change in momentum P can be determined as

$$\begin{aligned} P &= \int_0^t F(t)dt = M(1+m_{xx})(V - v_x)\sin\beta + M(1+m_{zz})v_z\cos\beta \\ &= VM(1+m_{xx})a_1 a_2 R^2 \frac{\tan\alpha}{x_\alpha D \cos\beta} \end{aligned} \tag{2.62}$$

The time variation of the force $F(t)$ associated with this impulse will depend on the hardness of the ground, the bow form, the bow structural response, and the flexibility of the ship hull. The impact is assumed to be fully plastic. Therefore, the grounding ship will lose some of the kinetic energy during this first phase of the impact. At the end of this phase, the kinetic energy can be expressed as

$$E_k^I = \frac{1}{2}M(1+m_{xx})v_x^2 + \frac{1}{2}M(1+m_{zz})v_z^2 + \frac{1}{2}MR^2(1+j_{yy})\dot{\theta}_y^{\,2}$$

or

$$E_k^I = E_0\left[\left(\frac{v_x}{V}\right)^2 + a_1\left(\frac{v_z}{V}\right)^2 + a_2\left(\frac{R\dot{\theta}_y}{V}\right)^2\right] \tag{2.63}$$

where

$$E_0 = \frac{1}{2}M(1+m_{xx})V^2$$

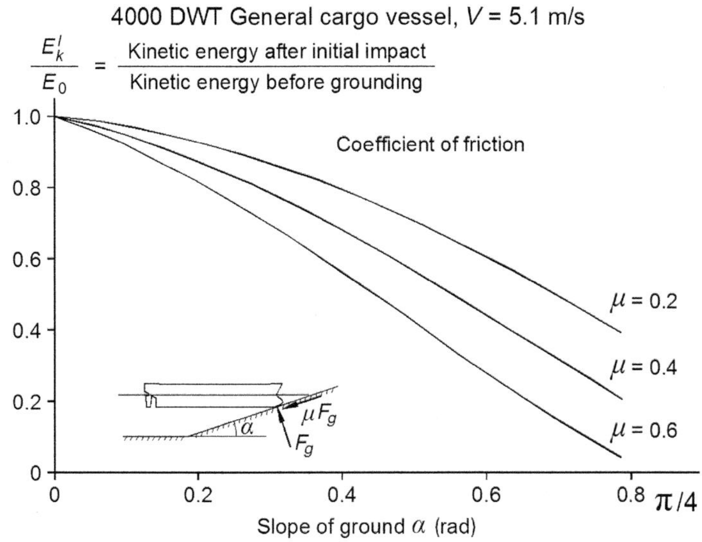

Fig. 2.31 Energy after initial impact as function of slope α. Added mass ratio $a_1 = a_2 = 1.82$. Based on Pedersen, P.T., 1994. Ship grounding and hull-girder strength. Mar. Struct. 7, 1–29.

In the second phase, this energy is assumed to be absorbed by friction, increase in potential energy local plastic deformation of the bow structure and the ground, and axial and vertical bending deformation of the ship hull.

From Fig. 2.31, it is seen that for slope angles α less than around 0.10 rad, the loss of kinetic energy during the initial impact phase can be neglected. But when the slope is around $\pi/4$, nearly all the kinetic energy is lost in this first phase. In this case, the impact is comparable with bow impact against a rigid wall.

2.6.1.2 Phase II. The sliding motion

For simplicity, we shall assume that after the initial impact phase, the ship bow will slide with decreasing velocity along a trajectory that is parallel with the sloping bottom.

In this phase II, it is of importance to determine the ground reaction force as the ship moves up the slope. To determine how far the ship hull will be lifted, the energy dissipation during the motion will be determined. This energy dissipation will then be compared with the initial kinetic energy of the ship reduced by the energy loss in phase I. When all the energy, Eq. (2.63), is dissipated, the ship will stop; this final situation represents the worst static loading condition for the ship hull.

It can be shown that even if this second phase of the grounding event also is a dynamic problem, the vertical acceleration forces in phase II are relatively modest for normal merchant ships. Hence, it is reasonable to neglect inertia forces associated with heave and pitch. In order to further simplify the problem and obtain analytical expressions, the hydrostatic forces and moments are approximated by linear expressions in the heave and pitch displacements, and only the effect of the vertical force component, F_z, is considered. With these assumptions, the relation between the vertical grounding force component F_z acting at a distance

$$l = (x_c - s) - x_F$$

from the centre of flotation, LCF, leads to a vertical displacement, w, of the centre of flotation that can be found from

$$F_z = \rho g A_z w$$

where A_z is the area of the waterplane. This vertical force component can also be expressed as

$$F_z = \rho g A_z \frac{z}{1 + l^2/r^2} \tag{2.64}$$

where z is the vertical displacement of the hull section at the location where the force F_z is acting and where

$$r = \left(\frac{\nabla \cdot GM_L}{A_z}\right)^{1/2}$$

is an equivalent radius of inertia. Here, ∇ is the displaced volume, and GM_L is the longitudinal metacentric height. If we introduce the waterplane area coefficient C_W and the block coefficient C_B, we also have

$$F_z = \frac{Mg}{T}\frac{C_W}{C_B}\frac{z}{1 + l^2/r^2} \tag{2.65}$$

As long as the first part of the bow has contact with the sloping bottom, then $s = 0$, and the distance l is a constant. But when the bow is pushed beyond the plane upper surface of the subsea island, Fig. 2.32, the distance l is a function of the position of the contact point.

The relation between the vertical contact force F and the vertical displacement z, Eq. (2.65), is only valid for 'small' values of z/T, where T is the draught. Therefore, unless the actual motion of the ship is confined to small perturbations of the initial draught, the linear expression (or first-order Taylor expansion) should be replaced by more accurate relations obtained, for example, from the ship's trim diagram.

To determine how far the ship will continue sliding along the surface of the ground, an energy balance shall be established. The kinetic energy of the ship before it strikes the ground less the energy lost in the initial impact in phase I must be equal to the work performed by the friction between the ship bottom and the ground plus the increase in

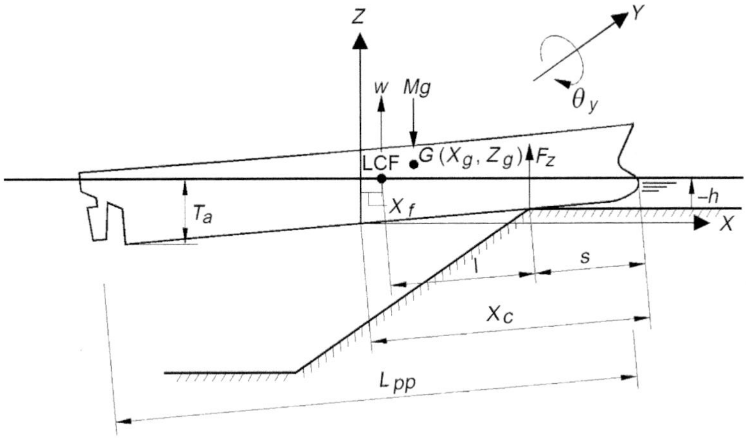

Fig. 2.32 Sliding on a subsea island.
Based on Pedersen, P.T., 1994. Ship grounding and hull-girder strength. Mar. Struct. 7, 1–29.

potential energy from the initial equilibrium state in the free-floating position to the end of the grounding. That is, the energy $E_k(t)$ during the second phase of the impact can be expressed as

$$E_k(t) = E_k^I - \int_0^{u(t)} F_f du - \int_0^{z(t)} F_z dz$$

where u is the distance the bow slides along the slope. This expression can be reduced to

$$E_k(t) = E_k^I - \int_0^{z(t)} \frac{\tan\beta}{\tan\alpha} F_z(z) dz \tag{2.66}$$

If the initial velocity is so large that the bow of the ship reaches the flat plateau of the ground, then the kinetic energy $E_k(t)$ will decrease as a function of the distance s, which the ship slides over the horizontal surface:

$$\begin{aligned} E_k(t) = E_k^I &- \int_0^{T_f+h} F_f(z)\frac{1}{\sin\alpha} dz - \int_0^{T_f+h} F_z(z) dz - \mu \int_0^s F_z(s_1) ds_1 \\ &- \int_0^s F_z \frac{T(s_1) + h - w(s_1)}{l(s_1)} ds_1 \end{aligned} \tag{2.67}$$

where T_f is the forward draught and h is the height of the horizontal plane above the sea surface; see Fig. 2.32.

It is the final resting situation that represents the worst loading case for the ship hull. In the simplified case where the ship does not reach the flat plateau of the ground and where the linear approximation (2.65) can be used for the ground reaction F_z as function of the lifted distance z, then Eq. (2.66) shows that when the ship comes to rest ($E_k(t)=0$), the bow will be lifted the distance

$$z_u = \left(\frac{2E_k^I C_B T (1 + l^2/r^2) \tan\alpha}{C_W M g \tan\beta} \right)^{1/2} \tag{2.68}$$

Since the kinetic energy E_k^I (see Eq. (2.63)) is proportional to the square of the initial velocity V, it is observed from Eq. (2.68) that z_u and hence also the sliding distance $x_{\max}$

$$x_{\max} = V\left((1 + l^2/r^2) \frac{(E_k^I/E_0) T C_B}{g C_W \tan\alpha \tan\beta} \right)^{1/2} \tag{2.69}$$

is proportional to V.

In addition, the vertical reaction force F_z (see Eq. 2.65) is proportional to z_u whereby the maximum vertical ground reaction at the contact point will, as a first approximation, be proportional to the initial velocity.

2.6.2 Model experiments and numerical examples

2.6.2.1 Model experiments

In Pedersen (1998), some test results were presented where ship models that were free to heave and pitch were forced to slide up along a sand slope. The tests were performed at the Danish Hydraulic Institute (Hansen Ottesen et al., 1994).

The experimental set-up is illustrated in Fig. 2.33, with a tanker model in scale 1:60 with a length of $L_{pp}=2.57$ m.

The bow of the ship model was connected to the rest of the hull by a dynamometer making it possible to measure the vertical shear force at this cross section of the ship. Based on the measured vertical shear force, the accelerations of the ship in the vertical plane, the heave and pitch motions, and the hydrostatic properties, the ground reaction forces F_z acting during the tests can be calculated. Fig. 2.34 shows typical results from such an experiment where the free-floating model is forced onto the slope with a constant speed $V=0.664$ m/s together with numerical predictions using the theory presented above.

It is immediately seen that the grounding tests qualitatively and quantitatively verify the mathematical model described in this section. That is, the grounding event is initiated by a short duration impact with relatively large forces followed by a gradually increasing ground reaction.

Besides forces, accelerations, heave and pitch motions, and position and speed of the ship model, also the furrow made by the ship in the sand slope was measured.

The measurements showed that for fine-grained, saturated, and drained sand, the furrow was quite shallow and that the coefficient of friction has a relatively constant value in the main part of the grounding event where the bow slides upwards on the

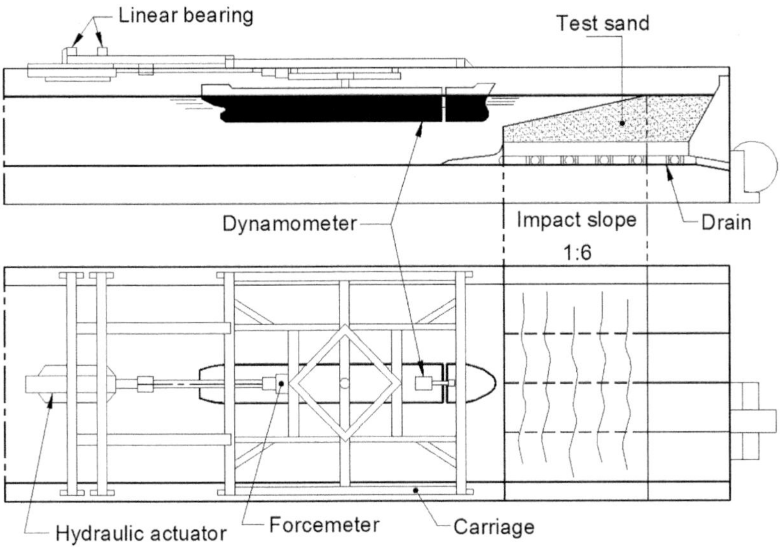

Fig. 2.33 Grounding of freely floating segmented ship model.
Based on Pedersen, P.T., 1994. Ship grounding and hull-girder strength. Mar. Struct. 7, 1–29.

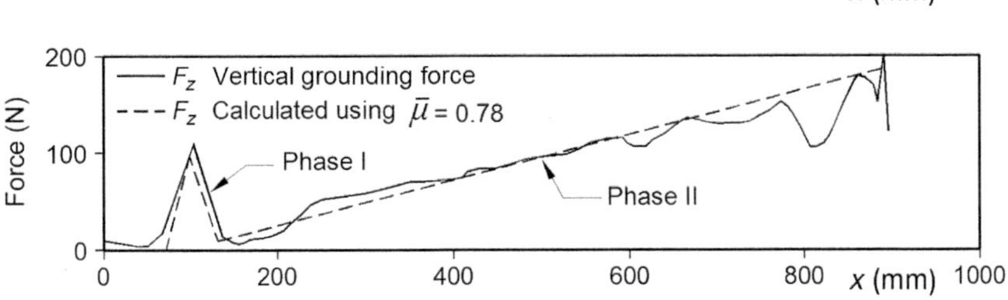

Fig. 2.34 Sample of measured and calculated longitudinal and vertical components of the grounding forces, for a ship model with bulbous bow, $V_{model} = 0.644\,\text{m/s}$.
Based on Pedersen, P.T., 1994. Ship grounding and hull-girder strength. Mar. Struct. 7, 1–29.

slope. Therefore, under these conditions, the experiments indicated that the mathematical model presented here is a reliable analysis model and the experiments served to provide realistic frictional coefficients.

The calculated empirical values of the coefficient of friction were relatively large, indicating that the coefficient μ in addition to simple friction also includes a contribution from the force used to deform the sloping surface.

For more coarse sand that in full scale would correspond to small stones or gravel, the test results showed that the furrow made in the bottom was so deep that this effect should be included in a realistic mathematical grounding model. In the next section, such an integrated soil/ship grounding mechanics model is briefly presented.

2.6.2.2 Numerical example: Grounding of a 4000 DWT ship in ballast

Fig. 2.35 shows the results of a nonlinear numerical calculation of the variation of the kinetic energy of a 4000 DWT ship in ballast with the following main characteristics:

$$L_{pp} = 119.5\,\text{m},\ B = 17.5\,\text{m},\ T_a = 4.16\,\text{m},\ T_f = 2.66\,\text{m},\ C_B = 0.65, C_W = 0.76,\ \text{and}\ V = 5.1\,\text{m/s}$$

The ship grounds on a slope 1:6.5, that is, $\alpha = 0.15$ rad, and the coefficient of friction is taken as $\mu = 0.6$. At sea level, the ground is horizontal, that is, $h = 0$. From Fig. 2.35, first, the drop in kinetic energy due to the initial impact is observed. After the impact, the kinetic energy is steadily decreasing while the forward part of the ship is sliding on the sloping bottom. This ship has so much kinetic energy that in the final resting position, the bow is lifted out of the water, and the bow stops behind the 'coastline'. In general, only relatively small ships in ballast conditions can in the final resting position have the bow lifted above the water surface.

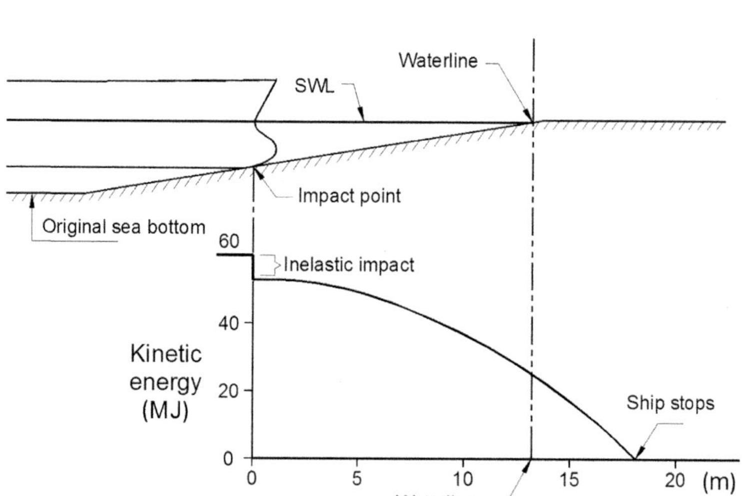

Fig. 2.35 Variation in kinetic energy with sliding distance.
Based on Pedersen, P.T., 1994. Ship grounding and hull-girder strength. Mar. Struct. 7, 1–29.

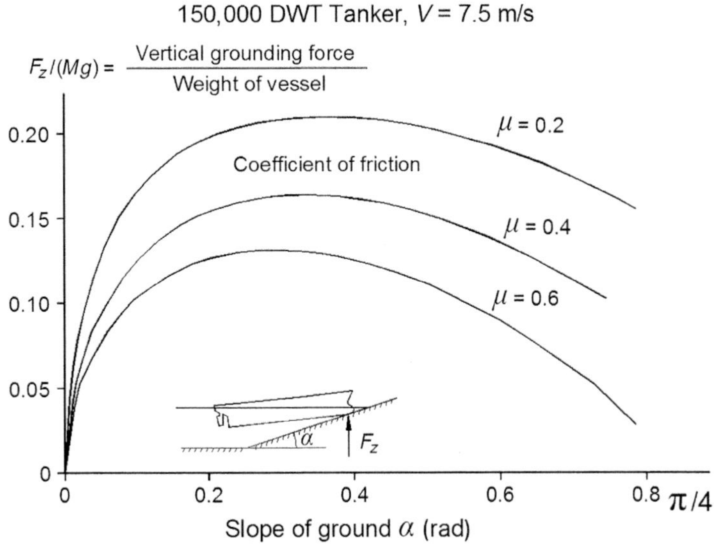

Fig. 2.36 Variation of the vertical ground reaction with slope α for different values of the coefficient of friction μ.
Based on Pedersen, P.T., 1994. Ship grounding and hull-girder strength. Mar. Struct. 7, 1–29.

2.6.2.3 Numerical example: 150,000 DWT tanker

Using the simplified analysis procedure, Fig. 2.36 shows the variation of the maximum ground reaction with slope α for a 150,000 DWT tanker grounding at high speed for various values of the coefficient of friction. The main characteristics of this ship are

$$L_{pp}=274.0\,\text{m},\ B=47.0\,\text{m},\ T_a=15.96\,\text{m},\ T_f=14.46\,\text{m},\ C_B=0.84, C_W=0.90,\ \text{and}\ V=7.5\,\text{m/s}$$

2.6.3 Sectional forces after grounding

It is often observed that when a ship grounds with relatively large speed, then the hull breaks either instantly, or it breaks after a short time during which it has been subjected to the action of tide and waves. This indicates that the sectional forces induced in the ship can become large compared with the strength of the ship hull.

In the following, a simplified analysis of the induced sectional forces will be presented for the case where the ship is at rest at its final position.

To obtain insight in the problem, the discussion will be based on an analytical approach where we utilize the linearized relation between ship motions and the associated forces.

Due to grounding, the changed buoyancy per unit length along the hull can be approximated as

$$q(x) = \rho g B(x)\left[w - \theta_y(x - x_F)\right]$$

or

$$q(x) = \rho g B(x)\frac{z}{1+(l/r)^2}\left[1 + \frac{l}{r^2}(x - x_F)\right]$$

The grounding-induced changes in the sectional shear force $Q(x)$ in the ship hull beam can be determined from

$$\frac{dQ}{dx} = -q$$

Thus, the shear force has an extreme value at a position $x = x_0$ where

$$x_Q = x_F - \frac{r^2}{l} \tag{2.70}$$

Integration of the line load $Q(x)$ leads to the shear force:

$$Q(x) = \frac{z\rho g}{1+(l/r)^2}\left[\left(1 - \frac{l x_F}{r^2}\right)\int_{-\frac{L}{2}}^{x} B(\xi)d\xi + \frac{l}{r^2}\int_{-\frac{L}{2}}^{x} \xi B(\xi)d\xi - H(x - x_c)A_z\right] \tag{2.71}$$

Here, $H(x - x_c)$ is the unit step function, which is zero for $x \le x_c$ and 1.0 for $x > x_c$, and $L = L_{pp}$ is the ship length.

If the ship breadth $B(x)$ is approximated by a mean value B_M, the following approximation for the variation of the sectional shear force is found:

$$Q(x) \approx \frac{z\rho g A_z}{1+(l/r)^2}\left[\left(1 + \frac{lL}{2r^2}\left(\frac{x}{L} - \frac{1}{2}\right)\right)\left(\frac{x}{L} - \frac{1}{2}\right) - H(x - x_c)\right] \tag{2.72}$$

Substitution of the x-coordinate of the extreme point for the grounding-induced shear force at $x = x_Q \approx -(r^2/l)$ $x = x_Q \approx -\frac{r^2}{l}x = x_Q \approx -\frac{r^2}{l}$ from Eq. (2.70) to Eq. (2.72) leads to

$$Q_{ext} \approx \frac{z\rho g A_z}{4\left(1+(l/r)^2\right)}\left(2 - \frac{lL}{2r^2} - \frac{2r^2}{lL}\right) \tag{2.73}$$

It is seen that when the support point is at the forward end of the ship, the maximum shear force will occur at the support point where the shear force equals the ground reaction force:

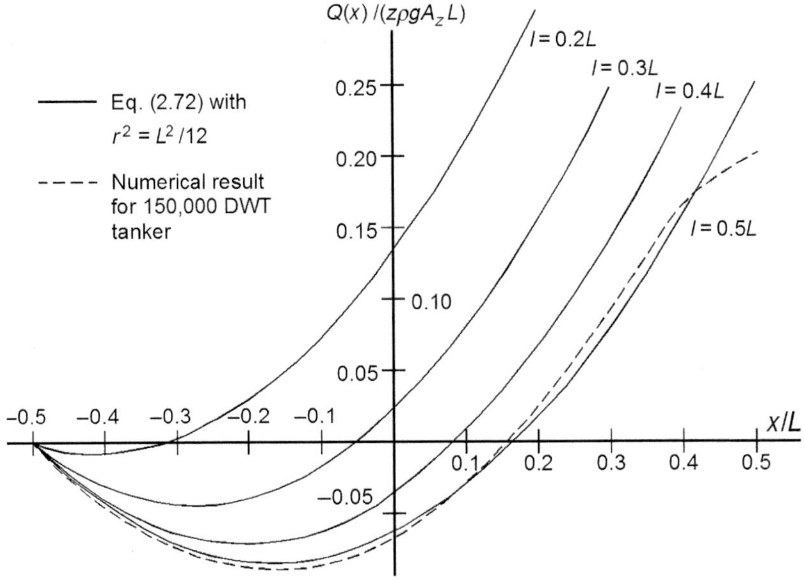

Fig. 2.37 Variation of the nondimensional shear force $Q(x)/(z\rho g A_z)$ along the hull beam. Based on Pedersen, P.T., 1994. Ship grounding and hull-girder strength. Mar. Struct. 7, 1–29.

$$Q_{\max} = \frac{\rho g A_z}{1 + (l/r)^2} z \quad > |Q_{ext}| \tag{2.74}$$

Fig. 2.37 shows how the nondimensional sectional shear force $Q(x)/(z\rho g A_z)$ caused by the final uplift z of the ship varies with the position l of the effective support point according to the approximation expressed by Eq. (2.72). It is also seen that the maximum shear force will occur at the support point. Included in Fig. 2.37 are results from an integration procedure for a 150,000 DWT tanker, which takes into account the actual variation of the waterline.

From the expression (2.72) for the sectional shear force, the additional vertical bending moment caused by the grounding can be determined by integration:

$$M(x) \approx \frac{z\rho g A_z L}{1 + (l/r)^2} \left[\left(\frac{x}{L} + \frac{1}{2} \right) \left[\frac{lL}{2r^2} \left(\frac{x}{L} \right) \left(\frac{x}{L} - \frac{1}{2} \right) + \frac{1}{2} \left(\frac{x}{L} + \frac{1}{2} \right) - \frac{4l}{L} \left(\frac{x^2}{L^2} - \frac{x}{2L} + \frac{1}{4} \right) \right] - \frac{x - x_c}{L} H(x - x_c) \right]$$

Due to the assumptions made, this expression should only be used for $r^2 \approx L^2/12$.

The extreme value of this grounding-induced bending moment occurs at the location along the hull beam where the corresponding shear force $Q(x)$ equals zero, that is, at

$$x_{m,ext} = \frac{L}{2} - \frac{2r^2}{l}$$

This leads to the following expression for the approximation to the maximum grounding-induced vertical bending moment in the ship hull:

$$M_{ext} \approx \frac{z\rho g A_z}{1+(l/r)^2}\left(1-\frac{2r^2}{lL}\right)\left[16\frac{r^4}{lL^3}-\frac{r^2}{L^2}\left(4+\frac{L}{l}\right)+\frac{l}{L}\right] \tag{2.75}$$

When the support point is close to the forward perpendicular, $l=0.5L$, this expression is nearly independent of practical values of r and has the value

$$M_{ext} \approx -\frac{1}{27}z\rho g A_z L \tag{2.76}$$

Fig. 2.38 shows the variation of the nondimensional vertical bending moment in the ship hull for various locations of the final effective support point. The approximation based on Eq. (2.75) is compared with the results from an accurate numerical integration procedure based on the exact hull shape for a 150,000 DWT tanker. It is seen that the approximation using an average constant breadth B_M is quite good for this ship type.

The expression for the maximum shear force (2.74) and for the maximum bending moment (2.75) can be used to estimate the hull loading in a given grounding scenario when the effective grounding force position l and uplifted distance z are known.

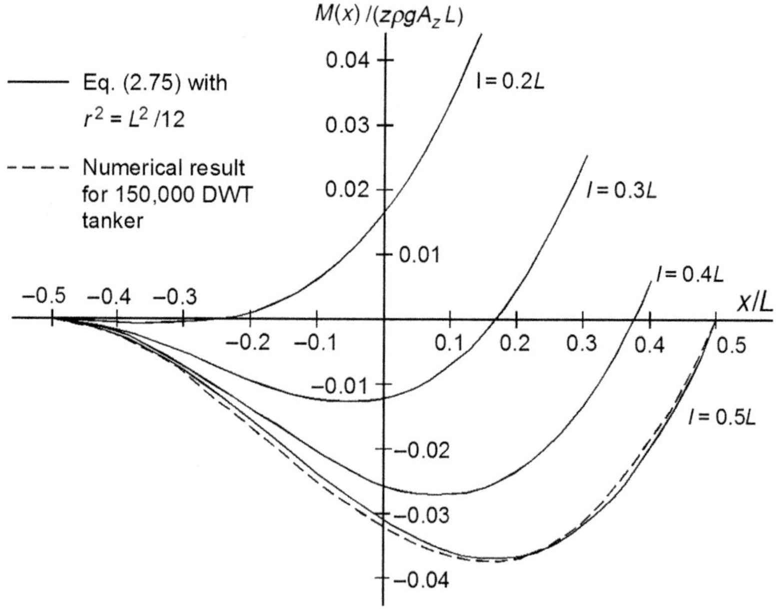

Fig. 2.38 Nondimensional grounding-induced vertical bending moment $M(x)/(z\rho g A_z L)$ for various positions of the effective support point.
Based on Pedersen, P.T., 1994. Ship grounding and hull-girder strength. Mar. Struct. 7, 1–29.

To evaluate the possible consequences of the grounding-induced sectional forces and moments, these generalized forces can be compared with the characteristic ultimate strength of the hull. The grounding-induced hull girder shear forces and bending moments will act together with the still water sectional forces. Since the ultimate strength capacity of the hull girder is related to the sum of the still water loads and the wave-induced loads, it seems appropriate to compare the grounding-induced hull girder sectional forces with wave-induced forces.

The International Association of Classification Societies (IACS) requires the maximum sagging wave-induced bending moment for ocean-going ships to be not less than

$$M_{IACS} = -110\left[10.75 - \left(\frac{300-L}{100}\right)^{1.5}\right]L^2B(C_B+0.7)$$

For the final resting situation of an event where a 150,000 tons tanker grounds on a plane hard surface with inclination α, Fig. 2.39 shows a comparison between the grounding-induced bending moments and the IACS requirements to the wave-induced sagging bending moments.

It is seen that for relatively high grounding speeds on a continuous sloping bottom, there is a high probability that the grounding-induced sagging bending moment will exceed the upper limit for the sagging wave-induced bending moments. In addition, the grounding bending moments can be increased in the stranded condition due to tide

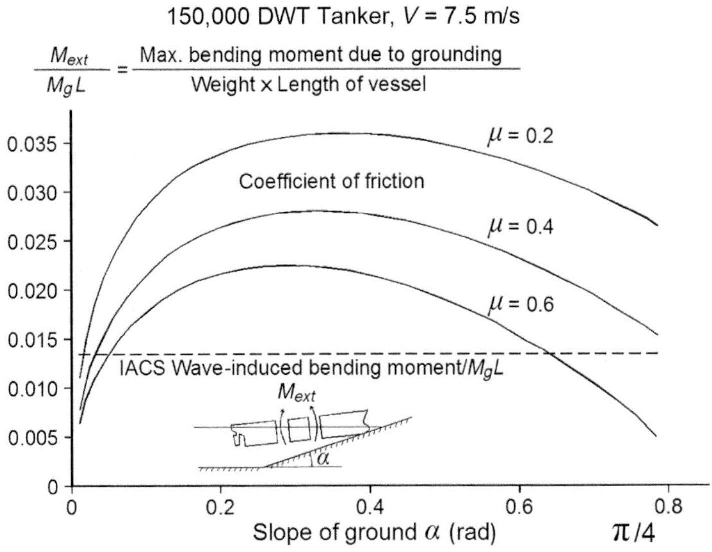

Fig. 2.39 Comparison of extreme values of grounding-induced sagging bending moment with IACS requirement to wave-induced bending moments for a 150,000 DWT tanker.
Based on Pedersen, P.T., 1994. Ship grounding and hull-girder strength. Mar. Struct. 7, 1–29.

and wave action. Similarly, large relative hull girder shear forces will be introduced in these situations.

Hand calculation exercise

A tanker has the following main dimensions: $L=322.7\,\text{m}$, $B=58.2\,\text{m}$, $T=20.8\,\text{m}$, $C_B=0.81$, $C_W=0.91$, and $r=\sqrt{\nabla GM_L/A_z}=85.6\,\text{m}$. This tanker has an initial velocity of 6.0 m/s when it runs aground on a hard sand slope with the constant inclination 1:6.

Assume that the effective coefficient of friction μ is 0.65 and use the simplified procedure to determine the following:

(a) The position when the ship is at rest
(b) The ground reaction force in the final position
(c) The maximum value of the bending moment in the final resting position

Solution to question (a)

The effective friction angle β can be determined from Eq. (2.61):

$$\tan\beta=\frac{\mu\cos\alpha+\sin\alpha}{-\mu\sin\alpha+\cos\alpha}\ \text{or}\ \tan\beta=\frac{0.65\cos 1/6+\sin 1/6}{-0.65\sin 1/6+\cos 1/6}\Rightarrow\tan\beta=0.984$$

Then, Eq. (2.69) can be used to determine the sliding distance:

$$x_{\max}=V\left[\left(1+\frac{l^2}{r^2}\right)\frac{(E_k^I/E_0)TC_B}{gC_W\tan\alpha\tan\beta}\right]^{\frac{1}{2}}$$

From Fig. 2.31, the ratio E_k^I/E_0 is estimated as 0.75. Thus,

$$x_{\max}=6\left[\left(1+\left(\frac{161}{85.6}\right)^2\right)((0.75\times 20.8\times 0.81\times 6)/(9.81\times 0.91\times 0.984))\right]^{0.5}=37.5\,\text{m}$$

and

$$z_{\max}=x_{\max}\tan\alpha\Rightarrow z_{\max}=6.25\,\text{m}$$

Solution to question (b)

The ground reaction force in this final position is found from Equation (2.65):

$$F_z=\frac{MgC_W}{T\ C_B}\frac{z}{(1+l^2/r^2)}$$

Here,

$$M=\rho\cdot C_B\cdot L\cdot T\cdot B$$

$$M=1025\cdot 0.81\cdot 322.7\cdot 20.8\cdot 58.2\Rightarrow M=324.34\cdot 10^6\,\text{kg}$$

That is,

$$F_z=\frac{324.34\cdot 9.81\cdot 0.91}{20.8\cdot 0.81}\cdot\frac{6.25}{1+(161/85.6)^2}\Rightarrow F_z=236.7\,\text{MN}$$

Solution to question (c)

In this case where the ground reaction force is acting at the fore body of the ship, Eq. (2.76) can be applied to estimate the maximum hull girder bending moment:

$$M_{ext} \approx -\frac{1}{27} z\rho g A_z L$$

$$M_{ext} \approx -\frac{1}{27} z\rho g C_W B L^2$$

$$M_{ext} \approx -\frac{1}{27} 6.25 \cdot 1025 \cdot 9.81 \cdot 0.91 \cdot 58.2 \cdot 322.72^2 \Rightarrow M_{ext} \approx -1284\,\text{MN}$$

2.6.4 Analysis of groundings on soft sea beds

As mentioned previously, the apparent coefficient of friction μ is a very important parameter. Therefore, in order to obtain a better basis for analysis of ships running aground on soft grounds, there is a need for a mathematical model for the interaction between ship and ground, which can be used in a consistent way to determine the detailed behaviour during the initial phases of the contact with the ground and, at the same time, to calculate the effective coefficient of friction in the second phase of the grounding event (see Pedersen and Simonsen, 1993).

The grounding forces on a ship bow arise when the bow is being pressed into the saturated soil. This generates a movement of the soil and of the pore water. In general, the dynamic behaviour of saturated soil is strongly influenced by the presence of pore water. The phenomenon of large dynamic deformations of saturated soil is extremely difficult to model numerically; see Hansen Ottesen et al. (1994). In the analysis by Pedersen and Simonsen (1993), it is assumed that the penetration of the ship bow generates a flow of pore water through the sand grain skeleton. This flow is governed by Darcy's law, and it is driven by the pressure in the pore water at the bow. In addition to this pore water pressure, the bow is subjected to the effective stresses in the grain skeleton at the bow surface. These stresses are determined by the theory of frictional soils in rupture. Frictional stresses on the bow surface are assumed to be related to the normal pressure by a simple Coulomb relation. The total soil reaction force is found by integration of normal pressure and frictional stresses over the surface of the bow.

Due to the special form of ship bows, the contact area between soil and bow can be approximated by a series of ellipsoidal forms and planes. This is used in the calculation of the soil and pore water kinematics in the reference given above.

The analysis has been verified through a comparison of calculated stopping lengths, effective coefficients of friction, and sea bed penetrations with corresponding experimental results obtained by model tests and full-scale tests where a condemned vessel with a V-shaped bow, a displacement of 300 metric tons, length between perpendiculars 29.55 m, moulded breadth 6.60 m, and a draught amidship equal to 2.67 m was run aground a number of times. See Sterndorff and Pedersen (1996).

As a numerical example of grounding on a soft sea beds, Simonsen and Pedersen (1993) considered a 220,000 DWT tanker with an initial velocity of 5.1 m/s. The main particulars of the ship were length, $L_{pp}=259.0$m; breadth $B=36.6$m; draught $T=13.2$m;

trim = 3.0 m; and mass = 1.07×10^8 kg. The sea bed consisted of coarse sand ($d = 2$ mm, $e = 0.6$, and frictional angle 37 degrees), and the inclination of the slope was 1:6 more than 7.0 m below the water surface and 1:20 above this level. For this case, the stopping distance was found to be 42.1 m and the maximum penetration to be approximately 1.0 m and the final bow lift 4.5 m. The maximum vertical grounding force F_z was found to be around 70 MN, and the effective coefficient of friction was approximately 0.8.

2.7 Concluding remarks

In this chapter, analytical methods have been presented for the energy released for the crushing and the impact impulse associated with ship–ship collisions, ship collisions with offshore platforms or wind farms, ship collisions with bridges or rigid walls, and ship grounding. The emphasis has been put on the derivation and validation of closed-form expressions for energy losses in any ship collision scenario. Comparisons of the present theory with previously published comprehensive simulation results and a large number of experimental results show that good agreement can be achieved.

References

Brach, R.M., 1993. Classical planar impact theory and the tip impact of a slender rod. Int. J. Impact Eng. 13 (1), 21–33.

Brown, A.J., 2002. Collision scenarios and probabilistic collision damage. Mar. Struct. 15, 335–365.

DNV (Det Norke Veritas), 2010. DNV-RP-C204: Design Against Accidental Loads.

Hansen Ottesen, N.-E., Simonsen, B.C., Sterndorf, M.J., 1994. The soil mechanics of ship beaching. In: 24th International Conference on Coastal Engineering, Kobe, Japan.

IACS (International Association of Classification Societies), 2016. I2 Structural Requirements for Polar Class Ships.

IMO (International Maritime Organization), 1974. International Maritime Organization Resolutions A265A.

Jonge, T.D., Laukeland, L., 2013. Collision between a spar platform and a tanker. In: Amdahl, J., Ehlers, S., Leira, B. (Eds.), Collision and Grounding of Ships and Offshore Structures. Taylor & Francis Group, London. ISBN: 978-1-138-00059-9.

Knott, M., Pruca, Z., 2000. Vessel collision design of bridges. In: Chen, W.F., Duan, L., Raton, B. (Eds.), Bridge Engineering Handbook. CRC Press.

Le Sourne, H., 2007. A ship collision analysis program based on super-element method coupled with large rotational ship movement analysis tool. In: Proceedings of the 4th International Conference on Collision and Grounding of Ships: Hamburg, Germanypp. 131–138.

Lee, G.J., 2014. Moment of inertia of liquid in a tank. Int. J. Nav. Archit. Ocean Eng. 6, 132–150.

Liu, Z., Amdahl, J., 2010. A new formulation of the impact mechanics of ship collisions and its application to a ship-iceberg collision. Mar. Struct. 23, 360–384.

Liu, B., Villavicencio, R., Zhang, S., Guedes Soares, C., 2017. Assessment of external dynamics and internal mechanics in ship collisions. Ocean Eng. 141, 326–336.

Lloyd's Register, 2014. Guidance Notes for Collision Analysis. Lloyd's Register Group Limited.

Minorsky, V.U., 1959. An analysis of ship collision with reference to protection of nuclear power ships. J. Ship Res. 3 (2), 1–4.

Motora, S., Fujino, M., Sugiura, M., Sugita, M., 1971. Equivalent added mass of ships in collisions. J. Soc. Nav. Archit. Jpn. 7, 138–148.

Pedersen, P.T., 1994. Ship grounding and hull-girder strength. Mar. Struct. 7, 1–29.

Pedersen, P.T., 1998. Gimsing, N.J. (Ed.), Ship Impact Against Protection Islands. The Storebælt Publications, East Bridge, pp. 58–64 (Chapter 3.7).

Pedersen, P.T., 2013. Ship collisions against wind turbines, quays and bridge piers. In: Amdahl, J., Ehlers, S., Leira, B. (Eds.), Collision and Grounding of Ships and Offshore Structures. Taylor & Francis Group, London. ISBN: 978-1-138-00059-9.

Pedersen, P.T., 2014. Risk assessment for ship collisions against offshore structures. In: Guedes Soares, C., Santos, T. (Eds.), Maritime Technology and Engineering. Taylor & Francis Group, London. ISBN: 978-1-138-02727-5.

Pedersen, P.T., Jensen, J.J., 1991. Ship impact analysis for bottom supported offshore structures. In: Advances in Marine Structures II. Elsevier Applied Sciences.

Pedersen, P.T., Li, Y., 2009. On the global ship hull bending energy in ship collisions. Mar. Struct. 22, 2–11.

Pedersen, P.T., Simonsen, B.C., 1993. Dynamics of ships running aground. J. Mar. Sci. Technol. 1 (1), 37–45.

Pedersen, P.T., Zhang, S., 1998. On impact mechanics in ship collisions. Mar. Struct. 11 (10), 429–449.

Petersen, M.J., 1982. Dynamics of ship collision. Ocean Eng. 9 (4), 295–329.

Petersen, M.J., Pedersen, P.T., 1981. Collision between ships and offshore platforms. In: Offshore Technology Conference, Houston, USA. Paper OTC 4134.

Pianc, 1984. Permanent International Association of Navigation Congress Report of the International Commission for Improving the Design of Fender Systems. Supplement to Bulletin No. 45 (1984), Brussels, Belgium.

Simonsen, B.C., Pedersen, P.T., 1993. Analysis of ship groundings on soft sea beds. J. Ship Ocean Technol. 1 (1), 35–47.

Sterndorff, M.J., Pedersen, P.T., 1996. Grounding experiments on soft bottoms. J. Mar. Sci. Technol. 1, 174–181.

Storheim, M., Amdahl, J., 2014. Design of offshore structures against accidental ship collisions. Mar. Struct. 37, 135–172.

Tabri, K., Määttänen, J., Ranta, J., 2008. Model-scale experiments of symmetric ship collisions. J. Mar. Sci. Technol. 13, 71–84.

Tabri, K., Matusiak, J., Varsta, P., 2009. Sloshing interaction in ship collisions—an experimental and numerical study. Ocean Eng. 36, 1366–1376.

Tabri, K., Varsta, P., Matusiak, J., 2010. Numerical and experimental motion simulations of non-symmetric ship collisions. J. Mar. Sci. Technol. 15, 87–101.

US-Guide, 1991. Guide specification and commentary for vessel collision design of highway bridges. vol. I. US Department of Transportation, US Federal Highway Administration, Publication No. FHWA-RD-006. Final Report.

Wevers, L.J., Vredeveldt, A.W., 1998. Full scale ship collision experiments. TNO Report 98-CMC-R1725. The Netherlands: Delft, p. 260.

Wolf, M., 2003. Full scale collision experiment, X-type sandwich side hull. EU Sandwich project report. Deliverable TRD448, p. 21.

Yu, Z., Amdahl, J., 2016. Full six degrees of freedom coupled dynamic simulation of ship collision and grounding accidents. Mar. Struct. 47, 1–22.

Zhang, S., 1999. The Mechanics of Ship Collisions. (PhD thesis). Technical University of Denmark, Denmark. ISBN: 87-89502-05-1.

Zhang, A., Suzuki, K., 2007. A comparative study of numerical simulations for fluid–structure interaction of liquid-filled tank during ship collision. Ocean Eng. 34, 645–652.

Zhang, S., Villavicencio, R., Zhu, L., Pedersen, P.T., 2017. Impact mechanics of ship collisions and validations with experimental results. Mar. Struct. 52, 69–81.

Internal mechanics of ship collision and grounding

3

3.1 Introduction and analysis methods

This chapter presents assessments of the structural response of ships during collisions and groundings. The main purpose is to establish simplified analytical methods for calculating the force–penetration curves and the absorbed energy–penetration relationships related to the analyses of ship–bridge collisions, ship–ship collisions, ship–platform collisions, and grounding events.

Ship structures are very complicated systems. Thus, the deformed, destroyed, and crushed modes of the structures due to collision or grounding are also very complex. A ship can be viewed as an assembly of welded plated structures where the shell plating, bottom plating, horizontal decks and bulkheads, etc. are constructed from various plates stiffened by stiffeners.

Observations from full-scale ship accidents and model experiments reveal that the primary energy-absorbing mechanisms of ship structures in collision and grounding are the following:

- Membrane deformation of the shell plating and attached stiffeners
- Folding and crushing of transverse web frames and longitudinal stringers
- Folding, cutting, and crushing of horizontal decks
- Cutting or crushing of ship bottoms
- Crushing of transverse bulkheads

The methods for the analysis of structural damage in ship collisions and grounding can be divided into four categories:

(1) Empirical methods
(2) Finite element methods
(3) Experimental methods
(4) Simplified analytical methods

3.1.1 Empirical methods

The first empirical method, commonly used even nowadays, was proposed by Minorsky (1959), who analysed 26 collision cases of full-scale ship accidents and developed the formula

$$E = 47.2R_T + 32.7,\ (\mathrm{MJ})$$

$$R_T = \sum P_N L_N t_N + \sum P_n L_n t_n,\ (\mathrm{m}^3)$$

Probability and Mechanics of Ship Collision and Grounding. https://doi.org/10.1016/B978-0-12-815022-1.00003-7

where R_T is the destroyed material volume of both the striking ship and the struck ship in m^3; E is the absorbed energy in MJ; P_N, P_n are the depths of damage in the Nth and the nth members of the striking ship and the struck ship, respectively; L_N, L_n are the lengths of damage in the Nth and the nth members of the striking ship and the struck ship; and t_N, t_n are the thicknesses of damage in the Nth and the nth members of the striking ship and the struck ship.

Fig. 3.1 presents the empirical curve of the energy–volume relationship derived by Minorsky (1959). It shows that the results for the 18 low-energy cases (shade area in Fig. 3.1), from the total of 26 collision scenarios, are considerably scattered. But the correlation of the other eight high-energy collisions is very good. Therefore, Minorsky's formula is generally considered valid for high-energy collisions.

Minorsky's formula is often used in the analyses of ship collisions due to its simplicity. When Minorsky tried to establish a function of absorbed energy with some related parameters, he also noted that the resistance of some members would be proportional to t^2 or some other function of the thickness. Moreover, several other factors were tried out, but they did not provide a better correlation than the volume factor. Therefore, Minorsky's formula indicates that the absorbed energy of a ship is simply proportional to the volume of destroyed material in high-energy collisions. This may not always be true, thus Chapter 4 next will provide alternative methods for design assessments.

The simplicity of Minorsky's formula has motivated investigators to extend it to low-energy collisions. Based on experiments, Woisin (1979) proposed the formula: $E=47.2R_T+0.5\sum(h\cdot t^2)$ (MJ), where R_T (in m^3) is the destroyed volume of materials, h (in m) is the height of broken or heavily deformed longitudinal members, and t (in cm) is the thickness of the members. Similarly, Vaughan (1980) established a formula that relates the absorbed energy and the destroyed volume and area: $E=93R_T+33A$

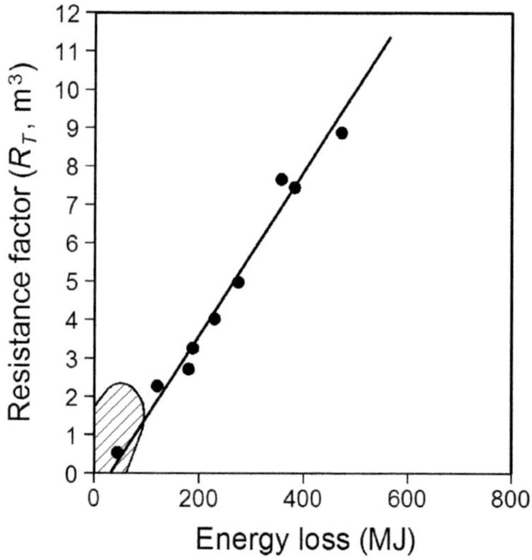

Fig. 3.1 Minorsky's original correlation.
Based on Minorsky, V.U., 1959. An analysis of ship collision with reference to protection of nuclear power ships. J. Ship Res. 3 (2), 1–4.

(MJ), where R_T is the destroyed volume of the materials (m^3) and A is area of tearing (m^2). Jones' (1979) formula for low-energy collision is based on the plastic theory of a beam subject to a point load. The formula is $E = 0.5 R_T \sigma_0 (2\delta/l)^2$, where R_T is the volume of the side shell involved in membrane deformation, σ_0 is the flow stress of the material, and $2\delta/l$ is the normalized final deflection over the span.

3.1.2 Finite element methods

The finite element method (FEM) is a popular means for the analysis of the structural response during collisions and groundings. For example, side collisions have been investigated by Sano et al. (1996), Kuroiwa (1996), Kitamura (1997), Lehmann and Peschmann (2002), Yamada (2006), Ehlers (2010), Martens (2014), Marinatos and Samuelides (2015), Storheim et al. (2015), Liu et al. (2017), and Ringsberg et al. (2018); grounding scenarios by Amdahl and Kavlie (1992) and Lemmen et al. (1996); and bow crushing by Lehmann and Yu (1998) and Yamada (2006).

It is often assumed that FEM calculations can give accurate results and may replace model experiments. However, as demonstrated by many investigators, for example, Storheim et al. (2015) and Ringsberg et al. (2018), this is not always the case. Even for experienced analysts, the FEM results can be very challenging in terms of accuracy. Chapter 5 of this book will specifically discuss the FEM techniques for ship collision analyses.

3.1.3 Experimental methods

Experiments are mainly carried out to verify theoretical procedures. Many experiments on ship collisions and grounding have been performed since the early 1960s. From 1962 to 1976, investigators in Italy, Germany, and Japan conducted a series of model tests. Some authors have given detailed reviews on these experiments, for example, Amdahl (1983), Jones (1979), Ellinas and Valsgard (1985), Samuelides and Frieze (1989), and Pedersen et al. (1993). The main purpose of the experiments carried out in Italy, Germany, and Japan was to design nuclear-powered ships to protect the nuclear reactor from collision damage.

In Italy, 24 model experiments were conducted to examine the efficiency of different types of side structures towards various types of striking ships. The models represented the structures of existing ships scaled to 1/15 and 1/10. The striking bow was set to run down along an inclined path, while the side structure model was mounted on a carriage that was free to move along the path. In order to account for the effect of the surrounding water, wings were attached to the carriage and immersed in water tanks on either side of it.

Twelve ship model tests were carried out in Germany (Woisin, 1979) during the period of 1967–76. The model scales were in the range between 1/12 and 1/7.5. An illustration of the test set-up and an example of a damaged bow after the test is shown in Fig. 3.2. It illustrates the striking bow running down an inclined railway path to impact a ship side model.

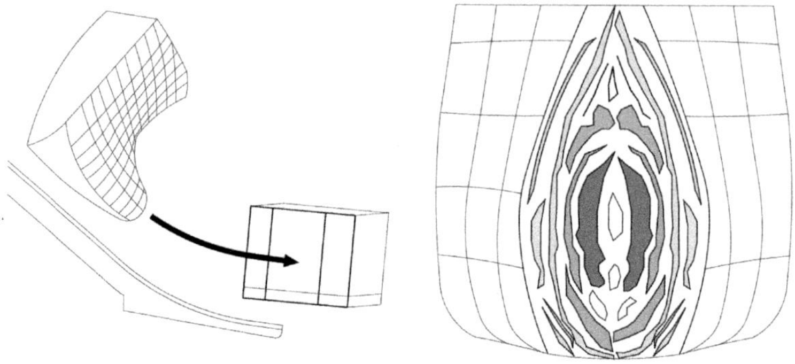

Fig. 3.2 Illustration of the test set-up for collision experiments in Germany.

During the period of 1966–70, a series of collision model tests were conducted in Japan. Both dynamic and static experiments were made that covered different aspects of the collision problem, such as the design of the side structure, the effect of the shape of the striking bow, and the effect of the added mass. A detailed summary of these experiments was given by Akita et al. (1972).

A 7-year project (from 1991 to 1997) on the prediction methodology of tanker structural failure and resulting oil spill was carried out in Japan (ASIS, 1998). The focus of the research project was on two main aspects: (1) the dynamic process of structural damage caused by collisions and grounding, and (2) the resulting process of oil spill and/or water ingress through the damaged hull. A series of full-scale ship collision experiments were carried out by TNO in the Netherlands in a joint project between Japan and the Netherlands in 1991. Two inland waterway tankers of 80 m length were used. Following the collision experiment in 1991, new ship–ship collision experiments were carried out as a joint project between Japan, the Netherlands, and Germany in the Netherlands in November 1997. A 1500 tonnes tanker collided with another 1500 tonnes tanker. The striking bow was a specially designed rigid bulb. The test section was installed in the middle of the struck ship, and it was full scale of a VLCC side of a conventional design and also a newly developed design. A photo from the full-scale collision experiments carried out in the Netherlands is shown in Fig. 3.3.

Many static tests of side collisions were carried out since static tests are easier to control than the dynamic ones. Ito et al. (1984) conducted a series of tests of double-sided hulls struck by a raked bow and a bulbous bow. The collision scenarios were grouped into five categories based on different locations of the impact. Amdahl and Kavlie (1992) performed model tests simulating a double hull indented by a rigid hexagonal body. The tests originally simulated stranding, but were also very useful for side collision analysis. Paik et al. (1999) carried out four collision model tests on rational design of double-hull tanker structures against collision. Wang et al. (2000) performed nine model tests to investigate the behaviour of a double hull in a variety of stranding or collision scenarios; Fig. 3.4 shows the test set-up and a damaged model following the test. Lehmann and Peschmann (2002) reported detailed results and analysis on the energy absorption of the ship structures in the collision experiments carried

Fig. 3.3 Full-scale collision experiments carried out by TNO in the Netherlands (ASIS, 1998).

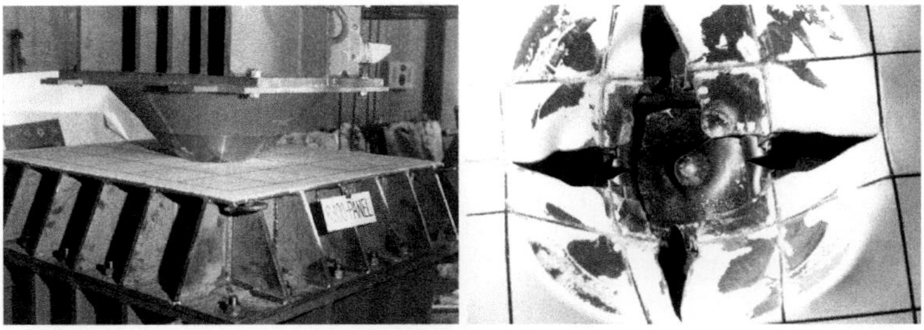

Fig. 3.4 Collision model test set-up and a damaged model after the tests by Wang et al. (2000). cf. Wang, G., Arita, K., Liu, D., 2000. Behavior of a double hull in a variety of stranding or collision scenarios. Mar. Struct. 13, 147–187.

out in the Netherlands. Tautz et al. (2013) conducted collision tests with rigid and deformable bulbous bows driven against double-hull side structures.

When a ship is involved in a head-on collision with relatively hard objects, such as rigid walls, the bow of the ship will be crushed. Amdahl (1983) carried out several model tests, crushing different bows, and provided valuable results. Fig. 3.5 shows one of the test models and an experimental result of the load–indentation curve. In Japan, several tests of bow crushing were also performed; for example, Yamada and Endo (2005) and Yamada (2006) carried out a series of model tests on bow crushing to a rigid wall; an example is shown in Fig. 3.6. Martens (2014) carried out bulbous bow crushing tests to investigate the effect of the rigidity of the bow on the damage to side structures of the struck ship (see Fig. 3.7).

3.1.4 Simplified analytical methods

Simplified analytical methods are based on the upper-bound theorem and assumptions concerning folding patterns from observations of accidental damages, experimental studies, or numerical simulations. Usually, the methods can give good predictions

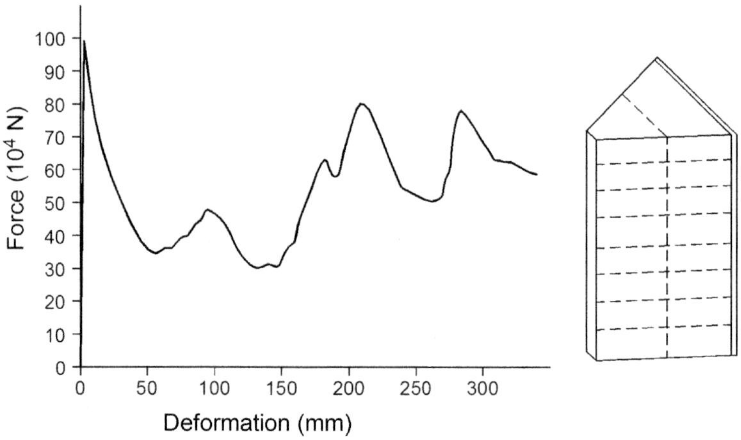

Fig. 3.5 A bow crushing test and the load–indentation curve.
Based on Amdahl, J., 1983. Energy Absorption in Ship-Platform Impacts. (Doctoral thesis). Norwegian Institute of Technology.

Fig. 3.6 A bow crushing test and a test model after the test.
cf. Yamada, Y., 2006. Bulbous Buffer Bows: A Measure to Reduce Oil Spills in Tanker Collision. (PhD thesis), Technical University of Denmark, Denmark; Yamada, Y., Endo, H., 2005. Collapse mechanism of the buffer bow structure on axial crushing. Int. J. Offshore Polar Eng. 15 (02).

through a fast simple analysis. Therefore, many authors have used simplified analytical methods for analysis of ship collisions and groundings.

A major assumption in the simplified analytical methods is that the different structural members, such as the side shell, decks, and web frames, do not interact, except at their joint boundaries, but contribute independently to the total collision resistance.

McDermott et al. (1974) developed a method for the analysis of tankers in minor collisions. The method is based on a number of basic assumptions. The mathematical model for the analysis procedure included bending and buckling of the stiffened hull plating, membrane tension of the hull plating and decks, and failure of the web frames.

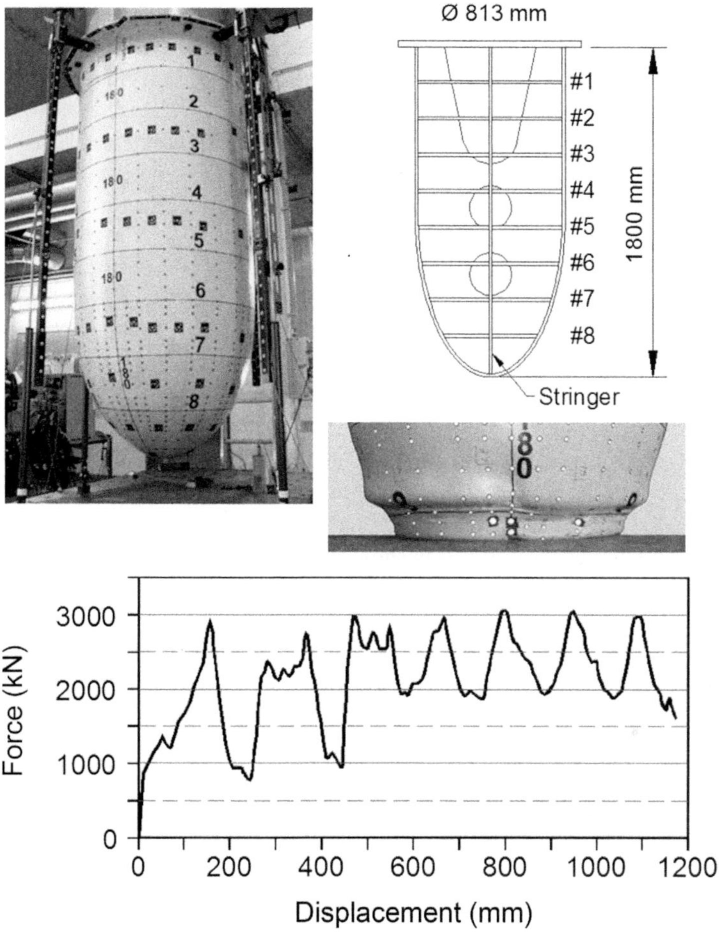

Fig. 3.7 A bow crushing test model and an example of crushing force curve.
Based on Martens, I., 2014. Konstruktive Aspekte beim Entwurf von Bugwülsten zur Verbesserung des Energieaufnahmevermogens bei Schiffskollisionen (in German). (PhD thesis), Schriftenreihe Schiffbau, Bericht 679. Technical University of Hamburg, Germany. ISBN 978-3-89220-679-8.

The results from the analysis indicated that most of the absorbed energy (about 70%–85%) in a collision is taken by membrane tension in the stiffened hull. Another part of the energy is absorbed by membrane tension in the stiffened deck and in-plane shearing of the frames. The bending energy in the stiffened hull is very small and can often be neglected.

Reckling (1983) proposed a method that takes into account the deformation of both the striking ship and the struck ship. The energy absorbed by the ships, up to the rupture of the struck ship hull, is computed by simple methods. A calculation example of two equal oil tankers of 141,000 deadweight (DWT, in tonnes) with bulbous bow was performed. The result showed that the absorbed energy under membrane tension of the side hull is only 18% of the total energy, while the energy absorbed by the web frames, decks, and bottom of the struck ship is 40% of the total energy. A remarkable 42% of the total energy is in this case absorbed by the striking bow structure.

Chang et al. (1980) presented a methodology for prediction of the structural response in collisions. The method is a synthesis of the FEM technique, the collapse theorem, and experimental data. The study showed that the collision force is a function of the thickness of the plate, the cross-sectional area of the strongest section, and the yield and ultimate strength of the materials. Similarly, the energy absorbed by the destroyed metal is a function of both the metal volume and the thickness–area ratio and the yield and ultimate strength of the materials. This study is limited to perpendicular collisions.

Research on bow crushing performed by Amdahl (1983), Yang and Caldwell (1988), and Kierkegaard (1993) is very valuable. The adopted methods are based on the assumption that a complex structure can be divided into fundamental elements, such as L-, T-, and X-elements. By adding the contributions by all basic components, the crushing behaviour of a bow can be determined. A typical calculation example performed by Kierkegaard (1993) is shown in Fig. 3.8.

Wierzbicki and Abramowicz (1983) carried out investigations of the axial crushing behaviour of basic structural elements. Lehmann and Yu (1995) developed a formula for the analysis of progressive folding of a conical shell, which can be used to predict the crushing load of a bulbous bow. Pedersen et al. (1993) studied the crushing load of ship bows. In this analysis, the formulae proposed by Gerard (1957), Amdahl (1983)

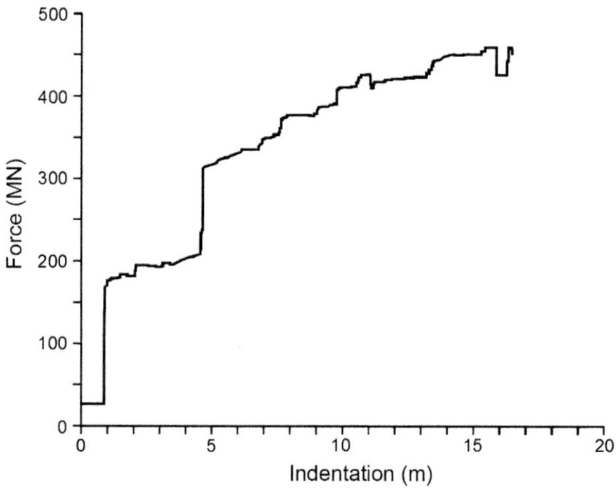

Fig. 3.8 Head-on crushing analysis of a ship using a simplified method (150,000 DWT bulk carrier with 7.72 m/s impact velocity).
Based on Kierkegaard, H., 1993. Ship bow response in high energy collisions. Mar. Struct. 6, 359–376.

and Yang and Caldwell (1988) are applied. A general review on the study of ship collisions and groundings can be found in Pedersen (2010) and Liu et al. (2018).

To perform damage analysis using the above-mentioned methods, the first step is to understand the material behaviour and its properties from elastic to large plastic deformation; this is discussed in the next section.

3.2 Materials modelling

3.2.1 Materials properties

Materials used for the construction, conversion, modification, or repair of ships, other marine structures and associated machinery are to be manufactured, tested, and inspected in accordance with Rules for the Manufacture, Testing and Certification of Materials; see, for example, Lloyd's Register (2018a). Eventually, all materials are to be supplied with the appropriate certification, as required by the Rules.

For standard mechanical **tensile tests**, the test specimen is designed with a gauge length of $L_0 = 5.65\sqrt{S_0}$ for plates and $L_0 = 5d$ for rolled bars, where S_0 is the cross-sectional area and d is the diameter of the specimen. The constant 5.65 was derived from the value $5\sqrt{4/\pi}$, which will give the same cross-sectional area as a cylindrical specimen where the gauge length is set at five times the diameter. The configurations of the specimens for standard tensile tests are shown in Fig. 3.9.

From the tensile tests, three mechanical properties are usually obtained: yield stress (σ_y), tensile strength (σ_u), and elongation (ε_f) at fracture. A typical engineering strain–stress curve is shown in Fig. 3.10.

For convenience in the industry, the determination of the engineering stress and engineering strain are based on the simple beam or rod theory:

$$\sigma = \frac{F}{S_0} \tag{3.1}$$

$$\varepsilon = \frac{L - L_0}{L_0} = \frac{\Delta L}{L_0} \tag{3.2}$$

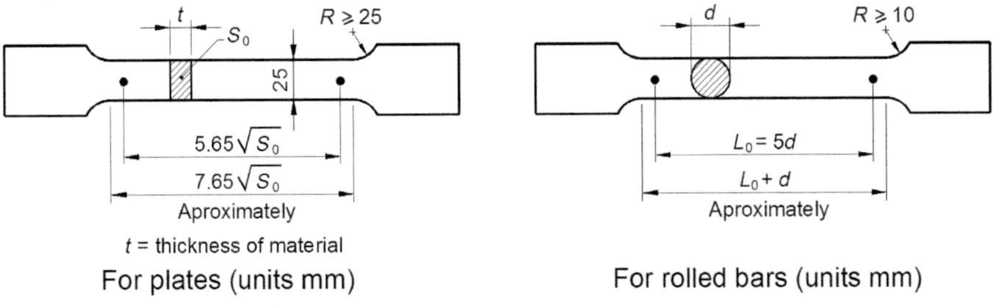

Fig. 3.9 Specimens the standard tensile tests.

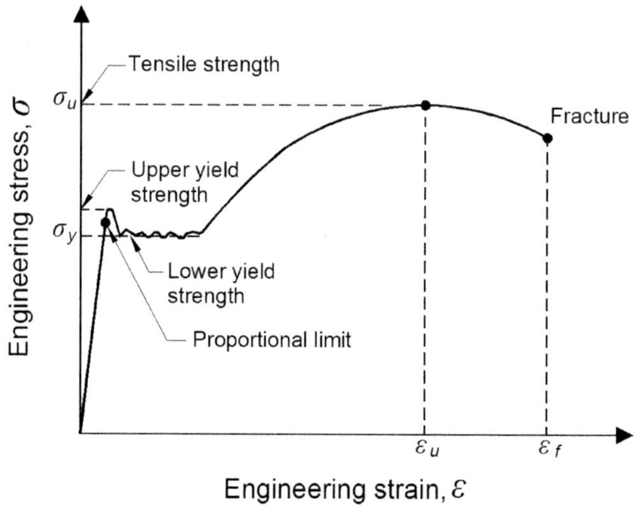

Fig. 3.10 Illustration of the engineering stress–strain curve.

where F is the instantaneous applied force and ΔL is the extension. This stress is a nominal stress over the original cross-sectional area S_0, and the elongation is the average strain over the gauge length L_0.

One of the key parameters used for analytical analysis of heavy collisions, that is, with large structural deformation and rupture, is the elongation ε_f at fracture which is obtained from the standard uniaxial tensile tests. It is well known that the engineering strain (elongation) depends on the gauge length. As mentioned, **the internationally accepted standard gauge length** for plates is $L_0 = 5.65\sqrt{S_0}$. Thus, if another gauge length is used, the measured elongation should be converted to the 'equivalent elongation' based on the standard gauge length $L_0 = 5.65\sqrt{S_0}$ by the following formula (BS EN ISO 2566-1, 1999; BS EN ISO 2566-2, 2001):

$$\varepsilon_f = 0.5\varepsilon_0\left(\frac{L}{\sqrt{S}}\right)^{0.40} \quad \text{for carbon steels} \tag{3.3a}$$

$$\varepsilon_f = 0.8\varepsilon_0\left(\frac{L}{\sqrt{S}}\right)^{0.127} \quad \text{for austenitic steels} \tag{3.3b}$$

where ε_0 is the actual measured elongation at fracture, L is the actual gauge length, and S is the original cross-sectional area. Table 3.1 shows the conversion factors for other commonly used gauge lengths.

A general relationship for the elongations between two different gauge lengths and cross-sectional areas can be written as

Table 3.1 Conversion factors for elongations of other commonly used gauge lengths to the standard gauge length of $5.65\sqrt{S_0}$.

Gauge length	$\mathbf{4.00\sqrt{S_0}}$	$\mathbf{5.65\sqrt{S_0}}$	$\mathbf{8.16\sqrt{S_0}}$	$\mathbf{11.30\sqrt{S_0}}$
Conversion factor (carbon steels)	0.870	1.000	1.158	1.317
Conversion factor (austenitic steels)	0.957	1.000	1.048	1.092

$$\varepsilon_{f2} = \varepsilon_{f1} \cdot \left(\frac{\left(\frac{L_1}{\sqrt{S_1}} \right)}{\left(\frac{L_2}{\sqrt{S_2}} \right)} \right)^{0.40} = \varepsilon_{f1} \cdot \left(\frac{L_1}{L_2} \right)^{0.40} \cdot \left(\frac{S_2}{S_1} \right)^{0.20} \quad \text{for carbon steels} \tag{3.4a}$$

$$\varepsilon_{f2} = \varepsilon_{f1} \cdot \left(\frac{\left(\frac{L_1}{\sqrt{S_1}} \right)}{\left(\frac{L_2}{\sqrt{S_2}} \right)} \right)^{0.127} = \varepsilon_{f1} \cdot \left(\frac{L_1}{L_2} \right)^{0.127} \cdot \left(\frac{S_2}{S_1} \right)^{0.064} \quad \text{for austenitic steels} \tag{3.4b}$$

Another well-known expression relating the elongation and the gauge length is the Barba's law: $\varepsilon_f = n + c\left(\sqrt{S}/L\right)$, where n and c are coefficients of the materials. For mild steels, the typical coefficients are $n=0.22$ and $c=0.7$, and this gives $\varepsilon_f=0.344$ using the standard gauge length of $L_0 = 5.65\sqrt{S_0}$.

For ship collision and grounding analyses, the **flow stress** is most often used, which, in this book, is defined as the average of the yield stress and the tensile strength:

$$\sigma_0 = \left(\sigma_y + \sigma_u\right)/2 \tag{3.5}$$

The material properties of the equivalent elongation (Eq. 3.3a, 3.3b) and the flow stress (Eq. 3.5) will be used in the simplified analytical methods for collision and grounding analyses, since the present methods are based on overall deformation mechanisms.

For shipbuilding steel materials, the Rules'minimum requirements for the material properties are shown in Table 3.2 (Lloyd's Register, 2018b). These are the minimum requirements as the actual values may be larger (see Section 3.2.4 Uncertainties in materials properties). Thus, it is important to get the actual values of the material properties from tensile tests in accordance with the internationally recognized test standards, such as BS EN 10002-1 (2001), ASTM (2004), and Lloyd's Register (2018b), particularly for calibrating theoretical procedures with experimental results.

Table 3.2 The minimum requirements for the material properties of steels (Lloyd's Register, 2018b).

Material	Yield stress (MPa)	Tensile strength (MPa)	Elongation on $5.65\sqrt{S_0}$
Mild steels	235	400	0.22
HT27 steels	265	400	0.22
HT32 steels	315	440	0.22
HT36 steels	355	490	0.21
HT40 steels	390	510	0.20
HT47 steels	460	570	0.17
Austenitic steels 316L	170 (0.2% Proof stress)	485	0.40

The material test temperatures: AH grade 0°C, DH grade −20°C, EH grade −40°C, and FH grade −60°C.

For more detailed structural analyses, such as the finite element analyses in Chapter 5, the engineering stress–strain curves are not adequate since the cross-sectional areas change during the deformations. Thus, the true stress and true strain should be used, which are determined from

$$\varepsilon_t = \int_{L_0}^{L} \frac{dL}{L} = \ln\left(\frac{L}{L_0}\right) = \ln\left(1 + \frac{\Delta L}{L_0}\right) = \ln(1+\varepsilon) \tag{3.6}$$

$$\sigma_t = \frac{F}{S} = \frac{F}{(L_0 S_0)/L} = \frac{F}{S_0}\frac{L}{L_0} = \frac{F}{S_0}\frac{L_0 + \Delta L}{L_0} = \sigma(1+\varepsilon) \tag{3.7}$$

Here, a constant material volume is assumed for the beam deformation by using $L_0 S_0 = LS$ in the derivation. This assumption is only valid up to the onset of necking (i.e. where the stress reaches the maximum in the engineering stress–strain curve; see Fig. 3.11). After this point, the deformations are not uniform and large local deformations in the necking region occur.

Beyond the onset of necking, the power law fit based on test data and proposed by Hollomon (1945) is commonly used

$$\sigma_t = K\varepsilon_t^{\,n} \tag{3.8}$$

where K is strength coefficient when $\varepsilon_t = 1.0$ and n is an exponent often called the strain hardening coefficient. The coefficients K and n can be determined from (Zhang et al., 2004a)

$$n = \ln(1 + A_g) \tag{3.9}$$

$$K = \sigma_u (2.718/n)^n \tag{3.10}$$

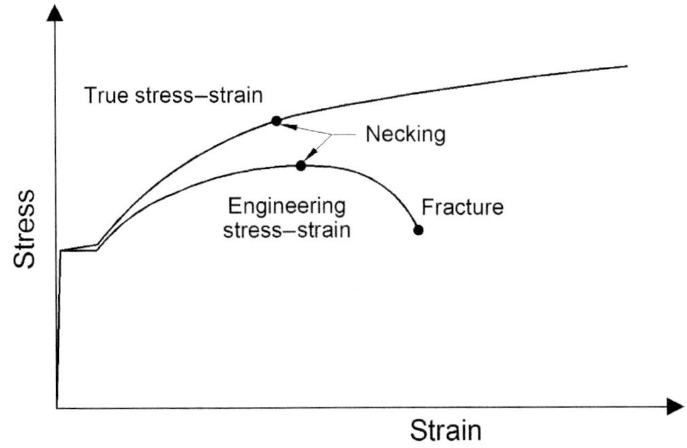

Fig. 3.11 Illustration of the engineering stress–strain curve and true stress–strain curve.

where A_g is the maximum uniform strain corresponding to the ultimate strength σ_u. If only the ultimate strength σ_u is available, A_g can be obtained from $A_g = 1/(0.24 + 0.01395\sigma_u)$.

Another approximate expression from Liu et al. (2017) is

$$\sigma_y = K \cdot 0.006^n \tag{3.11}$$

$$\sigma_u(1+n) = K \cdot (\ln(1+n))^n \tag{3.12}$$

More details on material modelling for nonlinear finite element analysis can be found in Chapter 5.

3.2.2 Temperature effects

Temperature will affect material properties; thus, tensile tests should be carried out at appropriate temperature to obtain the actual material properties. When such tensile properties are not available for the assessment temperatures, the following formula, based on data obtained at room temperature, can be applied to ferritic steels (BS 7910, 2015).

When the assessment temperature is below room temperature and above − 196°C, the yield stress σ_y and the tensile strength σ_u can be estimated from the room temperature values as

$$\sigma_{y(low\ temperature)} = \sigma_{y\ (room\ temperature)} + \frac{10^5}{(491 + 1.8T)} - 189 \tag{3.13}$$

$$\sigma_{u\,(low\ temperature)} = \sigma_{u\,(room\ temperature)} \cdot \left(0.7857 + 0.2423 \exp\left(-\frac{T}{170.646}\right)\right) \tag{3.14}$$

where T is the assessment temperature (°C).

When the assessment temperature is above the room temperature, the yield strength and the tensile strength decrease. They can be determined from

$$\sigma_{y,or\ u(high\ temperature)} = \sigma_{y,or\ u\,(room\ temperature)} - \Delta\sigma_T \tag{3.15}$$

where $\Delta\sigma_T$ is the amount that the strength decreases according to Table 3.3.

For the Young's modulus, the values given in Table 3.4 can be used, if reliable information is not available. For steels, Poisson's ratio is kept as 0.3.

3.2.3 Strain rate effects

It is well known that the yield stress of mild steels is sensitive to the strain rate. In particular, the yield stress of mild steels largely increases with the strain rate. Jones (2012) has discussed the strain rate effect on material properties. Various constitutive equations to describe the strain-rate-sensitive behaviour of materials have been proposed in the literature, and it has been highlighted that careful experimental work is required to generate the coefficients used in these equations. One of the most widely used empirical formulae to describe the strain rate effect of the flow stresses is that proposed by Cowper and Symonds (1957):

$$\frac{\sigma_{0d}}{\sigma_0} = 1 + \left(\frac{\dot{\varepsilon}}{D}\right)^{\frac{1}{q}} \tag{3.16}$$

where σ_{0d} is the dynamic flow stress, σ_0 is the static flow stress, and D and q are material constant. The coefficients D and q in the Cowper-Symonds equation have been also investigated and determined by various authors, and some of them are

Table 3.3 Yield stress and tensile strength reduction ($\Delta\sigma_T$) with the increase of temperature.

Temperature (°C)	Ferritic steels (MPa)	Austenitic steels (MPa)
25	0	8
50	0	40
75	15	65
100	30	90
125	40	105
150	50	120
175	60	130
200	70	140

Table 3.4 Young's modulus of steels at different temperatures.

Temperature (°C)	Ferritic steels (MPa)	Austenitic steels (MPa)
−200	219,000	208,000
−150	215,000	205,000
−100	212,000	203,000
−75	210,000	201,000
−50	209,000	200,000
0	**206,000**	**197,000**
25	205,000	195,000
50	203,000	193,000
100	201,000	190,000
150	198,000	187,000
200	195,000	183,000
250	191,000	179,000
300	187,000	176,000
350	182,000	172,000
400	177,000	169,000
450	171,000	165,000
500	164,000	160,000
550	156,000	156,000
600	146,000	151,000
650	135,000	146,000

summarized in Table 3.5. Fig. 3.12 shows a comparison between the Cowper–Symonds empirical equation and experimental results for mild steels and high-tensile steels (Paik, 2007).

For ship collisions, an estimated range of strain rates for shell membrane deformation can be from ~0.5 to ~5 (1/s). This is based on the following calculations. First, assume a high collision speed of 20 knots (=10.3 m/s) where the side shell is ruptured at 1.0 m deflection. This takes approximately 0.1 s (=1 m/10.3 m/s), assuming that the collision speed remains the same as at the beginning of the collision. Then, assuming that the elongation of steel materials is 0.4, the average strain rate is estimated as

Table 3.5 Coefficients of the Cowper–Symonds empirical formula for various materials.

Materials	D (1/s)	q (−)	Reference
Mild steels	40.4	5	Cowper and Symonds (1957)
Mild steels	500	4	Yang and Caldwell (1988)
High-tensile steels	6844	3.91	Abramowicz and Jones (1984)
High-tensile steels	3200	5	Paik and Thayamballi (2003)
Aluminium alloys	6500	4	Bodner and Symonds (1962)
Stainless steels	100	10	Forrestal and Sagartz (1978)

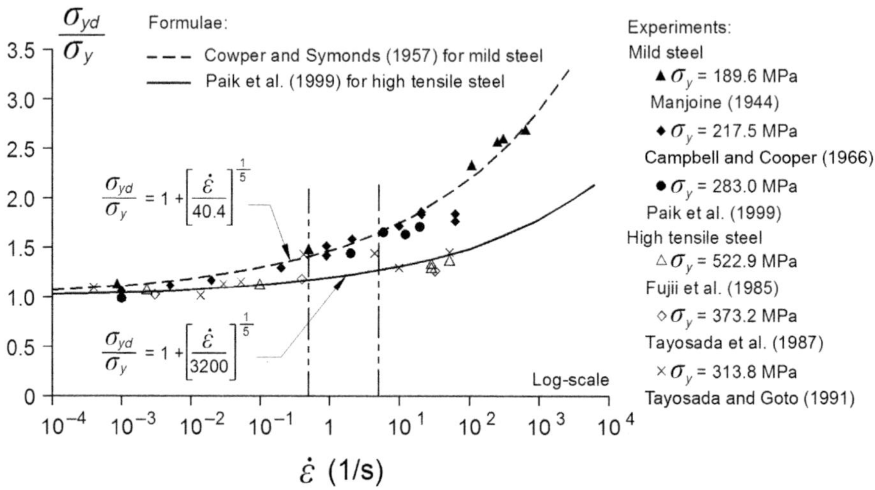

Fig. 3.12 Strain rate effect on flow stresses.
Based on Paik, J.K., 2007. Practical techniques for finite element modeling to simulate structural crashworthiness in ship collision and grounding (Part I: Theory). Ship Offshore Struct. 2 (1), 69–80. Reproduced with kind permission of Taylor and Francis Ltd. (journal web site: www.tandfonline.com).

4.0 (=0.4/0.1s). Similarly, if a low collision speed was set as 5.0knots, the average strain rate is estimated as 1.0 (1/s).

This may increase the dynamic flow stress by about 55% for mild steels and 22% for high-tensile steels using the theory described in this section. However, care is needed when using such increase of the flow stress for design assessments, and model tests using different strain rates are recommended if the dynamic effect is to be considered in the assessment.

Strain rates may also affect (reduce) the failure strains or the elongations. Using a similar approach as for the effect on flow stresses, the effect of strain rates on the elongations may be represented by the inverse of the Cowper–Symonds constitutive equation (Paik, 2007):

$$\frac{\varepsilon_{fd}}{\varepsilon_f} = \left[1 + \left(\frac{\dot{\varepsilon}}{D}\right)^{\frac{1}{q}}\right]^{-1} \tag{3.17}$$

where ε_{fd} is the dynamic elongation, ε_f is the static elongation, and D and q are constants. Eq. (3.17) gives a reasonable agreement with experimental data for mild steels, with yield stress of 217.5 MPa by using $D=10{,}000$ 1/s and $q=2$, and yield stress of 283 MPa using $D=6844$ 1/s and $q=3.91$, as shown in Fig. 3.13. It seems that the strain rate effect on the elongation at fracture is less sensitive in the range of the typical strain rates of 0.5–5 (1/s) in ship collisions.

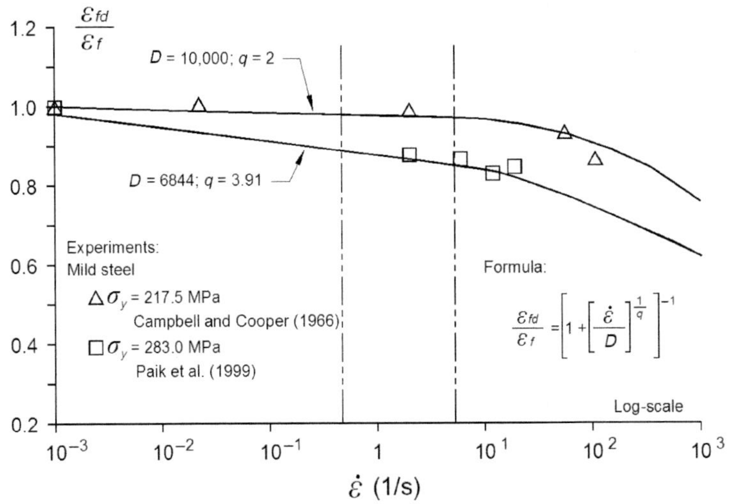

Fig. 3.13 Strain rate effect on elongations.
Based on Paik, J.K., 2007. Practical techniques for finite element modeling to simulate structural crashworthiness in ship collision and grounding (Part I: Theory). Ship Offshore Struct. 2 (1), 69–80. Reproduced with kind permission of Taylor and Francis Ltd. (journal web site: www.tandfonline.com).

3.2.4 Uncertainties in materials properties

The properties of materials are scattered, and even for the same material, the mechanical properties may not be exactly the same. Therefore, it may be necessary to consider the statistical variations in the experimental tensile properties.

Statistical analysis of the material properties of shipbuilding steels has been conducted for mild steels (Rules for ships' minimum requirement for the yield stress is 235 MPa) and HT36 steels (Rules for ships' minimum requirement for the yield stress is 355 MPa). The size of the samples is around 90 for each group, and the thicknesses of the steel plates vary from 8.0 to 50 mm, and the materials were produced from different mills. The results are presented in Table 3.6 together with some related data from BS 7910 (2015). It can be seen that: (1) the mean value of the yield stresses is much higher than the minimum requirements, about +20% for HT36 and +34% for mild steels; (2) the mean value of the tensile strength is also higher than the minimum requirements, about +12% for HT36 and +14% for mild steels; and (3) the mean value of the elongations is significantly higher than the minimum requirements, about +33% for HT36 and +50% for mild steels.

The coefficient of variation (COV = the ratio between the standard deviation and the mean value) for the yield stresses is 0.067 for HT36 steels and 0.091 for mild steels, which are close to the value 0.07 specified by BS 7910 (2015). The COV for tensile strength is 0.036 for HT36 steels and 0.044 for mild steels, which are close to the value 0.05, also specified by BS 7910 (2015). The COV for elongations is 0.1 for HT36 steels and 0.093 for mild steels, which are larger than the COV of the tensile strength; BS 7910 (2015) does not specify any value.

Table 3.6 Coefficient of variation (COV) for tensile properties of ferritic steels.

	Yield strength (MPa)			Tensile strength (MPa)			Elongation on $5.65\sqrt{S_0}$			Flow stress (MPa)	
	Rules req.	Statistics data (mean)	Statistics data (COV)	Rules req.	Statistics data (mean)	Statistics data (COV)	Rules req.	Statistics data (mean)	Statistics data (COV)	Statistics data (mean)	Statistics data (COV)
Mild steels	235	316	0.091	400	456	0.044	0.22	0.33	0.093	386	0.055
HT36 steels	355	426	0.067	490	550	0.036	0.21	0.28	0.100	488	0.041
BS 7910 for steels	–	–	0.070	–	–	0.050	–	–	–	–	–
BS 7910 for welds	–	–	0.100	–	–	0.100	–	–	–	–	–

It is evident that the elongation is strongly influenced by the tensile strength of the materials. The elongation at fracture decreases with the increase of the tensile strength for shipbuilding steels. Their relationship can approximately be expressed as

$$\varepsilon_f = 0.385 \cdot \left(\frac{400}{\sigma_u}\right) = \frac{154}{\sigma_u} \quad \text{for carbon steels} \tag{3.18a}$$

$$\varepsilon_f = 0.77 \cdot \left(\frac{400}{\sigma_u}\right) = \frac{308}{\sigma_u} \quad \text{for austenitic steels} \tag{3.18b}$$

where σ_u is the tensile strength of the material in MPa.

For nonlinear finite element analyses (Chapter 5), the material is often described by Eq. (3.8) and the simplified relationships in Eq. (3.10) or (3.12). In this case, Eq. (3.18a) can be expressed as

$$\varepsilon_f = \frac{154\left(\frac{2.718}{n}\right)^n}{K} \quad \text{for carbon steels} \tag{3.18c}$$

$$\varepsilon_f = \frac{154(n+1)}{K[\ln(n+1)]^n} \quad \text{for carbon steels} \tag{3.18d}$$

It must be noted that Eqs. (3.18c) and (3.18d) give almost identical result for given K and n values.

For design assessments, it normally uses either the lower bound or the mean minus two standard deviations, while the mean values are commonly used to conduct comparisons or validations between theoretical and experimental results. If the assessment is conducted using specified minimum values defined in a material specification, such as the Rules for ships' requirements (Lloyd's Register, 2018b), these values are considered as the lower bounds. Therefore, no further correction for the statistical variation is necessary.

3.3 Theorems for structural collapse analysis

3.3.1 Collapse analysis theorems

Simplified analytical methods based on perfectly rigid-plastic material behaviour are widely used in engineering analysis and design. It should be emphasized that the simplified analytical analysis is an approximate method. However, it has been proved that the method is valuable for estimating the collapse load of a structure subjected to extreme loads, and such collapse loads can be used as a basis for design. In order to describe the theory, we shall first give some related theorems. A basic assumption is that the material is perfectly plastic without strain hardening or softening.

3.3.1.1 The virtual work principle

The virtual work principle can be expressed in the following formula:

$$\int_A F_i u_i dA + \int_V T_i u_i dV = \int_V \sigma_{ij} \varepsilon_{ij} dV$$

where A is the whole surface area; V is the volume of the structure; F_i and T_i are the external force and the body force, respectively; σ_{ij} is a set of stresses; ε_{ij} is the strain field; and u_i is the displacement. The rate form of the virtual work equation is

$$\int_A F_i \dot{u}_i dA + \int_V T_i \dot{u}_i dV = \int_V \sigma_{ij} \dot{\varepsilon}_{ij} dV$$

3.3.1.2 Lower-bound theorem

If any system of generalized stresses can be found throughout a structure that is in equilibrium with the applied loads and nowhere violates the yield condition, then the structure will not collapse or be at the point of collapse.

3.3.1.3 Upper-bound theorem

If the work rate of a system of applied loads during any kinematically admissible collapse of the structure is equated to the corresponding internal energy dissipation rate, then the system of loads will cause collapse or be at the point of collapse.

Using the principle of virtual work, we can prove the lower-bound and the upper-bound theorems.

The two bound theorems can be used independently. If the calculated loads coincide in the two methods, the exact solution is found. But for large complex structures, it is difficult to find an exact solution. In this book, we use the upper-bound theorem to analyse the structural damage in ship collisions and groundings.

3.3.2 Formulation of the upper-bound method

The equilibrium for the external energy rate and the internal energy dissipation rate can be expressed as

$$F \cdot \dot{\delta} = \dot{E}_{int} \tag{3.19}$$

where F is the external force, $\dot{\delta}$ is the velocity at the force action point, and $\dot{E}_{int}$ is the internal energy rate.

For a general solid body, the internal energy rate $\dot{E}_{int}$ can be expressed as

$$\dot{E}_{int} = \iiint_V (\sigma_{ij}\dot{\varepsilon}_{ij})dV$$

where σ_{ij} is the set of stresses, $\dot{\varepsilon}_{ij}$ is the rate of the strain tensor, and V is the volume of the solid body. By the use of von Mises' flow theory, the rate of plastic energy dissipation is given as

$$\dot{E}_{int} = \iiint_V (\sigma_0\dot{\varepsilon}_e)dV$$

where $\dot{\varepsilon}_e = \sqrt{\frac{2}{3}(\dot{\varepsilon}_{ij}\dot{\varepsilon}_{ij})}$.

For a plane stress condition where $\sigma_{zz}=0$, the von Mises yield condition gives

$$F = \sigma_{xx}^2 + \sigma_{yy}^2 - \sigma_{xx}\sigma_{yy} + 3\sigma_{xy}^2 - \sigma_0^2 = 0$$

The strain rate can be determined from the associated flow rules (associated with the von Mises yield condition):

$$\dot{\varepsilon}_{xx} = \kappa\frac{\partial F}{\partial \sigma_{xx}} = \kappa(2\sigma_{xx} - \sigma_{yy})$$

$$\dot{\varepsilon}_{yy} = \kappa\frac{\partial F}{\partial \sigma_{yy}} = \kappa(2\sigma_{yy} - \sigma_{xx})$$

$$\dot{\varepsilon}_{xy} = \frac{1}{2}\kappa\frac{\partial F}{\partial \sigma_{xy}} = 3\kappa\sigma_{xy}$$

The stresses can then be expressed from the strain rates:

$$\sigma_{xx} = \frac{1}{3\kappa}(2\dot{\varepsilon}_{xx} + \dot{\varepsilon}_{yy})$$

$$\sigma_{yy} = \frac{1}{3\kappa}(2\dot{\varepsilon}_{yy} + \dot{\varepsilon}_{xx})$$

$$\sigma_{xy} = \frac{1}{3\kappa}(\dot{\varepsilon}_{xy})$$

Substituting the stresses into the von Mises criterion, the following expression is obtained for κ:

$$\kappa = \frac{1}{\sqrt{3}\sigma_0}\left(\dot{\varepsilon}_{xx}{}^2 + \dot{\varepsilon}_{yy}{}^2 + \dot{\varepsilon}_{xx}\dot{\varepsilon}_{yy} + \dot{\varepsilon}_{xy}{}^2\right)^{0.5}$$

For a deforming plate, the rate of internal plastic energy dissipation can be written as the sum of the bending and the membrane energy dissipation rate:

$$\dot{E}_{int} = \dot{E}_b + \dot{E}_m \tag{3.20}$$

The bending energy rate can be expressed as

$$\dot{E}_b = \iint_A \left(M_{\alpha\beta}\dot{k}_{\alpha\beta}\right)dA + \sum_{i=1}^{n}\left(M_{0i}\dot{\theta}_i l_i\right) \quad (\alpha, \beta = 1, 2) \tag{3.21}$$

where A is the plate area, $k_{\alpha\beta}$ is the curvature of the plating, and θ_i and l_i are the rotation and the length of the ith plastic hinge line, respectively. $M_{\alpha\beta}$ is the bending moment tensor, M_0 is the fully plastic bending moment $M_0 = \frac{2}{\sqrt{3}}\sigma_0^2 t/4$, and t is the plate thickness.

The fully plastic bending moment of a beam with unit width is $M_0 = \sigma_0^2 t/4$ where $\sigma_{xx} = \sigma_0$ and the term $2/\sqrt{3}$ is found from the von Mises yield criterion for plane strain where $\dot{\varepsilon}_{yy} = 0$:

$$\dot{\varepsilon}_{yy} = \kappa\left(2\sigma_{yy} - \sigma_{xx}\right) = 0 \Rightarrow \sigma_{yy} = \frac{1}{2}\sigma_{xx}$$

From $\sigma_{xx}^2 + \sigma_{yy}^2 - \sigma_{xx}\sigma_{yy} - \sigma_0^2 = 0$, $\sigma_{xx} = \frac{2}{\sqrt{3}}\sigma_0$ is obtained, and thus, $M_0 = \frac{2}{\sqrt{3}}\sigma_0^2 t/4$.

It is seen from expression (3.21) that the bending energy contains the continuous deformation field and the plastic hinge lines. For practical applications, simplified velocity fields are assumed so that only the plastic hinge lines are considered and the continuous deformation of the curvature is neglected. In this case, the bending energy is simplified as

$$\dot{E}_b = \sum_{i=1}^{n}\left(M_{0i}\dot{\theta}_i l_i\right) \tag{3.22}$$

The membrane energy rate of a deforming plate can be calculated from

$$\dot{E}_m = \iint_A \left(N_{\alpha\beta}\dot{\varepsilon}_{\alpha\beta}\right)dA, \quad (\alpha, \beta = 1, 2) \tag{3.23}$$

where $N_{\alpha\beta}$ is the membrane force tensor and $\dot{\varepsilon}_{\alpha\beta}$ is the strain rate tensor. By the use of von Mises' yield criterion, the membrane energy rate can be expressed as

$$\dot{E}_m = \frac{2}{\sqrt{3}}\sigma_0 t \iint\left(\sqrt{\dot{\varepsilon}_{xx}{}^2 + \dot{\varepsilon}_{yy}{}^2 + \dot{\varepsilon}_{xx}\dot{\varepsilon}_{yy} + \dot{\varepsilon}_{xy}{}^2}\right)dxdy \tag{3.24}$$

Therefore, the final expression on the rate of internal plastic energy dissipation for a deforming plate is the sum of the bending and the membrane energy dissipation rate:

$$\begin{aligned}\dot{E}_{int} &= \dot{E}_b + \dot{E}_m \\ &= \frac{2}{\sqrt{3}}\frac{\sigma_0{}^2 t}{4}\sum_{i=1}^{n}\left(\dot{\theta}_i l_i\right) + \frac{2}{\sqrt{3}}\sigma_0 t \iint\left(\sqrt{\dot{\varepsilon}_{xx}{}^2 + \dot{\varepsilon}_{yy}{}^2 + \dot{\varepsilon}_{xx}\dot{\varepsilon}_{yy} + \dot{\varepsilon}_{xy}{}^2}\right) dxdy \end{aligned} \tag{3.25}$$

In the simplified analytical method, a key element is to construct a kinematically admissible velocity and displacement field. This is primarily based on observations from experimental tests, full-scale accidents, existing analysis work, and analyser's experience and knowledge.

3.4 Lateral deformation of shell plating

3.4.1 Introduction

This section employs the upper-bound theorem to analyse the deformation of the side shell in ship collisions and to demonstrate how the force–deflection response and energy–deflection response are derived.

When a rigid striking bow collides with the side shell of a ship, the side structure of the struck ship can be deformed, destroyed, or penetrated. The deformation and damage to the struck ship may include tension of the shell plating, crushing of the transverse frames and the longitudinal stringers, and crushing of the decks and the bottom. By calculating the resistance of each component and adding all the component resistances together, the total resistance and dissipated energy can be determined. Fig. 3.14 shows an initial deformation of the side shell plating.

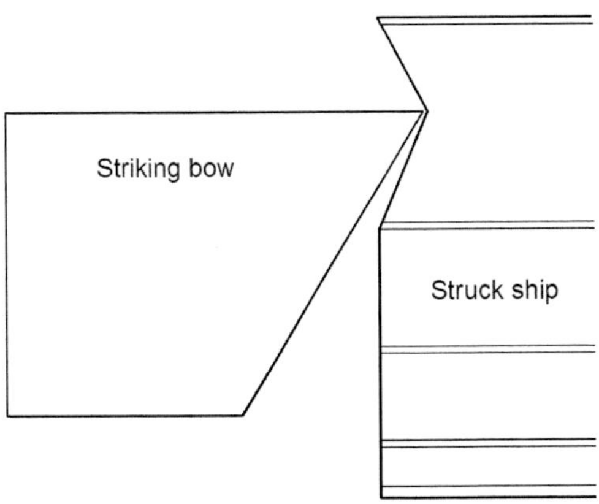

Fig. 3.14 A rigid bow strikes a side structure.

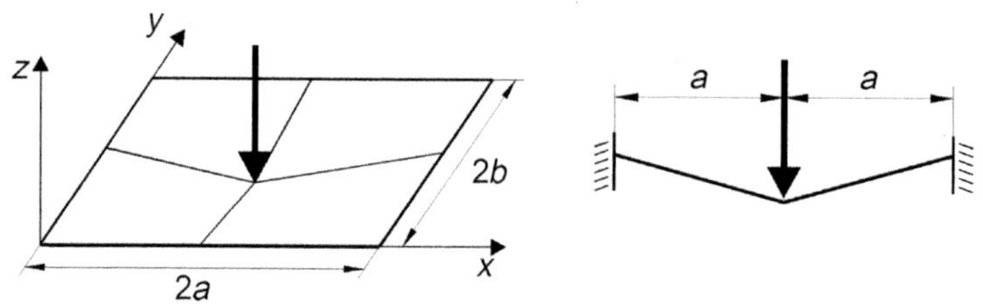

Fig. 3.15 Deformation of a rectangular plate subjected to a point load at its centre.

3.4.2 *Plates subjected to lateral point loads*

First, we consider an idealized case where a striking bow impacts the centre of a shell plate with boundaries at the adjacent decks and transverse frames. The distance between the two decks is $2b$, the spacing between the transverse frames is $2a$, and the thickness of the shell plate is t_p. The plate may suffer bending, shearing, and tension deformations. With the increase of the lateral deformation, the relative contribution of the bending force becomes weaker and can be neglected. Provided the plate edges are fixed, the membrane forces dominate the behaviour of the deformation. Therefore, the bending resistance is neglected, and only the moderate large normal deflection of the shell plate is considered. The xyz- coordinate for shell plate is presented in Fig. 3.15. The following function is used as the deformation of the shell plate:

$$w(x, y, t) = \delta(t) \cdot f(x, y) \tag{3.26}$$

where $\delta(t)$ is the time-dependent deflection at the centre of the plate and $f(x, y)$ is the deformation function.

Using Eq. (3.26), the nonlinear membrane strain and strain rate in the plate can be expressed as

$$\varepsilon_{xx} = \frac{1}{2}\left(\frac{\partial w}{\partial x}\right)^2 = \frac{1}{2}\delta^2\left(\frac{\partial f}{\partial x}\right)^2 \tag{3.27a}$$

$$\varepsilon_{yy} = \frac{1}{2}\left(\frac{\partial w}{\partial y}\right)^2 = \frac{1}{2}\delta^2\left(\frac{\partial f}{\partial y}\right)^2 \tag{3.27b}$$

$$\varepsilon_{xy} = \frac{1}{2}\frac{\partial w}{\partial x}\frac{\partial w}{\partial y} = \frac{1}{2}\delta^2\frac{\partial f}{\partial x}\frac{\partial f}{\partial y} \tag{3.27c}$$

$$\dot{\varepsilon}_{xx} = \delta\dot{\delta}\left(\frac{\partial f}{\partial x}\right)^2; \quad \dot{\varepsilon}_{yy} = \delta\dot{\delta}\left(\frac{\partial f}{\partial y}\right)^2; \quad \dot{\varepsilon}_{xy} = \delta\dot{\delta}\frac{\partial f}{\partial x}\frac{\partial f}{\partial y} \tag{3.28}$$

The equilibrium equation of the rate of external work and the rate of internal energy dissipation can be expressed as

$$F_p \cdot \dot{\delta} = \frac{2}{\sqrt{3}}\sigma_0 t_p \iint_A \left(\dot{\varepsilon}_{xx}^2 + \dot{\varepsilon}_{yy}^2 + \dot{\varepsilon}_{xx}\dot{\varepsilon}_{yy} + \dot{\varepsilon}_{xy}^2\right)^{\frac{1}{2}} dxdy \tag{3.29}$$

where A is the plate area and F_p is the resistance force. From Eqs (3.27a), (3.27b), (3.27c), (3.28), we have

$$\dot{\varepsilon}_{xx}^2 + \dot{\varepsilon}_{yy}^2 + \dot{\varepsilon}_{xx}\dot{\varepsilon}_{yy} + \dot{\varepsilon}_{xy}^2 = (\delta\dot{\delta})^2 \left(\left(\frac{\partial f}{\partial x}\right)^2 + \left(\frac{\partial f}{\partial y}\right)^2\right)^2 = (\dot{\varepsilon}_{xx} + \dot{\varepsilon}_{yy})^2$$

Then, Eq. (3.29) becomes

$$F_p \cdot \dot{\delta} = \frac{2}{\sqrt{3}}\sigma_0 t_p \iint_A (\dot{\varepsilon}_{xx} + \dot{\varepsilon}_{yy}) dxdy = \frac{2}{\sqrt{3}}\sigma_0 t_p \delta\dot{\delta} \iint_A \left(\left(\frac{\partial f}{\partial x}\right)^2 + \left(\frac{\partial f}{\partial y}\right)^2\right) dxdy$$

Therefore, the relationship between the resistance force and the deflection can be expressed in the general form

$$F_p = \frac{2}{\sqrt{3}}\sigma_0 t_p \delta \iint_A \left(\left(\frac{\partial f}{\partial x}\right)^2 + \left(\frac{\partial f}{\partial y}\right)^2\right) dxdy \tag{3.30}$$

In order to determine the collision resistance, a kinematically admissible deformation function should be proposed. Here, the following deformation function is used for one-quarter of the plate:

$$f(x, y) = \left(\frac{x}{a}\right)^n \left(\frac{y}{b}\right)^n, \quad n \geq 1$$

Then, we have $\frac{\partial f}{\partial x} = \frac{n}{a}\left(\frac{x}{a}\right)^{n-1}\left(\frac{y}{b}\right)^n$ and $\frac{\partial f}{\partial y} = \frac{n}{b}\left(\frac{x}{a}\right)^n\left(\frac{y}{b}\right)^{n-1}$, and for $n = 1$, the strains of the plate can be expressed as

$$\varepsilon_{xx} = \frac{1}{2}\delta^2\left(\frac{\partial f}{\partial x}\right)^2 = \frac{1}{2}\delta^2\left(\frac{1}{a}\frac{y}{b}\right)^2$$

$$\varepsilon_{yy} = \frac{1}{2}\delta^2\left(\frac{\partial f}{\partial y}\right)^2 = \frac{1}{2}\delta^2\left(\frac{1}{b}\frac{x}{a}\right)^2$$

$$\varepsilon_{xy} = \frac{1}{2}\delta^2\frac{\partial f}{\partial x}\frac{\partial f}{\partial y} = \frac{1}{2}\delta^2\left(\frac{1}{a}\frac{x}{b}\frac{y}{a}\frac{}{b}\right)$$

$$\left(\varepsilon_{xx}^2 + \varepsilon_{yy}^2 + \varepsilon_{xx}\varepsilon_{yy} + \varepsilon_{xy}^2\right)^{\frac{1}{2}} = \varepsilon_{xx} + \varepsilon_{yy} = \frac{1}{2}\delta^2\left[\left(\frac{1}{a}\frac{y}{b}\right)^2 + \left(\frac{1}{b}\frac{x}{a}\right)^2\right]$$

Integrating Eq. (3.30), we have the force–deflection relationship:

$$F_p = \frac{2}{\sqrt{3}}\frac{4n^2}{4n^2-1}\sigma_0 t\delta\left(\frac{b}{a}+\frac{a}{b}\right) = C\sigma_0 t\delta\left(\frac{b}{a}+\frac{a}{b}\right) \tag{3.31}$$

where

$C = \frac{2}{\sqrt{3}}\frac{4n^2}{4n^2-1}$

$n=1 \quad C=1.540$

$n=2 \quad C=1.232$

$n=3 \quad C=1.188$

$\vdots$

$n=\infty \quad C=1.155$

The energy–deflection relationship can be expressed as

$$E_p = \int_0^\delta F_p d\delta = \frac{1}{\sqrt{3}}\frac{4n^2}{4n^2-1}\sigma_0 t\delta^2\left(\frac{b}{a}+\frac{a}{b}\right) \tag{3.32}$$

It is seen from Eq. (3.31) that the effect of the power n of the deformation function $f(x,y)$ on the resistance force is not so significant. Observations from experiments, numerical simulations, and full-scale tests show that a linear function is a good approximation for ship collisions. For $n=1$, the relationship between the resistance and the deflection is

$$F_p = \frac{4}{3}\frac{2}{\sqrt{3}}\sigma_0 t_p\delta\left(\frac{b}{a}+\frac{a}{b}\right) = \frac{1}{3}\frac{2}{\sqrt{3}}\sigma_0 t_p A\delta\left(\frac{1}{a^2}+\frac{1}{b^2}\right) \tag{3.33}$$

where $A=2a\cdot 2b$ is the area of the plate. The energy absorbed by the plate can be calculated from

$$E_p = \int_0^\delta F_p d\delta = \frac{1}{3}\frac{1}{\sqrt{3}}\sigma_0 t_p A\delta^2\left(\frac{1}{a^2}+\frac{1}{b^2}\right) \tag{3.34}$$

When the impact position is not located at the centre of the shell plate, as illustrated in Fig. 3.16, we can follow the same procedure to derive the force–deflection relationship and the energy–deflection relationship. The derived expressions are

$$F_p = \frac{2}{\sqrt{3}}\frac{n^2}{4n^2-1}\sigma_0 t_p A\delta\left(\frac{1}{a_1a_2}+\frac{1}{b_1b_2}\right) \tag{3.35}$$

$$E_p = \frac{1}{\sqrt{3}}\frac{n^2}{4n^2-1}\sigma_0 t_p A\delta^2\left(\frac{1}{a_1a_2}+\frac{1}{b_1b_2}\right) \tag{3.36}$$

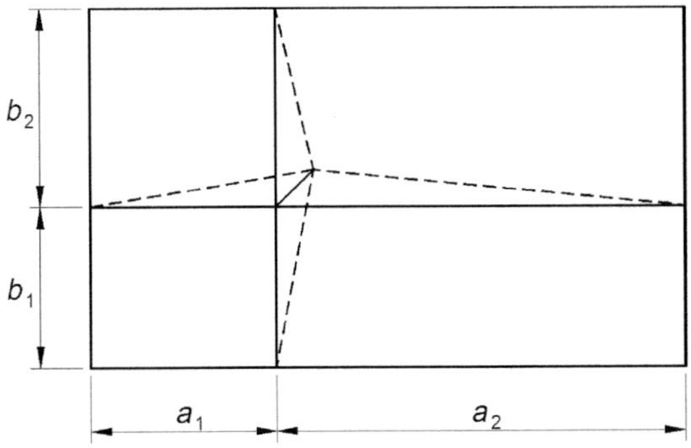

Fig. 3.16 Deformation of a rectangular plate subjected to an eccentric load.

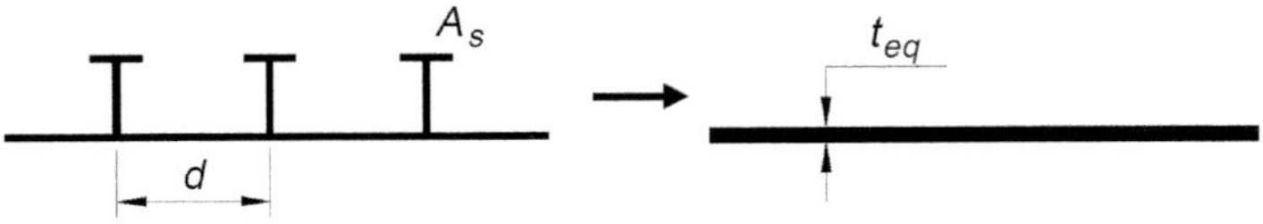

Fig. 3.17 Equivalent thickness of a stiffened plate.

The actual side shell is usually not a bare plate, and it is often stiffened by longitudinal stiffeners. In this case, the equivalent thickness method is employed to deal with the stiffeners. The idea of the equivalent method is to distribute the cross-sectional area of the stiffeners to the whole plate. That is,

$$t_{eq} = t_p + A_s/d \tag{3.37}$$

where t_{eq} is the equivalent thickness of the stiffened plate, t_p is the thickness of the shell plate, A_s is the sectional area of the stiffener, and d is the stiffener spacing (see Fig. 3.17).

For a stiffened panel, Eqs (3.35), (3.36) of the force–deflection relationship and the energy–deflection relationship become

$$F_p = \frac{2}{\sqrt{3}} \frac{n^2}{4n^2 - 1} \sigma_0 A \delta \left(\frac{t_{px}}{a_1 a_2} + \frac{t_{py}}{b_1 b_2} \right) \tag{3.38}$$

$$E_p = \frac{1}{\sqrt{3}} \frac{n^2}{4n^2 - 1} \sigma_0 A \delta^2 \left(\frac{t_{px}}{a_1 a_2} + \frac{t_{py}}{b_1 b_2} \right) \tag{3.39}$$

where t_{px} and t_{py} are the equivalent thicknesses of the stiffened plate in the x- and the y-directions, respectively. When the impact location is at the uppermost deck of the struck ship, as shown in Fig. 3.18, the contribution to the side shell to the force–deflection relationship and the energy–deflection relationship can be expressed as

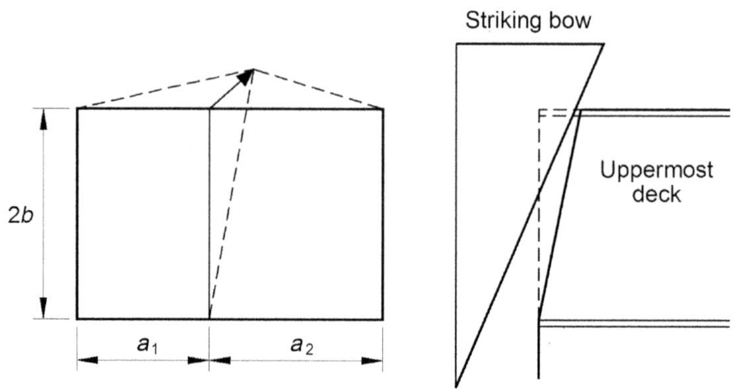

Fig. 3.18 Impact location at the uppermost deck of the struck ship.

$$F_p = \frac{2}{\sqrt{3}} \frac{n^2}{4n^2 - 1} \sigma_0 t_p A \delta \left(\frac{1}{a_1 a_2} + \frac{1}{(2b)(2b)} \right) \tag{3.40}$$

$$E_p = \frac{1}{\sqrt{3}} \frac{n^2}{4n^2 - 1} \sigma_0 t_p A \delta^2 \left(\frac{1}{a_1 a_2} + \frac{1}{(2b)(2b)} \right) \tag{3.41}$$

3.4.3 Plates subjected to lateral line loads or patch loads

3.4.3.1 Line loads

As illustrated in Fig. 3.19, a plate strip fixed at the ends impacted by a line object deforms uniformly. The force–deflection relationship and the energy–deflection relationship can be determined by

$$F_p = \frac{2}{\sqrt{3}} \frac{n^2}{2n - 1} \sigma_0 t_p \delta \left(\frac{2b}{a_1} + \frac{2b}{a_2} \right) \tag{3.42}$$

$$E_p = \frac{1}{\sqrt{3}} \frac{n^2}{2n - 1} \sigma_0 t_p \delta^2 \left(\frac{2b}{a_1} + \frac{2b}{a_2} \right) \tag{3.43}$$

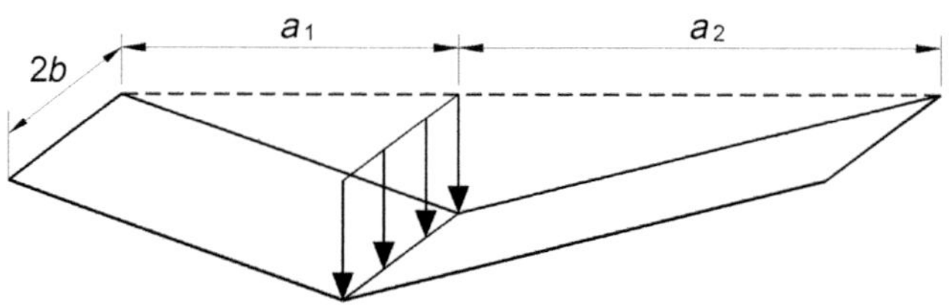

Fig. 3.19 Deformation of a plate strip subjected to a line load.

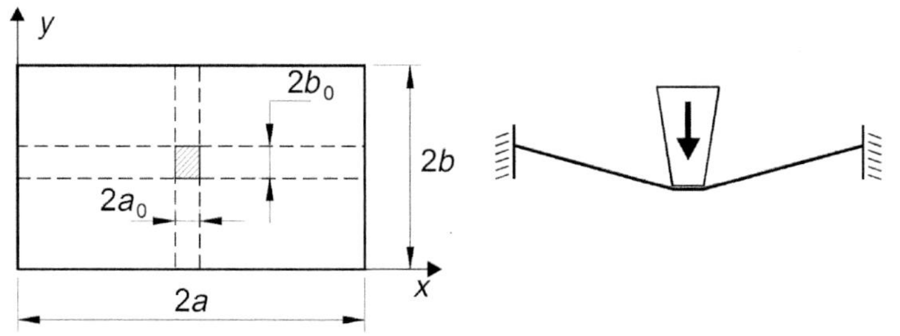

Fig. 3.20 Shell plate impacted by an object with rectangular impact section.

3.4.3.2 Patch loads (central impact)

As shown in Fig. 3.20, a blunt object with a flat face (dimension of $2a_0 \times 2b_0$) hits perpendicularly the centre of a fixed plate (dimension $2a \times 2b$). This situation may represent the case of dropped objects impacting decks.

The force–deflection relationship and the energy–deflection relationship of the blunt object impacting the centre of the plate can be expressed as

$$F_p = \frac{2}{\sqrt{3}}\frac{4n^2}{4n^2-1}\sigma_0 t_p \delta\left(\frac{b-b_0}{a-a_0}+\frac{a-a_0}{b-b_0}\right)+\frac{2}{\sqrt{3}}\frac{2n^2}{2n-1}\sigma_0 t_p \delta\left(\frac{b_0}{a-a_0}+\frac{a_0}{b-b_0}\right) \tag{3.44}$$

$$E_p = \frac{1}{\sqrt{3}}\frac{4n^2}{4n^2-1}\sigma_0 t_p \delta^2\left(\frac{b-b_0}{a-a_0}+\frac{a-a_0}{b-b_0}\right)+\frac{1}{\sqrt{3}}\frac{2n^2}{2n-1}\sigma_0 t_p \delta^2\left(\frac{b_0}{a-a_0}+\frac{a_0}{b-b_0}\right) \tag{3.45}$$

If $n=1$, $2a=2b$, and $2a_0=2b_0$, the force–deflection relationship and the energy–deflection relationship are simplified as

$$F_p = \frac{8}{3}\frac{2}{\sqrt{3}}\sigma_0 t_p \delta\left(1+\frac{3}{2}\frac{a_0}{a-a_0}\right) \tag{3.46}$$

$$E_p = \frac{4}{3}\frac{2}{\sqrt{3}}\sigma_0 t_p \delta^2\left(1+\frac{3}{2}\frac{a_0}{a-a_0}\right) \tag{3.47}$$

3.4.3.3 Patch loads (eccentric impact)

Similarly, for the eccentric impact, as shown in Fig. 3.21, the force–deflection relationship and the energy–deflection relationship are

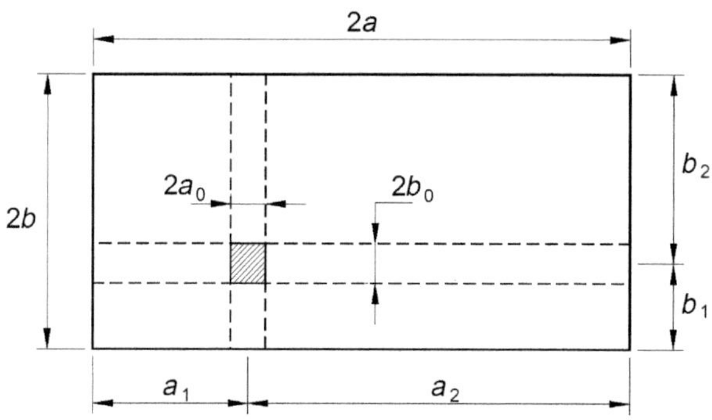

Fig. 3.21 Shell plate impacted eccentrically by an object with rectangular section.

$$F_p = \frac{2}{\sqrt{3}} \frac{4n^2}{4n^2-1} \sigma_0 t_p \delta \left(\frac{(a-a_0)(b-b_0)}{(a_1-a_0)(a_2-a_0)} + \frac{(a-a_0)(b-b_0)}{(b_1-b_0)(b_2-b_0)} \right) + \frac{2}{\sqrt{3}} \frac{2n^2}{2n-1} \sigma_0 t_p \delta \left(\frac{(a-a_0)b_0}{(a_1-a_0)(a_2-a_0)} + \frac{a_0(b-b_0)}{(b_1-b_0)(b_2-b_0)} \right) \tag{3.48}$$

$$E_p = \frac{1}{\sqrt{3}} \frac{4n^2}{4n^2-1} \sigma_0 t_p \delta^2 \left(\frac{(a-a_0)(b-b_0)}{(a_1-a_0)(a_2-a_0)} + \frac{(a-a_0)(b-b_0)}{(b_1-b_0)(b_2-b_0)} \right) + \frac{1}{\sqrt{3}} \frac{2n^2}{2n-1} \sigma_0 t_p \delta^2 \left(\frac{(a-a_0)b_0}{(a_1-a_0)(a_2-a_0)} + \frac{a_0(b-b_0)}{(b_1-b_0)(b_2-b_0)} \right) \tag{3.49}$$

3.4.4 Circular plate subjected to lateral patch loads

A patch load impacts the centre of a circular plate, such as a cylinder with a flat end hitting the centre of a circular plate fixed at the outer boundary, as illustrated in Fig. 3.22. The radius of the cylindrical impactor is a_0, and the radius of the circular plate is a. It is assumed that the deformation function is $f = \left(\frac{a-r}{a-a_0} \right)^n$. For $n=1$, the force–deflection relationship and the energy–deflection relationship can be expressed as

$$F_p = \frac{2}{\sqrt{3}} \pi \sigma_0 t_p \delta \left(1 + \frac{2a_0}{a-a_0} \right) \tag{3.50}$$

$$E_p = \frac{2}{\sqrt{3}} \frac{\pi}{2} \sigma_0 t_p \delta^2 \left(1 + \frac{2a_0}{a-a_0} \right) \tag{3.51}$$

For $n=2$, the force–deflection relationship and the energy–deflection relationship are expressed as

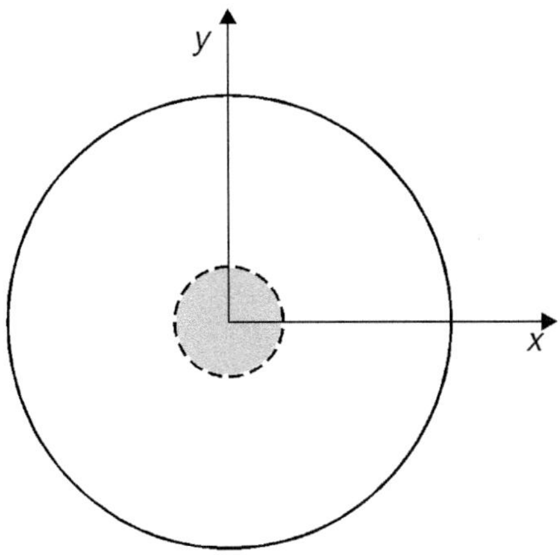

Fig. 3.22 Circular plate subjected to a patch load at its centre.

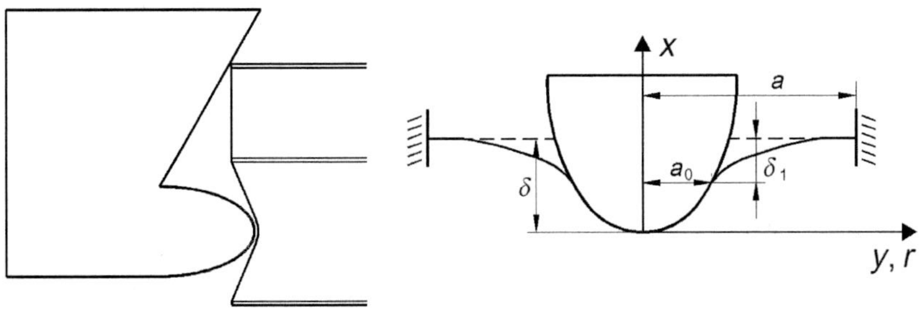

Fig. 3.23 Bulbous bow striking a side structure and the assumed deformation.

$$F_p = \frac{2}{\sqrt{3}} \frac{2\pi}{3} \sigma_0 t_p \delta \left(1 + \frac{4a_0}{a - a_0}\right) \tag{3.52}$$

$$E_p = \frac{2}{\sqrt{3}} \frac{\pi}{3} \sigma_0 t_p \delta^2 \left(1 + \frac{4a_0}{a - a_0}\right) \tag{3.53}$$

3.4.5 Bulbous bow impact on shell plating

Fig. 3.23 shows a bulb indenting into the side shell. To consider the effect of the radius of the bulb on the force–deflection and energy–deflection curves, first, we analyse a bulb indenting into the centre of a circular plate, which is also shown in Fig. 3.23, and after that, the method is extended to the case of a rectangular plate. In order to simplify the analysis and for convenience, the bulb has a simple elliptic form and can be represented by

$$w_b(r) = \frac{r^2}{R_b} \tag{3.54}$$

where $R_b = (R_V)^2/R_L$;R_V is the radius in the vertical direction and R_L is the radius in the longitudinal direction.

Based on the observations from the experiments and numerical simulations, the following appropriate function is used for the kinematically admissible plate deformation:

$$w = \begin{cases} \delta_1 \cdot f_1(\delta_1, r, \theta) = \delta_1 \cdot \left(\dfrac{a-r}{a-a_0}\right)^2, & a_0 \le r \le a \\ f_2(\delta, r, \theta) = \delta - \dfrac{r^2}{R_b}, & 0 \le r \le a_0 \end{cases} \tag{3.55}$$

where a is the radius of the plate, a_0 is the radius at the separating point between the bulb and the plate, δ is the deflection at the bulb tip, and δ_1 is the deflection at the separating point. The slopes at the separating point between the bulb and the plate are equal, which determines the separating radius a_0. That is, $\frac{\partial w_b}{\partial r} = \frac{\partial w}{\partial r} \Rightarrow \frac{a_0}{R_b} = \frac{\delta_1}{a-a_0}$. Then, the separating radius is

$$a_0 = \frac{R_b}{a}\delta \tag{3.56}$$

The deflection and the velocity at the separating point are expressed as

$$\delta_1 = \delta\left(1 - \frac{R_b\delta}{a^2}\right) \quad \text{and} \quad \dot{\delta}_1 = \dot{\delta}\left(1 - \frac{2R_b\delta}{a^2}\right) \tag{3.57}$$

The equilibrium of the external work rate and the internal energy dissipation rate is

$$F_p \cdot \dot{\delta} = \frac{2}{\sqrt{3}}\sigma_0 t \left(\iint\limits_{a_0 < r \le a} (\dot{\varepsilon}_{r1}) r dr d\theta + \frac{\partial}{\partial T} \iint\limits_{0 \le r \le a_0} (\varepsilon_{r2}) r dr d\theta \right)$$

where T is time. The strain and the strain rate are calculated from

$$\varepsilon_{r1} = \frac{1}{2}\left(\frac{\partial w}{\partial r}\right)^2 = 2\delta_1{}^2 \frac{(a-r)^2}{(a-a_0)^4}$$

$$\dot{\varepsilon}_{r1} = \left(\frac{4\delta_1\dot{\delta}_1}{(a-a_0)^4} + \frac{8\delta_1{}^2\dot{a}_0}{(a-a_0)^5} \right)(a-r)^2$$

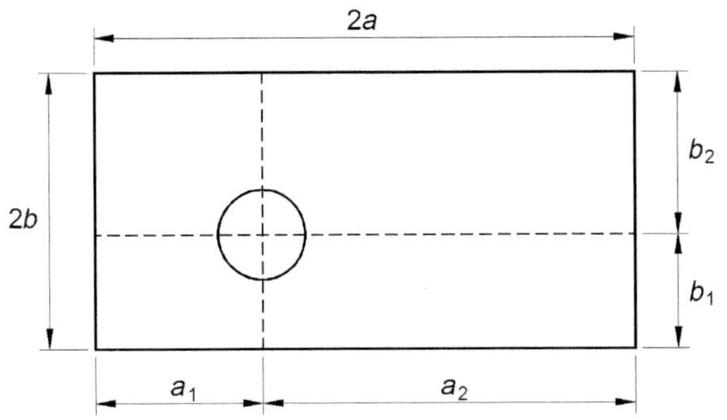

Fig. 3.24 A bulbous bow impacting shell plate eccentrically.

$$\varepsilon_{r2} = \frac{1}{2}\left(\frac{\partial f_2}{\partial r}\right)^2 = 2\left(\frac{r}{R_b}\right)^2$$

Through some derivations, the force–deflection relationship is written as

$$F_p = \frac{2\pi}{3}\frac{2}{\sqrt{3}}\sigma_0 t\delta\left(1 + 3\frac{R_b\delta}{a^2} + 6\left(\frac{R_b\delta}{a^2}\right)^2\right) \tag{3.58}$$

The energy–deflection relationship is

$$E_p = \frac{\pi}{3}\frac{2}{\sqrt{3}}\sigma_0 t\delta^2\left(1 + 2\frac{R_b\delta}{a^2} + 3\left(\frac{R_b\delta}{a^2}\right)^2\right) \tag{3.59}$$

The above method can be extended to get an approximate solution for the bulbous bow impacting the centre of a square plate (with a dimension of $2a \cdot 2a$). The force–deflection relationship is expressed as

$$F_p = \frac{32}{15}\frac{2}{\sqrt{3}}\sigma_0 t\delta\left(1 + 3\frac{R_b\delta}{a^2} + 6\left(\frac{R_b\delta}{a^2}\right)^2\right) \tag{3.60}$$

Similarly, the force–deflection relationship for a bulbous bow impacting the centre of a rectangular plate with a dimension of $2a \cdot 2b$ is expressed as

$$F_p = \frac{16}{15}\frac{2}{\sqrt{3}}\sigma_0 t\delta\left[\left(\frac{b}{a} + \frac{a}{b}\right) + 3\left(\frac{R_b\delta}{a^2} + \frac{R_b\delta}{b^2}\right) + 6\left(\left(\frac{R_b\delta}{a^2}\right)^2 + \left(\frac{R_b\delta}{b^2}\right)^2\right)\right] \tag{3.61}$$

As shown in Fig. 3.24, for a bulbous bow impacting a rectangular plate eccentrically, the force–deflection relationship is expressed as

$$F_p = 1.23\sigma_0 t\delta \left[ab\left(\frac{1}{a_1 a_2} + \frac{1}{b_1 b_2}\right) + \frac{3}{2}(R_b\delta)\left(\frac{1}{a_1^2} + \frac{1}{a_2^2} + \frac{1}{b_1^2} + \frac{1}{b_2^2}\right) + 3(R_b\delta)^2\left(\frac{1}{a_1^4} + \frac{1}{a_2^4} + \frac{1}{b_1^4} + \frac{1}{b_2^4}\right)\right] \quad (3.62)$$

3.4.6 Critical deflection of the shell plate at initial rupture

When the penetration or deflection is large enough, the shell plate will rupture. The critical deflection point of the shell at which rupture starts is one of the most challenging issues in collision damage assessments. This limit is also important to assess whether liquid or oil outflow occurs and the amount of oil that will flow out.

The fracture and rupture of shell structures in ship collisions are influenced by many factors, such as material, plate thickness, structural arrangement, welding, shape of indenter, and impact speed. It is an extremely complex problem to predict the rupture of structures accurately. Jones and Wierzbicki (1993) discussed the rupture criteria of ductile metal beams subjected to large dynamic loads. Three major failure criteria of the metal beams were discussed. The first is the **tensile tearing** failure mode, which occurs when the maximum tensile strain equals the critical rupture strain of the material, that is, $\varepsilon_{\max} = \varepsilon_f$. The second failure model is the **transverse shear failure** mode, which develops in a beam when large transverse shear deformations occur within a very short region of the plastic beam. When the total transverse shear displacement in a particular location equals a critical value, the beam or plate ruptures. The third failure criterion is the **energy density failure** mode. It is assumed that rupture occurs in a rigid-plastic structure when the absorption of plastic work per unit volume reaches a critical value.

The simplified analytical methods are based on overall deformation mechanisms. In this case, it is not possible to trace the strain history of material elements at a very detailed level. Therefore, as many authors did, for example, Wang (1995) and Paik and Pedersen (1995), we use the 'maximum strain failure criteria'. That is, when the maximum strain in a structure reaches a critical strain, the structure ruptures.

When the critical rupture strain is known, the critical deflection or penetration of the shell plating can be determined. For example, consider a point load acting in the middle of a plate strip with a span $2b$, as shown in Fig. 3.25. In this case, the strain in the plate strip due to the lateral deflection can be calculated from

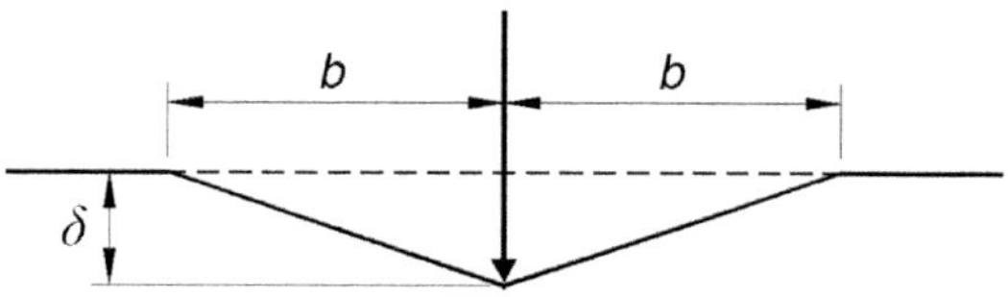

Fig. 3.25 Lateral deflection of a plate strip.

$$\varepsilon = \sqrt{1 + \left(\frac{\delta}{b}\right)^2} - 1 \approx \frac{1}{2}\left(\frac{\delta}{b}\right)^2$$

where δ is the deflection at the middle point. When the deflection is large enough, the strain in the plate strip reaches the critical rupture value. The critical rupture deflection or penetration is then determined from

$$\delta_f = b \cdot \sqrt{2\varepsilon_f} \tag{3.63}$$

A typical critical rupture strain of mild steels from standard uniaxial tensile tests is around 0.35. If this critical strain $\varepsilon_f = 0.35$ is used, then the critical deflection or penetration is $\delta_f = 0.837b$.

Amdahl (1995) pointed out that due to scale effects, bending strains and material imperfections, the critical strain is far too large when assessing full-scale collisions. The critical strain value he suggested for side collisions is between 5% and 10%, if the simple beam theory is used.

In the minor collision analyses performed by McDermott et al. (1974), the critical rupture strain for mild steel material in side collisions is evaluated from $\varepsilon_{cr} = 0.10 \cdot (\varepsilon_f / 0.32)$. The main reason of having to use such small value of critical rupture strain is that the theoretical model used to estimate critical deflection is too simple. Thus, to get a good agreement with the experimental data, the critical rupture strain has to be tuned and reduced.

Through finite element analyses and experimental studies, Simonsen and Lauridsen (2000), Lee et al. (2004), and Liu et al. (2014) have investigated the critical defection of shell plate initial rupture. It has been found that the radius of the indenter and the plate size have a significant influence on the critical deflection. The expressions from their studies are presented below:

(1) Simonsen and Lauridsen (2000):

$$\delta_f = 1.41 n^{0.33} b^{0.48} R_b^{\ 0.52} \tag{3.64}$$

(2) Lee et al. (2004):

$$\delta_f = (2D_c)^{0.35} b^{0.62} R_b^{\ 0.38} \tag{3.65}$$

(3) Liu et al. (2014):

$$\delta_f = 1.15 n^{0.18} b^{0.5} R_b^{\ 0.5} \tag{3.66}$$

where n is the material strain hardening exponent, b is the radius of a circular plate or half width of a square/rectangular plate short side if the impact is at the centre or the plate width if the impact is at the edge (i.e. the shortest distance to the support edge), and R_b is the radius of the indenter; see Fig. 3.26. D_c is the material critical damage value for ductile fracture, defined as the product of the failure strain and the average stress triaxiality, which can be taken as 0.27 for punch indentations.

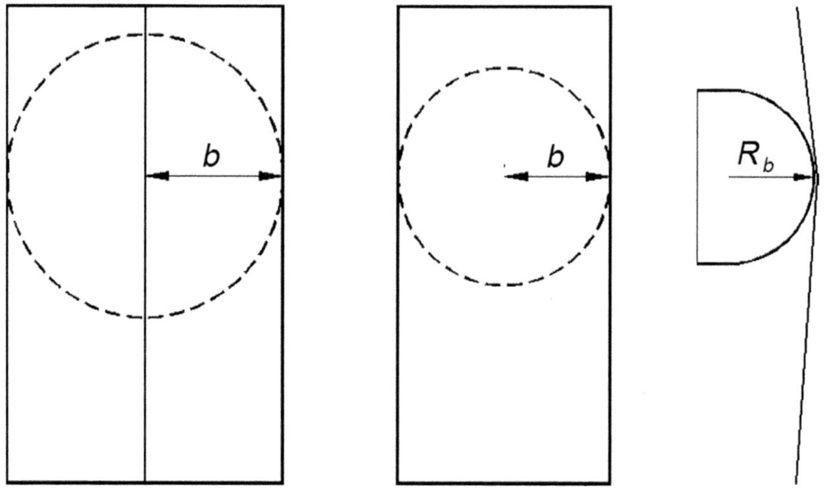

Fig. 3.26 Illustration of a bulb indenting into plates.

Each method has been validated by the authors' own experimental data using four edge-clamped plates (or circular plates) indented by rigid spheres. Following the same principle and collecting a wide range of experimental results (a total of 37 experimental data) and introducing the fracture strain as an input parameter, a new empirical formula for determining the critical deflection of shell plate at initial rupture is proposed here:

$$\delta_f = {\varepsilon_f}^{0.4} b^{0.4} {R_b}^{0.6} \tag{3.67}$$

A comparison using the various formulae for calculating the critical deflections of the 37 experiments is presented in Fig. 3.27 and Table 3.7. It can be seen that the methods proposed by Simonsen and Lauridsen (2000), Lee et al. (2004), and Liu et al. (2014) give a similar level of good agreement for the plate models (circle symbols in Fig. 3.27; 19 experimental data) as these methods were validated using plate models by the authors. However, for the double-hull models (diamond symbols in Fig. 3.27; 18 experimental data), Eqs. (3.64), (3.65), (3.66) overestimate the results. The new formula (Eq. 3.67) generally underestimates the results for the plate models, but it gives reasonable predictions for the double-hull models. It should be noted that for plate models, the ideal boundary condition of fully clamped edges may be difficult to achieve in laboratory model tests, as in-plane displacements occur which increases the experimentally observed critical deflection. This could be one of the reasons that such calibrated methods give higher results on double-hull models.

The last example in the comparison in Table 3.7 is a collision accident of a large LNG ship struck by an oil tanker. The side shell of the LNG ship was ruptured by the bulb of the striking oil tanker, and the inner hull of the LNG ship was only dented without rupture. The maximum penetration into the side shell was about 3.0m. The critical rupture deflection is estimated between 2.3 and 2.6m for the collision incident. While the methods of Simonsen and Lauridsen (2000), Lee et al. (2004), and Liu et al. (2014) predict more than 3.3m, the present method, Eq. (3.67), gives an estimate of 2.58m.

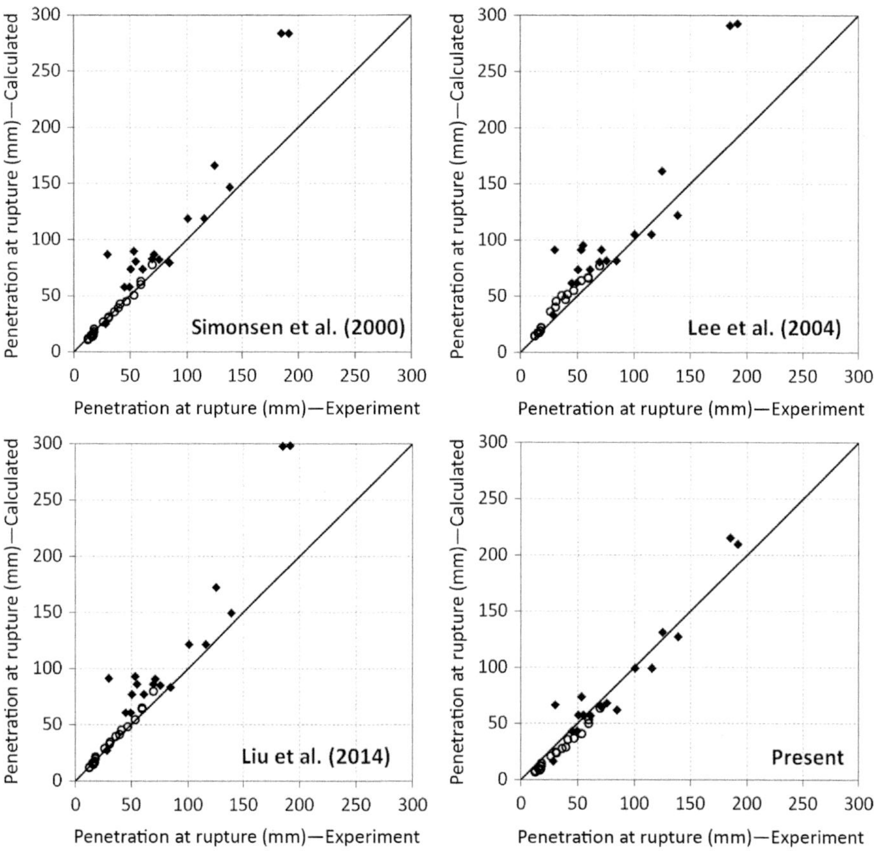

Fig. 3.27 Comparisons between calculated and experimental results on critical deflections (note: Eq. (3.67) is used in the last figure labelled with 'Present').

3.4.7 Calculation examples

3.4.7.1 Example 1

Simonsen and Lauridsen (2000) carried out model tests of plates indented by rigid cones with different radii. The test set-up is illustrated in Fig. 3.28. The specimen was made of mild steel plate with a thickness of 1.0 mm. The properties of the material are as follows: yield stress 185 MPa and tensile strength 298 MPa. The case with the largest cone radius ($R_b = 50$ mm) is selected for the analysis. The comparison between the model test results and the analytical results (Eq. 3.58) is presented in Table 3.8 and Fig. 3.28; the overall agreement is good. To investigate the effect of the rigid cone radius, the analytical results for $R_b = 0$ are calculated and also presented in Table 3.8. It is seen that the radius of the indenter has significant effect on the load–deflection curves.

Table 3.7 Critical deflections at rupture of shell plates. Comparisons between hand calculations and experimental data.

Reference	Model type	Model ID	b (mm)	Thickness (mm)	R_b (mm)	ε_f	n	Experimental δ_f (mm)	Simonsen δ_f (mm)	Lee δ_f (mm)	Liu δ_f (mm)	Present, Eq. (3.67) δ_f (mm)
Arita and Aoki (1985)	Double hull (D.H.)	S-I	100	3.2	80	0.35	0.20	51.0	73.8	74.0	77.0	57.5
	D.H.	S-II	140	4.5	80	0.34	0.20	71.4	86.8	91.2	91.1	65.0
	D.H.	S-III	100	4.5	80	0.34	0.20	61.2	73.8	74.0	77.0	56.8
Wang et al. (2000)	D.H.	P-10	100	2.3	10	0.35	0.20	28.5	25.0	33.6	27.2	16.5
	D.H.	P-50	100	2.3	50	0.35	0.20	45.0	57.8	61.9	60.9	43.4
	D.H.	P-100	100	2.3	100	0.35	0.20	69.9	82.9	80.6	86.1	65.7
	D.H.	P-200	100	2.3	200	0.35	0.20	115.7	118.9	104.9	121.7	99.6
	D.H.	P-300	100	2.3	300	0.35	0.20	139.2	146.8	122.4	149.1	127.0
	D.H.	W-50	100	2.3	50	0.35	0.20	49.8	57.8	61.9	60.9	43.4
	D.H.	W-200	100	2.3	200	0.35	0.20	100.5	118.9	104.9	121.7	99.6
	D.H.	C-50	200	2.3	50	0.35	0.20	55.1	80.6	95.2	86.1	57.2
	D.H.	C-200	200	2.3	200	0.35	0.20	125.1	165.8	161.2	172.2	131.4
Tautz et al. (2013)	D.H.	CE-1	400	4.0	305	0.32	0.19	185.0	283.2	290.8	297.9	215.5
Lehmann and Peschmann (2002)	D.H.	Lehmann	408	5.0	300	0.30	0.19	191.7	283.4	292.6	298.4	209.6
Paik et al. (1999)	D.H.	ST-3-BW	141.0	2.8	80	0.46	0.22	53.5	89.8	91.6	93.0	73.6
	D.H.	ST-4-BW	141.0	4.0	80	0.36	0.20	30.1	87.1	91.6	91.4	66.7
	D.H.	ST-3-OW	117.5	2.8	80	0.46	0.22	75.5	82.3	81.8	84.9	68.4
	D.H.	ST-4-OW	117.5	4.0	80	0.36	0.20	84.6	79.8	81.8	83.5	62.0

Simonsen and Lauridsen (2000)	Circular plate (C.P.)	1	111.4	1.0	10	0.52	0.19	26.8	25.9	35.9	28.5	20.1
	C.P.	2	111.4	1.0	13	0.52	0.19	31.2	29.7	39.7	32.5	23.6
	C.P.	3	111.4	1.0	25	0.52	0.19	41.8	41.7	50.9	45.0	34.9
	C.P.	4	111.4	1.0	50	0.52	0.19	60.0	59.9	66.2	63.7	52.9
	Square plate (S.P.)	1	160	1.0	10	0.52	0.19	31.7	30.8	45.0	34.1	23.3
	S.P.	2	160	1.0	13	0.52	0.19	37.0	35.4	49.7	38.9	27.2
	S.P	3	160	1.0	25	0.52	0.19	54.2	49.7	63.7	53.9	40.3
	Rectangular plate (R.P.)	1	125	1.0	25	0.52	0.19	47.5	44.1	54.7	47.7	36.5
Lee et al. (2004)	S.P.	1	110	0.9	20	0.44	0.22	40.0	38.8	46.4	41.1	28.5
	S.P.	2	110	1.14	50	0.44	0.22	60.0	62.5	65.7	64.9	49.4
	S.P.	3	110	1.4	75	0.44	0.22	70.0	77.1	76.7	79.5	62.9
Liu et al. (2014)	R.P.	ST-D10	38.1	1.4	5	0.353	0.21	12.9	11.2	14.2	12.0	7.4
	R.P	ST-D16	38.1	1.4	8	0.353	0.21	16.0	14.3	17.0	15.2	9.8
	R.P.	ST-D20	38.1	1.4	10	0.353	0.21	17.8	16.0	18.5	16.9	11.3
	R.P.	ST-D30	38.1	1.4	15	0.353	0.21	18.7	19.8	21.5	20.8	14.4
	R.P.	Al-D10	38.1	2	5	0.216	0.15	13.1	10.0	14.2	11.3	6.1
	R.P.	Al-D16	38.1	2	8	0.216	0.15	16.9	12.8	17.0	14.3	8.1
	R.P.	Al-D20	38.1	2	10	0.216	0.15	18.1	14.3	18.5	16.0	9.2
	R.P.	Al-D30	38.1	2	15	0.216	0.15	18.7	17.7	21.5	19.5	11.8
Test case 1	Ferry incident	Steel	1125	8.5	785	0.300	0.19	500–600	760	791	802	560
Test case 2	LNG ship incident	Steel	4020	18	4284	0.300	0.19	2300–2600	3387	3319	3539	2580

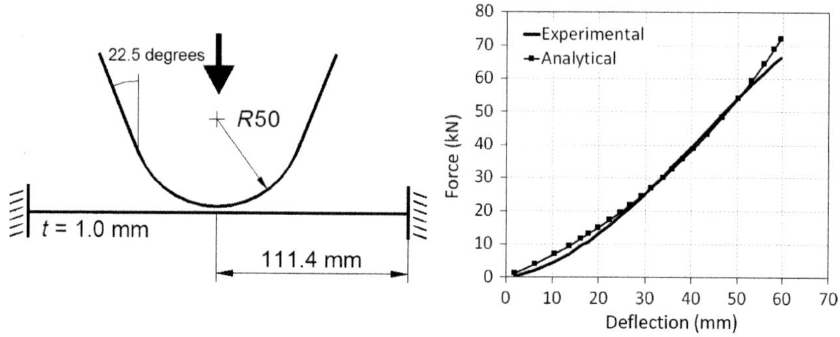

Fig. 3.28 A bulbous bow indenting a circular plate and comparison between the results from the model test and the theoretical analysis based on Eq. (3.58).

Table 3.8 Comparison between the results from the model test and the theoretical analysis for a bulbous bow indenting a circular plate (bow radius $R_b = 50$ mm).

Deflection (mm)	Force (kN, experiment) $R_b = 50$ mm	Force (kN, analytical) $R_b = 50$ mm	Force (kN, analytical) $R_b = 0$ mm
1.9	0.2	1.1	1.1
6.4	2.1	4.0	3.7
10.5	4.7	7.0	6.1
13.8	7.1	9.5	8.0
16.3	9.6	11.6	9.5
17.9	10.8	13.1	10.5
20.0	13.2	15.0	11.7
22.4	15.7	17.3	13.1
24.7	18.7	19.6	14.4
26.8	21.0	21.8	15.6
29.3	24.1	24.6	17.1
31.4	27.1	27.0	18.3
33.9	30.4	30.1	19.8
35.8	33.6	32.6	20.9
38.1	36.7	35.7	22.3
40.4	40.0	38.9	23.6
43.4	44.4	43.3	25.4
46.7	49.4	48.4	27.2
50.1	54.2	54.1	29.2
53.1	58.4	59.4	31.0
55.9	61.8	64.7	32.7
58.1	64.8	68.9	33.9
59.7	66.5	72.1	34.8

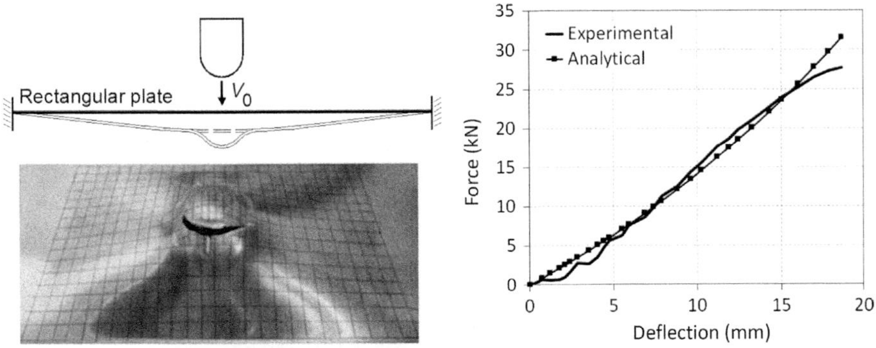

Fig. 3.29 A bulbous bow impacting a rectangular plate and results comparison between the model test and the theoretical analysis using Eq. (3.61).

3.4.7.2 Example 2

Liu et al. (2014) carried out model tests of rectangular plates impacted by rigid spherical indenters with different radius. The test set-up is illustrated in Fig. 3.29. The specimen was made of mild steel rectangular plate with dimensions $127 \times 76.2 \times 1.4$ (unit: mm). The properties of the material are as follows: yield stress 228 MPa and tensile strength 364 MPa. The case with the largest cone radius ($R_b = 15$ mm) is selected for the analysis. The comparison between the model test results and the analytical results (Eq. 3.61) is presented in Fig. 3.29. It is seen that the agreement is good, except near the peak load where the plate is just about to rupture. The analytical result is higher than the experimental result at this point, and this may be caused by the local necking of the plate. Similar observations also apply to Example 1.

3.4.7.3 Example 3

Zhang et al. (2018) carried out model tests of rectangular plates indented by rigid flat rectangular indenters. A test model and its deformation are illustrated in Fig. 3.30. The specimen was made of a steel rectangular plate with dimensions of $600 \times 400 \times 1.5$

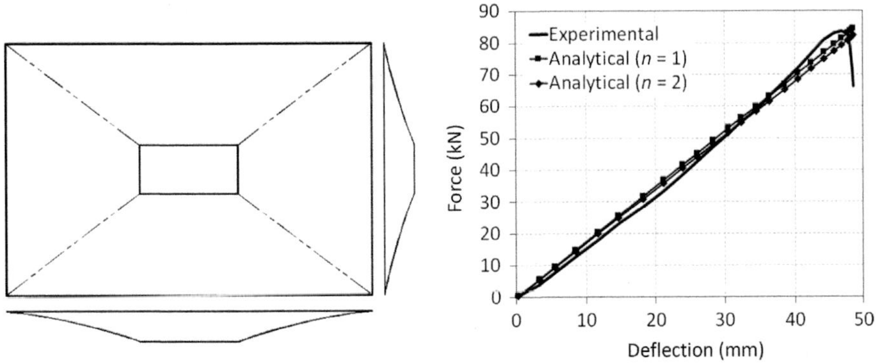

Fig. 3.30 A rectangular plate indented by a rigid flat rectangular indenter.

(unit: mm). The rigid flat rectangular indenter with dimensions of 160×80 (unit: mm) acts at the centre of the plate. The properties of the material are as follows: yield stress 183 MPa and tensile strength 298.4 MPa; the flow stress is (183+298.4)/2=240.75 MPa. The comparison between the model test results and the analytical result (Eq. 3.44) is presented in Fig. 3.30, where the deformation exponents $n=1$ and $n=2$ are used in the analytical calculations. It is seen that both exponents ($n=1$ and $n=2$) give good agreement with the experimental results and that the analytical results are not sensitive to the deformation exponent.

3.5 Crushing of web frames and stringers

3.5.1 Introduction

Similar to the above section, this section employs the upper-bound theorem to analyse the crushing deformation of side frames and stringers in ship collisions and to demonstrate how the force–deflection curve and the energy–deflection curve are derived.

3.5.2 Analysis method

Web frames may suffer in-plane distortions if the webs are well reinforced with horizontal stiffeners or suffer folding when the webs are not reinforced with stiffeners. McDermott et al. (1974) studied the in-plane distortions of webs. The resistance of the web frames can be approximated by the shear forces leading to yielding: $F_w = \left(1/\sqrt{3}\right)\sigma_0 t\delta$ when $\delta < H_{web}$, and $F_w = \left(1/\sqrt{3}\right)\sigma_0 A_w$ when $\delta \geq H_{web}$, where H_{web} is the web height and A_w is the section area of the web frame.

For a deep web, the most often observed deformation mode in ship accidents and experiments is folding or crushing when an external load is in the web plane. A typical picture of a model test of web crushing and the assumed deformation mode for the present analysis is shown in Fig. 3.31.

The folding deformation mode of the web frame includes bending and membrane tension. The equilibrium of the principle of virtual work is used to determine the crushing force: $F\dot{\delta} = \dot{E}_b + \dot{E}_m$where F is the crushing force, δ is the indentation, $\dot{E}_b$ is the bending energy rate, and $\dot{E}_m$ is the membrane energy rate.

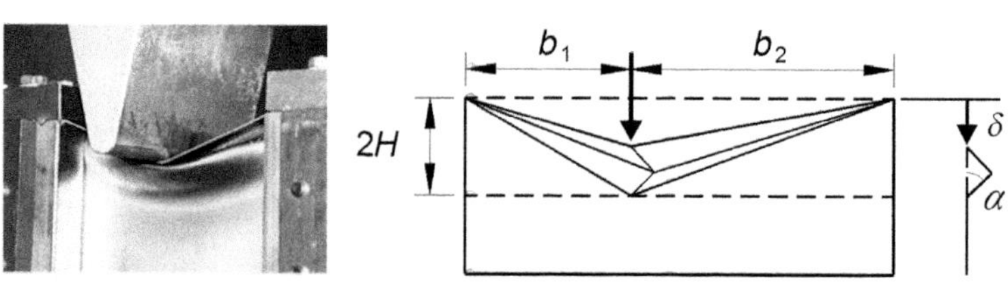

Fig. 3.31 Web crushing test and the assumed deformation mode for web crushing analysis.

3.5.2.1 Bending energy rate

Three bending plastic hinge lines are formed in the folding or crushing processes. The upper plastic hinge line is created between the web and its attached shell plating. The middle hinge moves out of its original plane, and the lower hinge remains in the web plane. The lengths of the hinge lines are assumed to be equal, which is the length of the web ($=b_1+b_2$, as shown in Fig. 3.31). Therefore, the rate of the bending energy can be determined from, $\dot{E}_b=4M_0\dot{\alpha}\cdot(b_1+b_2)$ where $M_0=\frac{2}{\sqrt{3}}\frac{\sigma_0 t^2}{4}$ and α is the bending angle (see Fig. 3.31). The relationship between the bending angle α and the indentation δ can be found from a geometric analysis. That is, $\delta=2H(1-\cos\alpha)\approx H\alpha^2$ $\dot{\alpha}=\frac{\dot{\delta}}{2\sqrt{H\delta}}$, where $2H$ is the folding length. Thus, the rate of the bending energy is

$$\dot{E}_b=\frac{1}{\sqrt{3}}\sigma_0 t^2\frac{(b_1+b_2)}{\sqrt{H\delta}}\cdot\dot{\delta} \tag{3.68}$$

When the first fold is completed, the total bending energy is expressed as

$$E_b=2\pi M_0(b_1+b_2)=\frac{\pi}{\sqrt{3}}\sigma_0 t^2(b_1+b_2) \tag{3.69}$$

3.5.2.2 Membrane energy rate

The main contribution to the membrane energy is the in-plane stretching. It can be written as

$$\dot{E}_m=\iint\limits_{S1}\frac{2}{\sqrt{3}}\sigma_0 t\dot{\varepsilon}_1 dS+\iint\limits_{S2}\frac{2}{\sqrt{3}}\sigma_0 t\dot{\varepsilon}_2 dS$$

where $S_1=2H\cdot b_1$ and $S_2=2H\cdot b_2$ are the deformed area of the web. For the upper fibre in the web (such as the part b_1), the strain and the strain rate can be calculated from $\varepsilon_0=\frac{1}{2}\left(\frac{\delta}{b_1}\right)^2$ and $\dot{\varepsilon}_0=\frac{\delta\dot{\delta}}{{b_1}^2}$. For the lower fibre in the deformed part of the web, the indentation is assumed to be linear with the level of the fibre, that is, $\delta_Z(z)=(z/(2H))\delta$ and $z=[0,\ 2H]$. So, the strain rate at each level becomes $\dot{\varepsilon}_1=\left(\frac{z}{2H}\right)^2\frac{\delta\dot{\delta}}{{b_1}^2}$. The average strain rate in the whole area $S_1=2H\cdot b_1$ can be obtained by integrating the strain rate over the area:

$$\dot{\varepsilon}_{av}=\frac{1}{S_1}\iint\limits_{S_1}\dot{\varepsilon}_1 ds=\frac{1}{2H\cdot b_1}\iint\limits_{S1}\left(\frac{z}{2H}\right)^2\frac{\delta\dot{\delta}}{{b_1}^2}ds=\frac{1}{3}\frac{\delta\dot{\delta}}{{b_1}^2}=\frac{1}{3}\dot{\varepsilon}_0 \tag{3.70}$$

It is seen that the average strain rate is only 1/3 of the strain rate in the upper fibre. By multiplication of the area and the average strain, the rate of the total membrane energy is

$$\dot{E}_m = \frac{2}{3}\frac{2}{\sqrt{3}}\sigma_0 tH\left(\frac{\delta}{b_1}+\frac{\delta}{b_2}\right)\dot{\delta} \tag{3.71}$$

When the first fold is completed, $\delta = 2H$, the total membrane energy can be determined from

$$E_m = \frac{1}{3}\frac{2}{\sqrt{3}}\sigma_0 tH\delta^2\left(\frac{1}{b_1}+\frac{1}{b_2}\right) = \frac{4}{3}\frac{2}{\sqrt{3}}\sigma_0 tH^3\left(\frac{1}{b_1}+\frac{1}{b_2}\right) \tag{3.72}$$

3.5.2.3 The total energy and crushing force

By adding the contributions of bending and membrane, the total energy rate can be expressed as

$$\dot{E} = \dot{E}_b + \dot{E}_m = \frac{1}{\sqrt{3}}\sigma_0 t^2\frac{(b_1+b_2)}{\sqrt{H\delta}}\cdot\dot{\delta} + \frac{2}{3}\frac{2}{\sqrt{3}}\sigma_0 tH\left(\frac{\delta}{b_1}+\frac{\delta}{b_2}\right)\dot{\delta}$$

$$E = E_b + E_m = \frac{\pi}{\sqrt{3}}\sigma_0 t^2(b_1+b_2) + \frac{4}{3}\frac{2}{\sqrt{3}}\sigma_0 tH^3\left(\frac{1}{b_1}+\frac{1}{b_2}\right)$$

The instantaneous crushing force and the mean crushing force can be expressed as

$$F = \frac{\dot{E}}{\dot{\delta}} = \frac{1}{\sqrt{3}}\sigma_0 t^2\frac{(b_1+b_2)}{\sqrt{H\delta}} + \frac{2}{3}\frac{2}{\sqrt{3}}\sigma_0 tH\left(\frac{\delta}{b_1}+\frac{\delta}{b_2}\right) \tag{3.73}$$

$$F_m = \frac{E}{2H} = \frac{\pi}{2\sqrt{3}}\sigma_0 t^2(b_1+b_2)\frac{1}{H} + \frac{2}{3}\frac{2}{\sqrt{3}}\sigma_0 tH^2\left(\frac{1}{b_1}+\frac{1}{b_2}\right) \tag{3.74}$$

By minimizing the mean crushing force, we get the optimum folding length:

$$\frac{\partial F_m}{\partial H} = 0 \Rightarrow : H = \left(\frac{3\pi}{16}b_1 b_2 t\right)^{\frac{1}{3}} = 0.8383(b_1 b_2 t)^{\frac{1}{3}} \tag{3.75}$$

Then, the instantaneous crushing force and the mean crushing force can be simplified as

$$F = 0.631\sigma_0 t^{1.83}\frac{(b_1+b_2)}{(b_1 b_2)^{0.17}\delta^{0.5}} + 0.645\sigma_0 t^{1.33}\delta\frac{(b_1+b_2)}{(b_1 b_2)^{0.67}} \tag{3.76}$$

$$F_m = (1.082 + 0.541)\sigma_0 t^{\frac{5}{3}} \frac{(b_1 + b_2)}{(b_1 b_2)^{\frac{1}{3}}} = 1.623\sigma_0 t^{\frac{5}{3}} \frac{(b_1 + b_2)}{(b_1 b_2)^{\frac{1}{3}}} \tag{3.77}$$

It is seen that the bending contributes to 2/3 of the mean crushing force and the remaining 1/3 is from membrane deformation. When $b_1 = b_2 = b$, that is, the impact point is located in the middle of the web, the formulae are simplified to

$$F = 1.262\sigma_0 t^{1.83} b^{0.67} \frac{1}{\delta^{0.5}} + 1.290\sigma_0 t^{1.33} \delta \frac{1}{b^{0.33}}; \quad F_m = 3.25\sigma_0 t^{1.67} b^{0.33} \tag{3.78}$$

where $\delta \leq 2H$; $H = 0.838\ t^{0.33} b^{0.67}$.

In the above derivation, the crushing distance is $2H$ when the first fold is completed. However, the web plate cannot be totally compressed and the shortened distance is smaller than $2H$. Normally, the effective crushing length is about 73% of the folding length $2H$ for bare plate (Abramowicz, 1983). Here, we take 75%, and the effective mean crushing force after considering the effective crushing length is then expressed as

$$F_m = 4.33\sigma_0 t^{1.67} b^{0.33} \tag{3.79}$$

Assuming different folding patterns, various formulae have been derived by a number of authors, for example, Wang and Ohtsubo (1997), Simonsen (1997a), Simonsen and Ocakli (1999), Hong and Amdahl (2008), and Liu (2015). Some of the existing formulae are summarized in Fig. 3.32. Such difference is due to the fact that the upper-bound theorem is an approximate method and the results depend on the construction of the deformation patterns.

3.5.2.4 Subsequent folding

After the first fold is completed, the second fold can be formed at a further indentation. The crushing mode is illustrated in Fig. 3.33. It is assumed that the second folding pattern is similar to the first fold. In the subsequent folding, the first folded web is compressed. The behaviour is like that of a plate strip indented centrally by a transverse load. The resistance force of the first folded web can be determined from

$$F_1 = \frac{8}{\sqrt{3}}\sigma_0 t H \left(\frac{\delta - 2H}{b}\right) = 4.62\sigma_0 t H \left(\frac{\delta - 2H}{b}\right) \tag{3.80}$$

Eq. (3.79) is still valid for calculating the resistance of the second fold. By taking into account the resistance of the first folded web, the crushing force for the second folding process can be expressed as

$$F = 4.62\sigma_0 t H \left(\frac{\delta - 2H}{b}\right) + 4.33\sigma_0 t^{1.67} b^{0.33} \tag{3.81}$$

Authors	Folding pattern	Half-folding length (H)	Effective mean crushing force (F_m)
Wang and Ohtsubo (1997)	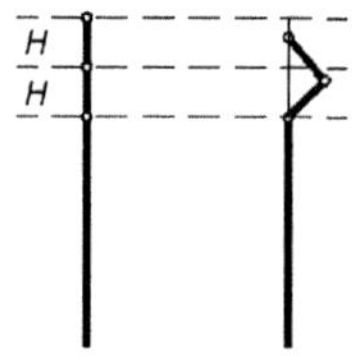	$H = 0.811\, t^{0.33} b^{0.67}$	$F_m = 4.36\, \sigma_0 t^{1.67} b^{0.33}$
Simonsen (1997)		$H = 0.671\, t^{0.33} b^{0.67}$	$F_m = 4.26\, \sigma_0 t^{1.67} b^{0.33}$
Zhang (1999)		$H = 0.838\, t^{0.33} b^{0.67}$	$F_m = 4.33\, \sigma_0 t^{1.67} b^{0.33}$
Liu (2015)		$H = 0.443\, t^{0.33} b^{0.67}$	$F_m = 4.72\, \sigma_0 t^{1.67} b^{0.33}$
Simonsen and Ocakli (1999)	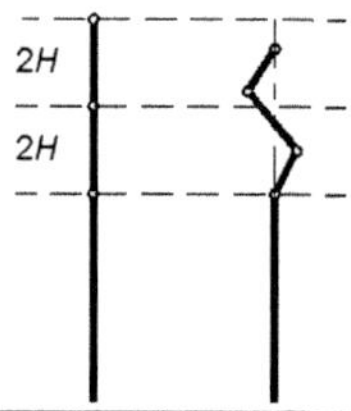	$H = 0.377\, t^{0.33} b^{0.67}$	$F_m = 5.40\, \sigma_0 t^{1.67} b^{0.33}$
Hong and Amdahl (2008)	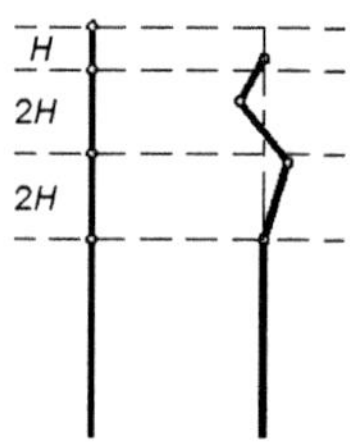	$H = 0.395\, t^{0.33} b^{0.67}$	$F_m = 5.82\, \sigma_0 t^{1.67} b^{0.33}$

Fig. 3.32 A summary of the existing formulae for web crushing analysis.

In a similar way, the crushing force in the third fold, the fourth fold, etc. can be determined, until the first folded web ruptures. In summary, the resistance force of web crushing can be determined as follows

$$F = \begin{cases} 4.33\sigma_0 t^{1.67} b^{0.33}, & \delta \leq 2H \\ 4.33\sigma_0 t^{1.67} b^{0.33} + 4.62\sigma_0 tH\left(\dfrac{\delta - 2H}{b}\right), & 2H < \delta \leq 4H \\ \vdots & \\ 4.33\sigma_0 t^{1.67} b^{0.33} + 4.62\sigma_0 tH(n-1)\left(\dfrac{\delta - n \cdot H}{b}\right), & (n-1)(2H) < \delta \leq n(2H) \end{cases} \tag{3.82}$$

3.5.2.5 Concertina folding/tearing

After the first folded web has ruptured, the folding mode changes to concertina folding or tearing (assume the web is deep enough). Wierzbicki (1994) studied concertina tearing and found that the concertina tearing force can be expressed as

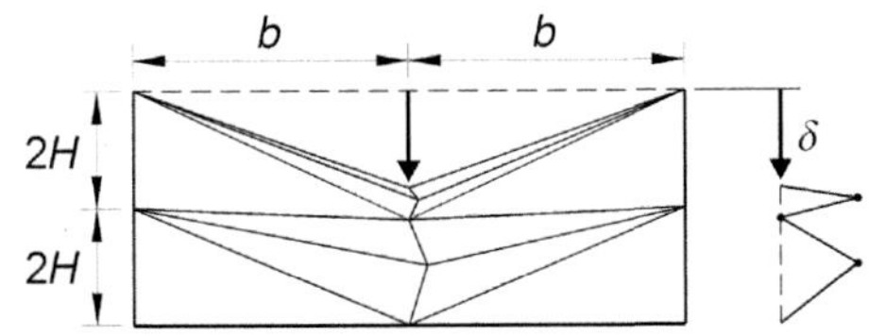

Fig. 3.33 Subsequent folding of web crushing.

$$F_{con} = 5.04\sigma_0 t^{1.67} b^{0.33} + \frac{8}{3} R_c \cdot t \tag{3.83}$$

where R_c is the specific work of fracture. Wierzbicki gave the range of the specific work of fracture for mild steel: $R_c = 300$–1000 (N/mm). In fact, when Wierzbicki derived the formula for the concertina tearing force, he only considered the deformation within onefold. The deformation may extend to more than one fold as discussed above.

Observations from many experiments show that the rupture of the webs occurs during the second fold. Thus, it is assumed that when the indentation is equal to $\delta = 3H$, the upper fibre of the web starts to rupture. With further indentation, the concertina folding forms. Thus, the mean crushing force in the second lobe may represent the concertina tearing force. The dissipated energy in the first folded lobe before rupture can be calculated from

$$E_{m1} = \int_{2H}^{3H} (4.62\sigma_0 tH(\delta - 2H)/b) d\delta = 2.31\sigma_0 tH^3/b$$

After rupture has started, the resistance from the first folded lobe does not vanish completely. The lower-level fibre in the first folded lobe still offers resistance, and it is assumed that the resistance keeps constant. Thus, the dissipated energy during this indentation can be estimated from

$$E_{m2} = [4.62\sigma_0 tH(3H - 2H)/b] \cdot H = 4.62\sigma_0 tH^3/b$$

The mean crushing resistance from the first folded lobe during the second folding process is calculated from

$$F_{m1} = (E_{m1} + E_{m2})/(2H) = 3.465\sigma_0 tH^2/b = 2.435\sigma_0 t^{1.67} b^{0.33}$$

By adding the contributions from the first folded lobe and the second newly crushed lobe, the mean crushing force for concertina tearing is expressed as

$$F_c = (2.435 + 4.33)\sigma_0 t^{1.67} b^{0.33} = 6.77\sigma_0 t^{1.67} b^{0.33} \tag{3.84}$$

3.5.3 Bulbous bow impact on web frames or stringers

The collision scenario of a bulbous bow impacting the side of a ship in a stringer position is illustrated in Fig. 3.34. When the bulbous bow indents into the side of a ship, the shell plate will stretch, and the stringer web will fold and crush. The simplified analysis model is presented in Fig. 3.35. The length of the stringer is $2b$, which is the transverse web frame spacing. It is assumed that the transverse web frames do not deform unless the ship bow hits the web frame directly. The width of the shell plate is d. The thickness of the shell plate is t, and the thickness of the stringer web is t_w.

3.5.3.1 Analysis formulas

The dissipated energy of a bulbous bow impacting the side stringer is composed of the tension of the side plate and the crushing of the stringer web: $E = E_{plate} + E_{web}$. As shown in Fig. 3.35, the deformation of the side plate can be assumed to be

$$w_1(x,y,t) = \delta_1(t) \cdot f_1(x,y) = \delta_1 \left(\frac{b-x}{b-b_0} \right)^2, \quad b_0 \le x \le b$$

where δ_1 is the deflection at the separating point between the bulb and the plate and b_0 is the radius at the separating point. Following similar procedures as in Section 3.4.4, the collision resistance due to the side shell stretching is expressed as

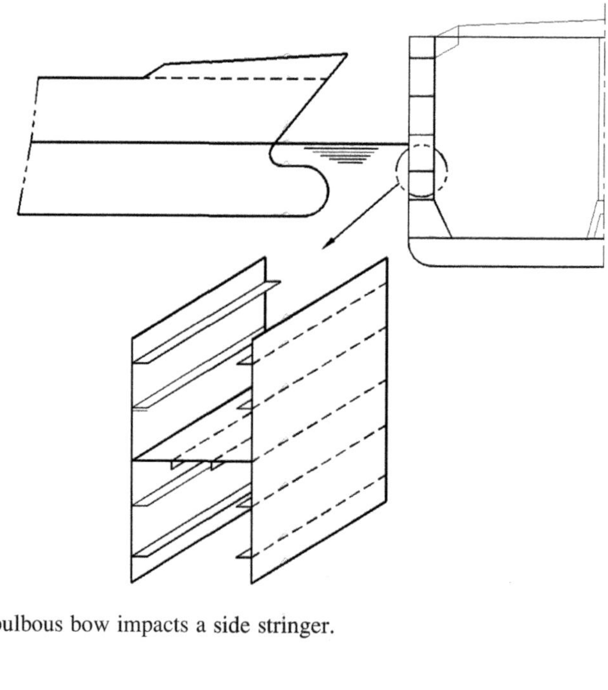

Fig. 3.34 A bulbous bow impacts a side stringer.

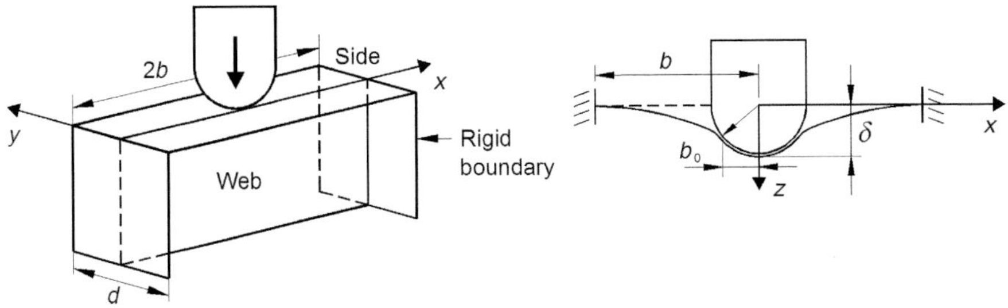

Fig. 3.35 Analysis model for a bulbous bow impacting a stringer.

$$F_{plate} = \frac{8}{3}\frac{2}{\sqrt{3}}\sigma_0 t\delta\left(1 + \frac{3}{2}\frac{R_b\delta}{b^2}\right)\frac{d}{b} \tag{3.85}$$

By integrating Eq. (3.85) with respect to the indentation δ, the dissipated energy of the shell plate is obtained as

$$E_{plate} = \frac{4}{3}\frac{2}{\sqrt{3}}\sigma_0 t\delta^2\left(1 + \frac{R_b\delta}{b^2}\right)\frac{d}{b} \tag{3.86}$$

Eq. (3.78) for the web crushing force is used here for stringer web crushing; thus, we rewrite the formula

$$F_{web} = 1.262\sigma_0 t_w{}^{1.83} b^{0.67}\frac{1}{\delta^{0.5}} + 1.290\sigma_0 t_w{}^{1.33}\delta\frac{1}{b^{0.33}}$$

$$H = 0.838 t_w{}^{0.33} b^{0.67} \text{ (Half} - \text{fold length)}$$

The dissipated energy of the stringer web during crushing can be expressed as

$$E_{web} = 2.524\sigma_0 t_w{}^{1.83} b^{0.67}\delta^{0.5} + 0.645\sigma_0 t_w{}^{1.33}\delta^2\frac{1}{b^{0.33}}$$

By adding the contributions from the shell plate and the stringer web, the total collision resistance and absorbed energy can be written as

$$\begin{aligned} F &= F_{plate} + F_{web} \\ &= 3.08\sigma_0 t\delta\left(1 + 1.5\frac{R_b\delta}{b^2}\right)\frac{d}{b} + 1.262\sigma_0 t_w{}^{1.83} b^{0.67}\frac{1}{\delta^{0.5}} + 1.290\sigma_0 t_w{}^{1.33}\delta\frac{1}{b^{0.33}} \end{aligned} \tag{3.87}$$

$$\begin{aligned} E &= E_{plate} + E_{web} \\ &= 1.54\sigma_0 t\delta^2\left(1 + \frac{R_b\delta}{b^2}\right)\frac{d}{b} + 2.522\sigma_0 t_w{}^{1.83} b^{0.67}\delta^{0.5} + 0.645\sigma_0 t_w{}^{1.33}\delta^2\frac{1}{b^{0.33}} \end{aligned} \tag{3.88}$$

3.5.4 Calculation examples

3.5.4.1 DTU crushing test

At the Technical University of Denmark (DTU), Simonsen and Ocakli (1999) performed a series of model tests for quasistatic web crushing. A rigid wedge indents into the web at the midspan. The length of the web is $2b = 150$ mm, and the thickness is $t = 1.0$ mm. The width of the attached flange is 50 mm, and its thickness is 1.0 mm. The flow stress of the material is $\sigma_0 = 223.2$ MPa. The comparison of the present calculation using Eqs (3.84), (3.87) with the experimental result is presented in Fig. 3.36; it is seen that the agreement is good. When the penetration is equal to the critical rupture value (31.8 mm), the web and the attached flange rupture, and the crushing resistance drops immediately. After that, the crushing of the web is represented as concertina tearing with a constant force.

3.5.4.2 MIT crushing experiment

Another comparison is performed with a web crushing test carried out at MIT (Simonsen, 1997a). The test model is a web with a flange, and the two sides of the web are clamped. A rigid object indented the model at the middle of the web length (similar to the above DTU test).

The basic data of the model are the following: the length of the web is $2b = 166.8$ mm, and the thickness is $t = 0.737$ mm; the width of the attached flange is 41.7 mm, and its thickness is 0.737 mm. The flow stress of the material is $\sigma_0 = 236$ MPa.

The half-folding length calculated by Eq. (3.78) is $H = 0.838t^{0.33}b^{0.67} = 14.5$ mm, while the model test result is 13.9 mm. The comparison of the crushing force obtained by Eq. (3.87) and the experimental result is shown in Fig. 3.37; very good agreement is found. In this example, rupture did not happen.

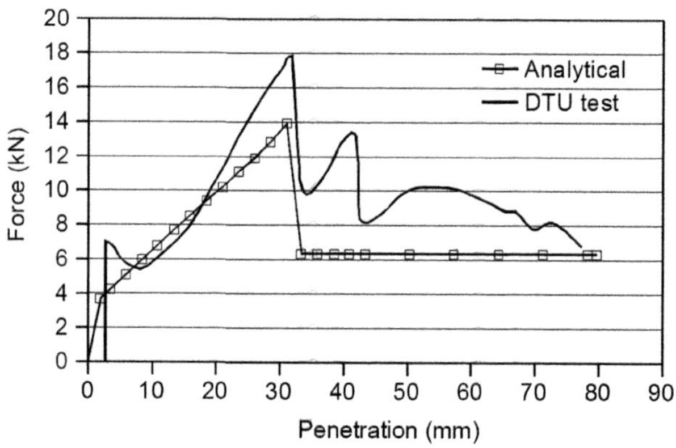

Fig. 3.36 Comparison between experimental (DTU web crushing test) and analysis results using Eq. (3.87) before rupture and Eq. (3.84) after rupture.

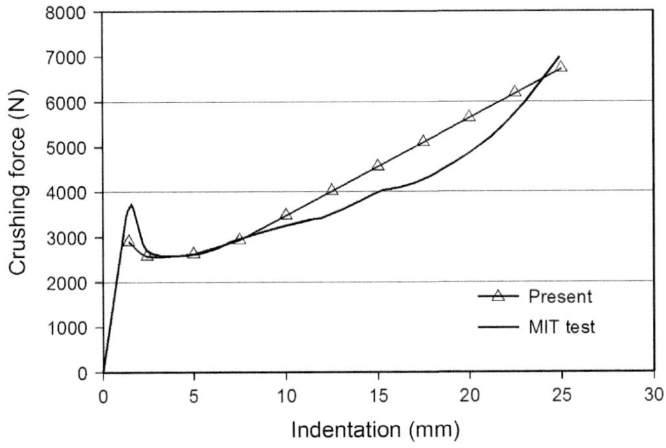

Fig. 3.37 Crushing force comparison between the present result analysis and MIT test result.

3.5.4.3 Concertina folding tests

An example of concertina folding is the model test performed by Yahiaoui et al. (1994). The main parameters of the model are the following: $\sigma_0 = 330\,\text{MPa}$, $2b = 50\,\text{mm}$, and $t = 1.14\,\text{mm}$. The comparison of the mean crushing forces obtained by the present Eq. (3.84), the experimental result, and Wierzbicki's formula are 8073, 8500, and 6760 N, respectively, and it is presented in Fig. 3.38. Reasonable agreement is found between Eq. (3.84) and the experimental result.

3.5.4.4 Drop impacting experiment

Qvist et al. (1995) carried out a large-scale impact test at the B&W Shipyard in Denmark. The model was used to simulate part of the side structure of a 40,000 DWT tanker on a full scale and a large tanker (>100,000 DWT) on the scale 1/2. The test models before and after testing are shown in Fig. 3.39. A steel ball ($M = 2750\,\text{kg}$) dropped from a height of 5.0 m colliding with the stringer model at the middle. This means that the impact velocity is $V = 10\,\text{m/s}$. Thus, the energy to be dissipated by destroying the stringer model is $E = 0.5MV^2 = 137.5$ (kJ).

The detailed parameters of the model are the following: the spacing of the transverse frames is $2b = 1350\,\text{mm}$, and the thickness is $t = 8.0\,\text{mm}$; the width of the attached shell plate is $d = 450\,\text{mm}$, and the thickness is 8.0 mm. The flow stress of the material is $\sigma_0 = 317.5\,\text{MPa}$. The radius of the steel ball is $R_b = 500\,\text{mm}$.

The result calculated by Eq. (3.88) for the indentation at the end of the impact and the test results are shown in Table 3.9. The calculation results of the energy contribution from the shell plating and the stringer web at the end of the collision are 66% and 34%. From the comparison, it is seen that good agreement is achieved. It should be also noted that the shell plating absorbs much more energy than the stringer web.

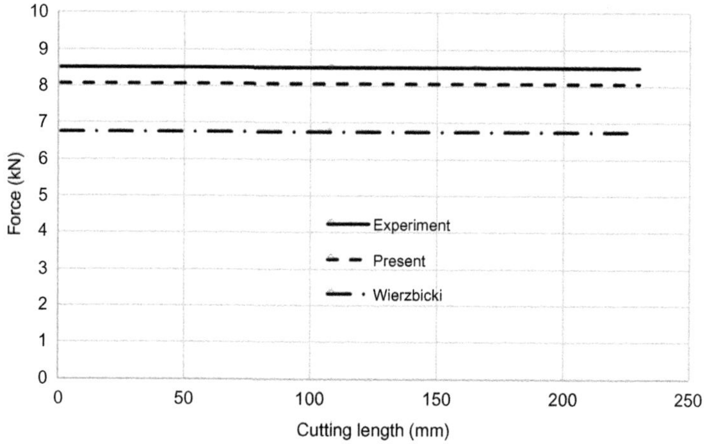

Fig. 3.38 Concertina folding of a plate test and concertina tearing force comparison between the present result analysis, Eq. (3.84), and test result.

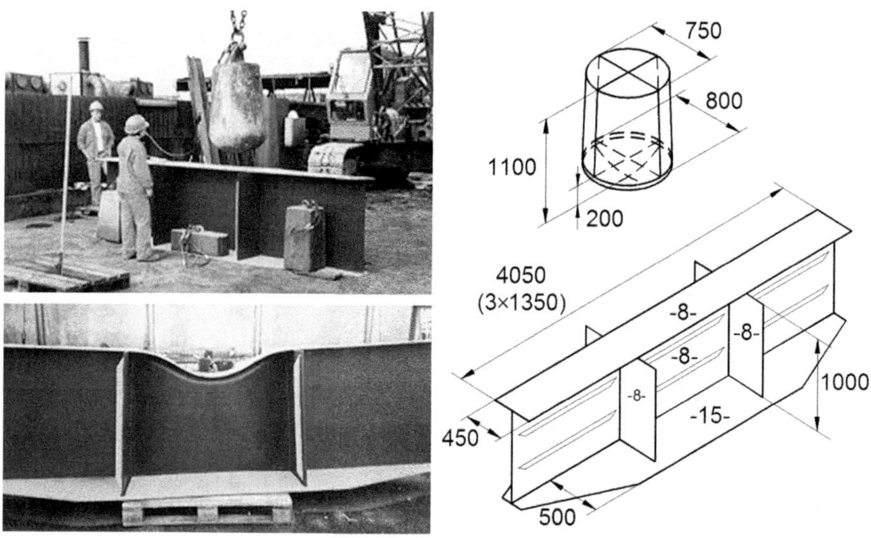

Fig. 3.39 Experiment of a steel ball impacting a side-stringer model.

Table 3.9 Comparison of the calculation result and the test result of the drop impact test by Qvist et al. (1995).

	Experiment, first model	**Experiment, second model**	**Eq. (3.88)**
Indentation (mm)	190	175	171

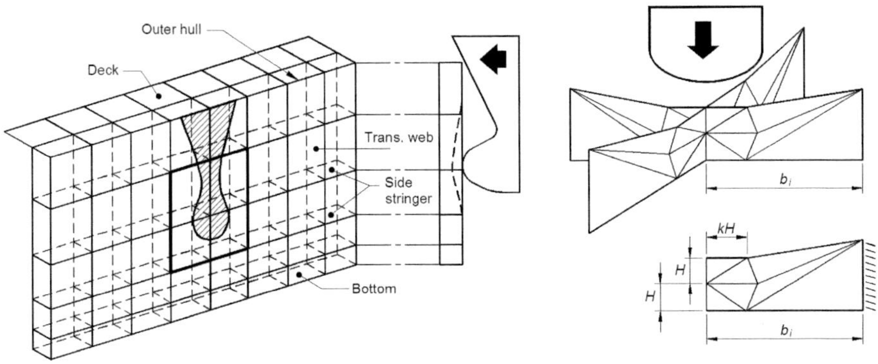

Fig. 3.40 A striking bow indents into the side of a ship at the intersection of a transverse web and a longitudinal stringer and the analysis model.

3.6 Crushing of web and stringer intersections

3.6.1 Introduction

This section considers a collision scenario as that shown in Fig. 3.40; a striking bow impacts the side structure of a ship at an intersection of a transverse web frame and a longitudinal stringer. The intersection is dented and crushed by the striking bow. The analysis model for the damage mode of the intersection is also depicted in Fig. 3.40.

3.6.2 Analysis method for the mean crushing force

First, we derive the formula for the crushing of an individual plate unit as shown in the lower part of Fig. 3.40. When the first fold has been completed, the total dissipated energy can be obtained by adding the bending energy and the in-plane deformation energy: $E = E_b + E_{m1} + E_{m2}$, where E_b is the bending energy along the plastic hinge lines, E_{m1} is the in-plane deformation energy in the area of $kH \cdot 2H$, and E_{m2} is the in-plane deformation energy in the area of $(b - kH) \cdot 2H$.

For the bending energy, three horizontal hinge lines are considered, while the inclined hinge lines are neglected because the dissipated energy in those inclined hinge lines is small (Paik and Pedersen, 1995). Therefore, the bending energy can be expressed as

$$E_b = 4M_0 b\frac{\pi}{2} = 1.8138\sigma_0 t^2 b \tag{3.89}$$

Amdahl (1983) and Paik and Pedersen (1995) studied the dissipated energy in the area of $(kH \cdot 2H)$. It is expressed as

$$E_{m1} = \frac{2}{\sqrt{3}} N_0 H^2 \left(\sqrt{k^2 + \frac{1}{4}} \cdot \arcsin \frac{1}{\sqrt{4k^2+1}} + k \right) = 1.4896\sigma_0 t H^2 \tag{3.90}$$

where $k = 0.5733$ (Amdahl, 1983) and $N_0 = (2/\sqrt{3})\sigma_0 t$. The membrane energy in the area of $(b - kH) \cdot 2H$ is determined from

$$E_{m2} = \frac{4}{3}\frac{2}{\sqrt{3}}\sigma_0 t H^3 \frac{1}{b - kH} = 1.5396\sigma_0 t H^3 \frac{1}{b - kH} \tag{3.91}$$

Thus, the total dissipated energy can be expressed as

$$E = 1.8138\sigma_0 t^2 b + 1.4896\sigma_0 t H^2 + 1.5396\sigma_0 t H^3 \frac{1}{b - kH} \tag{3.92}$$

When the total dissipated energy is known, the mean crushing force can be calculated from

$$F_m = \frac{E}{2H} = \sigma_0 t \left(0.9069tb\frac{1}{H} + 0.7448H + \frac{0.7698H^2}{b - kH} \right) \tag{3.93}$$

where $2H$ is the folding length. For simplicity, we use the first two terms to minimize the mean crushing force:

$$\frac{\partial F_m}{\partial H} = 0 \Rightarrow : H = 1.103(tb)^{0.5} \tag{3.94}$$

It is now necessary to verify the accuracy of the folding length. Several examples have been studied. One of the examples is shown in Fig. 3.41, which gives the mean crushing force with various folding lengths H. The data of the example are $b = 1.2\,\text{m}$, $t = 8\,\text{mm}$, and $\sigma_0 = 300$ MPa. By simulation, we get the optimum half-folding length $H = 99$ mm, and the minimum mean crushing force is $F_m = 403.9$ (kN). By application of Eq. (3.94), we obtain the half-folding length $H_1 = 108$ mm, and the related mean crushing force is $F_1 = 405.5$ (kN). The relative error of the mean crushing force is only 0.4%. Therefore, using Eq. (3.94) to calculate the optimum

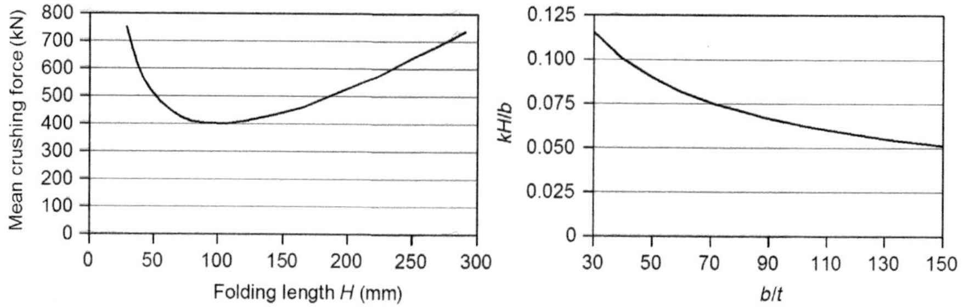

Fig. 3.41 Mean crushing force with various folding lengths (left figure) and value of kH/b with various b/t (right figure).

folding length, and thus obtain the mean crushing force, has resulted in a good accuracy. By substitution of Eq. (3.94) into Eq. (3.93), the mean crushing force is expressed as

$$F_m = 1.6437\sigma_0 t^{1.5} b^{0.5} + 0.9365\sigma_0 t^2 \left(\frac{b}{b-kH}\right) \tag{3.95}$$

To further simplify Eq. (3.95), we calculate the ratio of kH/b. The result of kH/b with various b/t is shown in Fig. 3.41. It is seen that, for a wide range of the ratio $b/t=30$–150, the value of $kH/b=11\%$–5%. If we assume that $kH/b=7.5\%$, Eq. (3.95) becomes

$$\begin{aligned} F_m &= 1.6437\sigma_0 t^{1.5} b^{0.5} + 1.012\sigma_0 t^2 \\ &= 1.6437\sigma_0 t^{1.5} b^{0.5} \left(1 + 0.6159\left(\frac{t}{b}\right)^{0.5}\right) \end{aligned} \tag{3.96}$$

The second term of Eq. (3.96) is the contribution by the tension deformation of the area $(b-kH)\cdot 2H$. This indicates that the contribution from the tension in the area $(b-kH)\cdot 2H$ is small. For a typical example, when $b/t=55$, the mean crushing force is $F_m=1.6437\sigma_0 t^{1.5} b^{0.5}(1+0.083)$. Thus, the contribution from the tension deformation in the area of $(b-kH)\cdot 2H$ is less than 8%.

The indenting and crushing mechanism of the L-, T-, and X-intersection can be generated by the individual plate unit. Thus, the mean crushing force of the L-, T-, and X-sections can be expressed as shown in Fig. 3.42. If an effective crushing length is taken as 75% of the folding length, the formulae in Fig. 3.42 need to be divided by a coefficient of 0.75.

3.6.3 Subsequent crushing

When the first fold has been completed, subsequent folds can form before the web plate ruptures. It is assumed that the subsequent folds have the same geometry as the first, as shown in Fig. 3.43.

For the second folding process, the crushing resistance is composed of two parts. One is the newly crushed second fold, and the other is the completely compressed first fold. That is,

$$F = F_{fold1} + F_{fold2}$$

Eq. (3.96) is still valid for the mean crushing resistance of the newly crushed second fold $F_{fold2}=1.6437\sigma_0 t^{1.5} b^{0.5} + 1.012\sigma_0 t^2$. The fully compressed first fold can be approximated as a plate strip, and the resistance can be determined from $F_1=2.31\sigma_0 tH(\delta-2H)/b_1$, where $b_1=b-kH\approx 0.925b$. Then, the mean crushing resistance of the first folded lobe is

L-section	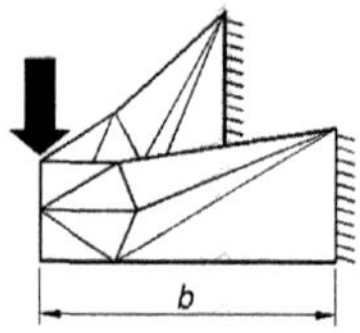	$F_L = 2.3245\,\sigma_0 t^{1.5} c^{0.5} + 2.025\,\sigma_0 t^2$ $c = b + b$
T-section	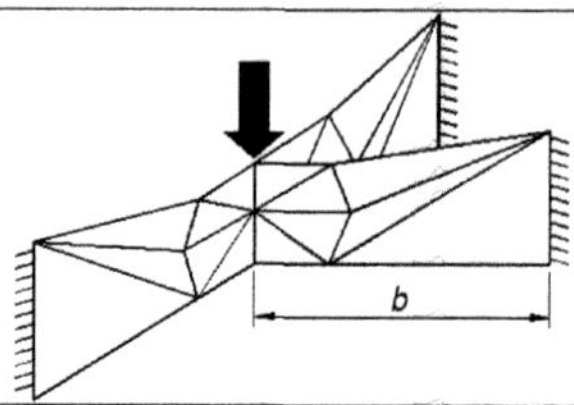	$F_T = 2.8470\,\sigma_0 t^{1.5} c^{0.5} + 3.036\,\sigma_0 t^2$ $c = b + b + b$
X-section	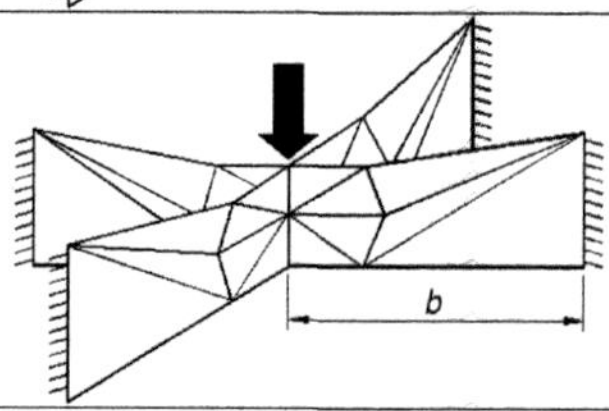	$F_X = 3.2874\,\sigma_0 t^{1.5} c^{0.5} + 4.048\,\sigma_0 t^2$ $c = b + b + b + b$

Fig. 3.42 The mean denting force of the L-, T-, and X-sections (note: the formulae need to be divided by a coefficient of 0.75 if an effective crushing length is taken as 75% of the folding length).

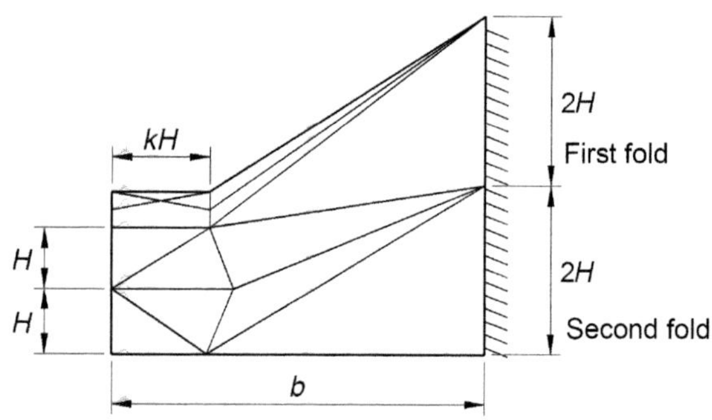

Fig. 3.43 The subsequent denting and crushing mode.

$$F_{fold1} = \left(\int_{2H}^{4H} F_1 d\delta \right) / (2H) = \left(6.704\sigma_0 t^{2.5} b^{0.5}\right) / (2H) = 3.04\sigma_0 t^2$$

Therefore, during the second folding, the mean crushing force is

$$\begin{aligned} F &= 3.04\sigma_0 t^2 + 1.6437\sigma_0 t^{1.5} b^{0.5} + 1.012\sigma_0 t^2 \\ &= 1.6437\sigma_0 t^{1.5} b^{0.5} + (1+3)1.012\sigma_0 t^2 \end{aligned} \tag{3.97}$$

Similarly, the mean crushing force of the Nth fold can be expressed as

$$\begin{aligned} F_N &= 1.6437\sigma_0 t^{1.5} b^{0.5} + [1+3(N-1)]1.012\sigma_0 t^2 \\ &= 1.6437\sigma_0 t^{1.5} b^{0.5} + (3N-2)1.012\sigma_0 t^2 \end{aligned} \tag{3.98}$$

After several folds and when the penetration reaches the critical penetration, the web plate ruptures. After rupture (assume in N_0 folds), the mean crushing force is assumed to be constant, and it is calculated from

$$F_{N_0} = 1.6437\sigma_0 t^{1.5} b^{0.5} + (3N_0-2)1.012\sigma_0 t^2 \tag{3.99}$$

By combining the individual plate units, the mean crushing force of the subsequent Nth fold for the L-, T-, and X-sections can be expressed as

$$\text{L}-\text{section}: \begin{array}{l} F_L = 2.3245\sigma_0 t^{1.5} c^{0.5} + (3N-2)2.025\sigma_0 t^2 \\ c = b+b \end{array} \tag{3.100}$$

$$\text{T}-\text{section}: \begin{array}{l} F_T = 2.8470\sigma_0 t^{1.5} c^{0.5} + (3N-2)3.036\sigma_0 t^2 \\ c = b+b+b \end{array} \tag{3.101}$$

$$\text{X}-\text{section}: \begin{array}{l} F_X = 3.2874\sigma_0 t^{1.5} c^{0.5} + (3N-2)4.048\sigma_0 t^2 \\ c = b+b+b+b \end{array} \tag{3.102}$$

It should be also noted that if an effective crushing length is taken as 75% of the folding length, the above formulae need to be divided by a coefficient of 0.75.

3.6.4 Calculation examples

Amdahl and Kavlie (1992) studied crushing of a cruciform by finite element simulations using different mesh sizes. All edges of the model were restrained to the displacement in the axial direction only. All vertical side edges were fully clamped. The calculation model and the related data are shown in Fig. 3.44.

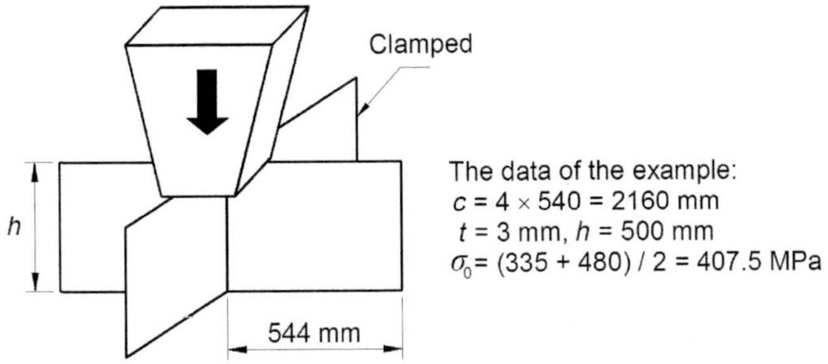

Fig. 3.44 Crushing of an X-section and the related data.

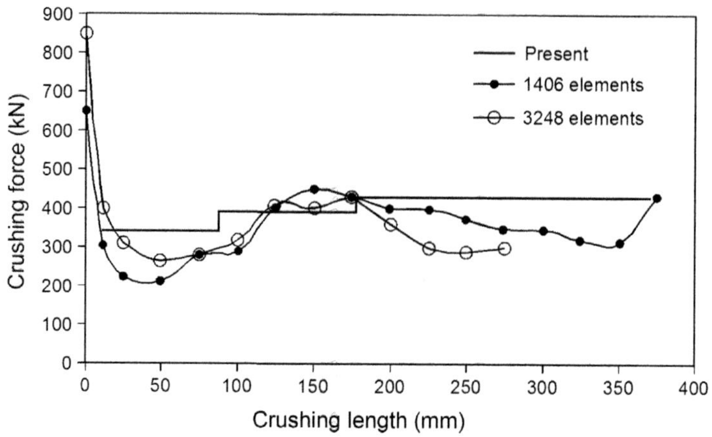

Fig. 3.45 Comparison of the crushing force of the X-section obtained by the present method, Eq. (3.102), and the finite element simulation.

Using Eq. (3.94), we get the folding length $2H=88.8$ mm, and by application of Eq. (3.102), we get the mean crushing force $F_X(N=1)=323.31+14.85=338.2$ (kN). For the following Nth fold, the mean crushing force is $F_X(N)=323.31+(3N-2)14.85$ (kN).

The comparison of the present results with Amdahl and Kavlie's simulation results is shown in Fig. 3.45. Good agreement is found. In the analytical calculations, it is assumed that the rupture occurs in the fold $N_0=3$. It is noted that both the FE analysis and the present analytical results do not consider the effective crushing length in this example.

3.7 Axial crushing of structural elements

3.7.1 Introduction

This section investigates the axial crushing of structural elements in ship collisions that form the foundation for ship bow crushing analysis.

3.7.2 Axial crushing analysis of cylindrical shells

As a classical example, we first look at the axial crushing analysis of cylindrical shells, which was firstly carried out by Alexander (1960). Fig. 3.46 shows typical folding patterns of this type that are given by Jones (2012). The folding pattern can be axisymmetric (the left-hand picture, also called regular folding) or nonaxisymmetric (the centre and right-hand pictures, also called irregular folding).

The crushing with the regular folding pattern, as illustrated in Fig. 3.47, is analysed next to demonstrate the analysis process. The diameter of the cylinder shell is D, and

Fig. 3.46 Axial crushing of cylindrical shells.
cf. Jones, N., 2012. Structural Impact. Cambridge University Press.

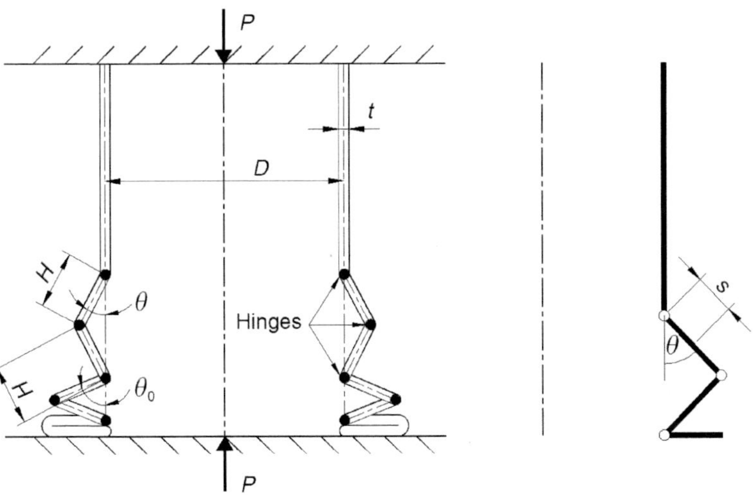

Fig. 3.47 Analysis models for axial crushing of cylindrical shells with a regular folding pattern.

the wall thickness is t. For one complete fold, three plastic hinges are formed. The bending energy at the upper and lower hinges can be calculated from

$$E_1 = 2\left(M_0 \cdot \pi D \cdot \frac{\pi}{2}\right)$$

The bending energy at the middle hinge, where the diameter increases from D to $D + 2H\sin\theta$, is calculated from

$$E_2 = 2M_0 \int_0^{\pi/2} \pi(D + 2H\sin\theta)d\theta = 2M_0\pi\left(D \cdot \frac{\pi}{2} + 2H\right)$$

Therefore, the total bending energy at the three hinges is

$$E_b = E_1 + E_2 = 2\pi M_0(\pi D + 2H) \tag{3.103}$$

The radial deformation at the distance s (see Fig. 3.47) can be determined from

$$\varepsilon_r = \frac{\pi(D + 2\sin(\theta)s) - \pi D}{\pi D} = \frac{2\sin(\theta)s}{D}$$

The energy due to stretching deformation can be determined by different ways; one of the simple methods is to calculate the change of the area between hinges:

$$\Delta A = 2\left(\pi(D + 2H)^2/4 - \pi D^2/4\right) - 2\pi DH = 2\pi H^2$$

Thus, the energy due to stretching deformation is

$$E_m = \sigma_0 t \Delta A = 2\pi\sigma_0 t H^2 \tag{3.104}$$

The external work has to be dissipated by plastic energies in bending and stretching:

$$F_m \cdot 2H = E_b + E_m$$

Therefore, the mean crushing force is expressed as

$$F_m = \frac{\pi\sigma_0 t^2}{\sqrt{3}}\left(\frac{\pi D}{2H} + 1\right) + \pi\sigma_0 t H$$

The folding length is unknown, but by minimizing the crushing force, we find

$$\frac{\partial F_m}{\partial H} = 0 \Rightarrow H = \sqrt{\frac{\pi t D}{2\sqrt{3}}} \approx 0.9523\sqrt{tD}$$

Finally, the mean crushing force is

$$F_m = 6\sigma_0 t\sqrt{tD} + 1.8\sigma_0 t^2 \tag{3.105}$$

The above analysis assumes that the deformation is outwards. If the materials deform inwards, a similar analysis leads to

$$F_m = 6\sigma_0 t\sqrt{tD} - 1.8\sigma_0 t^2$$

In reality, the cylindrical shell may deform partially outwards and partially inwards, so the average can be taken as

$$F_m = 6\sigma_0 t^{1.5} D^{0.5} \tag{3.106}$$

This was obtained by Alexander (1960). If the effective crushing length is taken as 75% of the folding length, the effective mean crushing force is

$$F_m = (6/0.75)\sigma_0 t^{1.5} D^{0.5} = 8\sigma_0 t^{1.5} D^{0.5} \tag{3.107}$$

Using the equilibrium of internal energy dissipation rate and external energy dissipation rate, the instantaneous crushing force can be derived (Amdahl, 1983). The total bending energy rate at the three hinges is

$$\dot{E}_b = 4\pi M_0 (D + H\sin\theta)\dot{\theta}$$

The energy rate of stretching of the tube in the circumferential direction can be determined from

$$\dot{E}_m = 2\pi\sigma_0 t\left(H^2 \cos\theta\right)\dot{\theta}$$

The external work rate is

$$\dot{E}_{ext} = F \cdot (2H\sin\theta)\dot{\theta}$$

Therefore, the instantaneous crushing force can be expressed as

$$F = \frac{\pi\sigma_0 t^2}{\sqrt{3}\sin\theta} \cdot \left(\frac{D}{H} + \sin\theta + \sqrt{3}\frac{H}{t}\cos\theta\right) \tag{3.108}$$

3.7.3 Axial crushing of conical shells

For crushing analysis of conical shells, Lehmann and Yu (1995) developed analytical solutions. The theory is based on three straight folding elements that are connected through plastic hinges. The analytical formulae can be reduced to formulae similar to that derived by Alexander (1960) for cylindrical shells. For application to transversely framed bulbous bow crushing analysis, the effective mean crushing force can be determined from

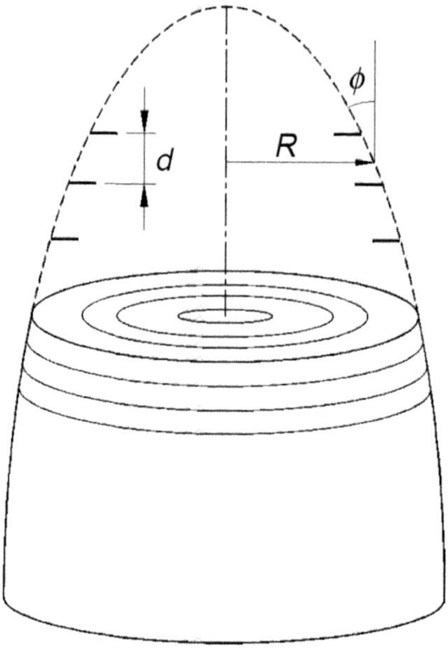

Fig. 3.48 Analysis models for axial crushing of conical shells.

$$F_m = 2.09\sigma_0 t^2 \left(\frac{2\pi R}{d} + \frac{d}{t} + (\pi + 2\phi)\tan\phi + 1 \right) \tag{3.109}$$

where t is the shell thickness, d is the frame spacing, R is the average radius at the concerned section, and ϕ is the conic angle (see Fig. 3.48). In this analysis, it is assumed that the folding length ($2H$) is confined between two adjacent frames; thus, it can be calculated from $2H = d/\cos\phi$.

3.7.4 Axial crushing of L-, T-, and X-structural elements

A fundamental theoretical analysis on axial crushing of basic elements, such as L-, T-, and X-elements, was made by Wierzbicki and Abramowicz (1983) where the basic folding mechanism was constructed and developed. Later, Kierkegaard (1993) further investigated the axial crushing behaviours of these basic elements.

When a ductile plated-shell structure, for example, a boxlike structure, is subjected to axial loads, the structure will usually be deformed in a folding pattern. The collapse mode is a progressive folding normally starting at the contact and gradually extending until all kinetic energy is absorbed by the structure or the applied external loads are removed.

Experimental tests were carried out by DiPaolo and Tom (2006) on axial crush configuration response of thin-walled, steel box components under quasistatic testing conditions. The folding deformation was a symmetric axial crush mode. Crush characteristics were evaluated, and for each material type, the observed differences were

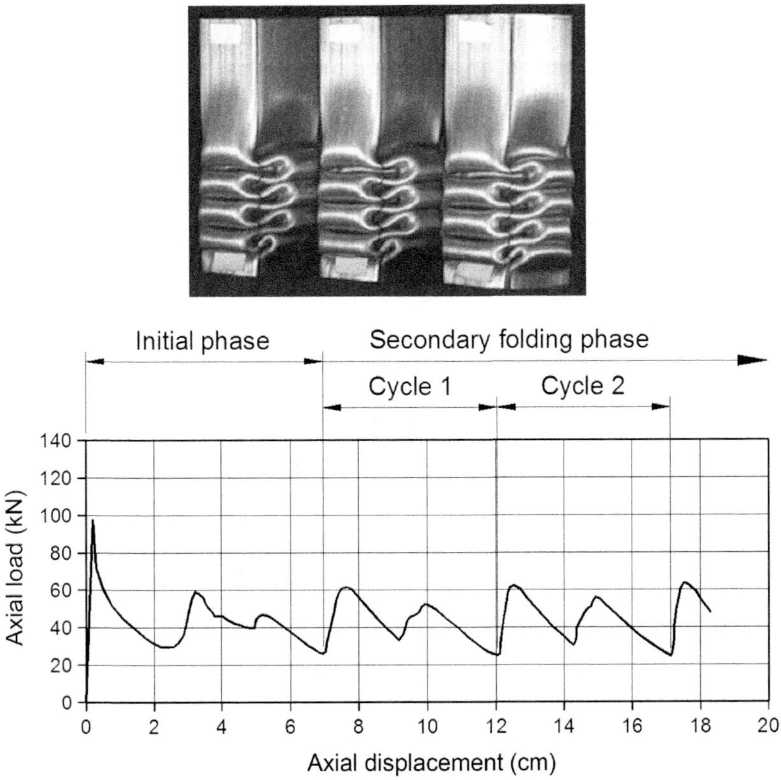

Fig. 3.49 Crushing pattern of a box structure under axial loads.
cf. DiPaolo, B.P., Tom, J.G., 2006. A study on an axial crush configuration response of thin-wall, steel box components: the quasi-static experiments. Int. J. Solids Struct. 43, 7752–7775.

less than 7% for maximum and minimum load magnitudes and less than 2% for the energy absorption, displacement, and mean load quantities in both the initial phase and the secondary folding phase cycles. The damage pattern and the load–deformation curve for a box-shaped structure under axial loads are shown in Fig. 3.49.

During crushing of a structural element, large plastic deformations will occur provided the material has sufficient ductility and no fractures were assumed. The energy is dissipated by the structure in mainly two ways: (1) by bending in the plastic hinge lines and (2) by in-plane stretching of the plate fields. The energy absorption capability depends on the structural arrangement and the material. As shown in Fig. 3.49, the crushing forces vary with the crushing distance. This reflects the progressive formation of folds during crushing. For the ship collision analysis, the mean crushing force is a useful parameter. A number of analytical formulae are available to predict the mean crushing force, for example, Wierzbicki and Abramowicz (1983), Amdahl and Kavlie (1992), Wierzbicki et al. (1993), Abramowicz (1994), Paik and Pedersen (1995), and Wang and Ohtsubo (1997). Formulae for the effective mean axial crushing forces are summarized in Fig. 3.50 where the half-folding length is determined from $H = C^{2/3}t^{1/3}$,

L-section	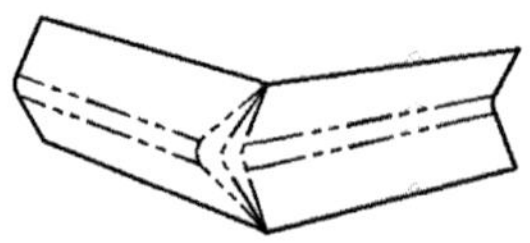	$F_L = 3.263\,\sigma_0 t^{\frac{5}{3}} c^{\frac{1}{3}}$ $c = b + b$
T-section	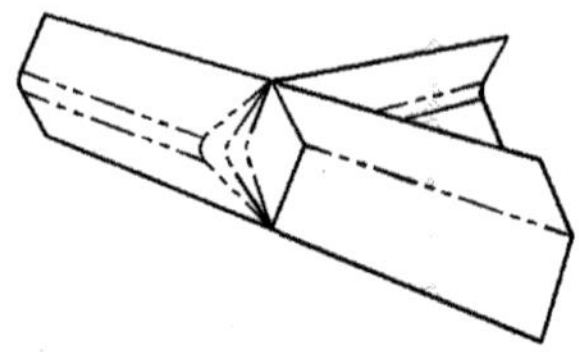	$F_T = 3.421\,\sigma_0 t^{\frac{5}{3}} c^{\frac{1}{3}}$ $c = b + b + b$
X-section	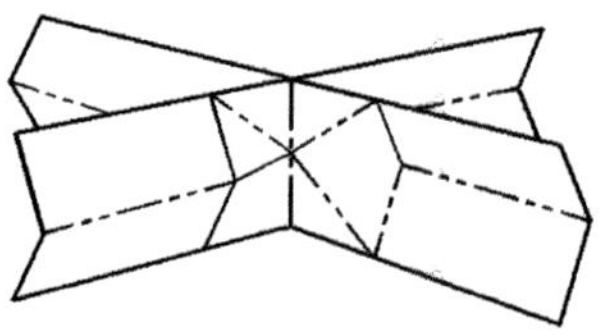	$F_X = 5.439\,\sigma_0 t^{\frac{5}{3}} c^{\frac{1}{3}}$ $c = b + b + b + b$

Fig. 3.50 The effective mean crushing force (after taking into account the effective mean crushing distance coefficient of 0.75) of the L-, T-, and X-sections.

where C is the flange length. Detailed derivations of the folding mechanisms can be found in references such as Wierzbicki and Abramowicz (1983). They show that two-thirds of the plastic energy is dissipated through deformations at stationary and moving plastic hinge lines, and the remaining one-thirds is dissipated by tension deformations that are confined in relative small areas.

3.7.5 Experimental crushing tests and analytical calculation examples

Significant number of model tests on axial crushing of structures has been carried out in the past, for example, Abramowicz and Jones (1984), Ku et al. (2001), Tarigopula et al. (2006), and DiPaolo and Tom (2006). A number of model tests by DiPaolo and Tom (2006) and Tarigopula et al. (2006) are selected here for analyses and comparisons.

3.7.5.1 Quasistatic axial crushing tests by DiPaolo and Tom (2006)

A series of quasistatic axial crushing tests were carried out by DiPaolo and Tom (2006). The test model is a square tube with a cross-sectional dimension of $50 \times 50\,\text{mm}$ with plate thickness $1.4\,\text{mm}$ (Fig. 3.51). The material properties for the models C2-3 and C4-3 are presented in Table 3.10 that was obtained from uniaxial

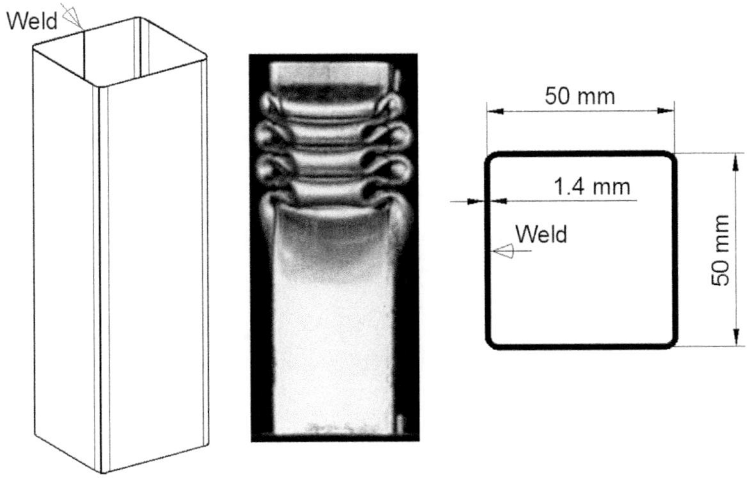

Fig. 3.51 Model illustration of the axial quasistatic crushing tests by DiPaolo and Tom (2006).

Table 3.10 Material properties and result comparison of the axial crushing tests (DiPaolo and Tom, 2006).

Model ID	**Material**	**Yield stress (MPa)**	**Ultimate strength (MPa)**	**Flow stress (MPa)**
C2-3	A36 steel	337	365	351
C4-3	A36 steel	340	364	352

Model ID	**Experimental result**		**Analytical result**	
	Folding length 2*H* (mm)	**Mean force (kN)**	**Folding length 2*H* (mm)**	**Mean force (kN)**
C2-3	44.4	23.9	29.8	21.1
C4-3	44.8	23.7	29.8	21.2

tensile tests. The analytical results and the experimental results are also presented in Table 3.10. The analytical solution is

$$F_L = 4 \times 3.263\sigma_0 t^{5/3} c^{1/3} = 4 \times 3.263 \times 351 \times 1.4^{5/3} \times 50^{1/3}/1000 = 21.1\ (\text{kN})$$

It is seen that the analytical method underestimates the mean crushing force by −12% and the folding length (2*H*) by −33%.

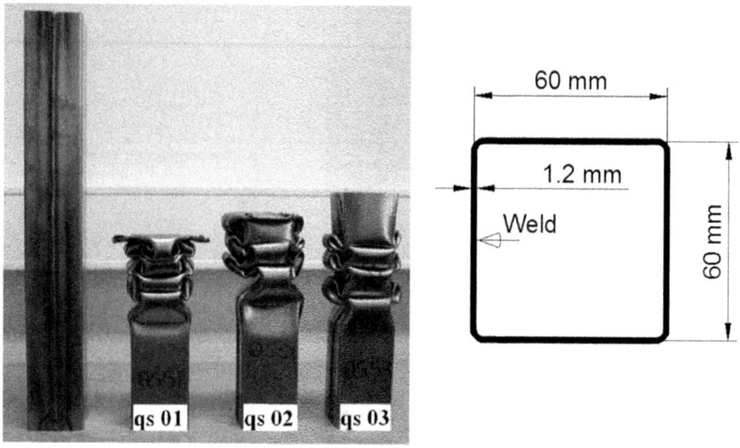

Fig. 3.52 Model illustration of the axial quasistatic crushing tests by Tarigopula et al. (2006).

3.7.5.2 Quasistatic axial crushing tests by Tarigopula et al. (2006)

Similar quasistatic axial crushing tests on square tubes were carried out by Tarigopula et al. (2006). The specimens have a nominal square cross section with rounded corners, and the average dimensions of the specimens, with respect to the centreline of the plate, are $60 \times 60 \times 1.2$ mm. The square sections were continuously welded as shown in Fig. 3.52. The properties of the materials were obtained from uniaxial tensile tests (Table 3.11). The mean crushing force is calculated from

$$F_L = 4 \times 3.263 \sigma_0 t^{5/3} c^{1/3} = 4 \times 3.263 \times 640 \times 1.2^{5/3} \times 60^{1/3}/1000 = 36.9\ (\text{kN})$$

Table 3.11 Material properties and result comparison of axial crushing tests (Tarigopula et al., 2006).

Model ID	Yield stress (MPa)	Ultimate strength (MPa)	Flow stress (MPa)
qs01, qs02, qs03	532	748	640

	Experimental result		Analytical result	
Model ID	**Folding length 2*H* (mm)**	**Mean force (kN)**	**Folding length 2*H* (mm)**	**Mean force (kN)**
qs01	–	36.2	32.0	36.9
qs02	–	34.2	32.0	36.9
qs03	–	36.4	32.0	36.9

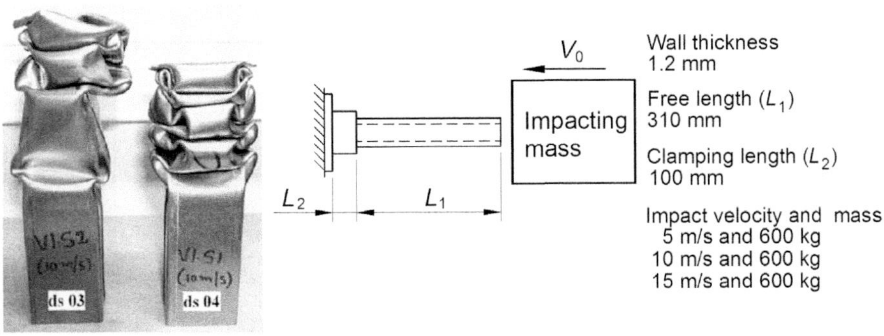

Fig. 3.53 Model illustrations of axial dynamic crushing tests by Tarigopula et al. (2006).

The comparison between the analytical and experimental results is presented in Table 3.11; very good agreement is achieved.

3.7.5.3 *Dynamic axial crushing tests by Tarigopula et al. (2006)*

Dynamic axial crushing tests on square tubes were also carried out by Tarigopula et al. (2006). The specimens are similar to the above quasistatic crushing tests with a nominal square cross section of $60 \times 60 \times 1.2$ mm. The specimens were impacted by a 600 kg mass at three speeds of 5, 10, and 15 m/s. The folding patterns, which are more complicated than the static cases, are shown in Fig. 3.53.

While it is difficult to estimate an accurate value of the strain rate in a square tube crushing, Abramowicz and Jones (1984) have proposed that the average strain rate can be approximated as

$$\dot{\varepsilon} = \frac{1}{3}\frac{V_0}{b} \tag{3.110}$$

where V_0 is the initial impact speed and b is the width of the cross section. They also estimated the coefficients $D = 6844$ (1/s) and $q = 3.91$ in the Cowper–Symonds empirical formula for steels. Then, we can estimate the dynamic effect by

$$\frac{\sigma_{0d}}{\sigma_0} = 1 + \left(\frac{V_0}{3b \times 6844}\right)^{\frac{1}{3.91}} \tag{3.111}$$

A comparison of the dynamic effects on the mean crushing force for square tubes between the experimental results (Tarigopula et al., 2006) and analytical results by Eq. (3.111) is presented in Table 3.12; reasonably good agreement is found.

Table 3.12 Dynamic effect on the axial crushing force normalized to quasistatic crushing force (Tarigopula et al., 2006).

Impact speed (m/s)	Experimental	Analytical
0 (static)	1.00	1.00
5	1.14	1.24
10	1.21	1.29
15	1.37	1.32

3.8 Crushing of ship bows against rigid walls

3.8.1 Introduction

This section is to study bow crushing analysis methods and maximum crushing forces of a ship colliding to rigid walls such as bridge pylons or piers.

3.8.2 Analysis procedure for bow crushing

The most important task for ship bow crushing analysis is to determine the crushing force, absorbed energy, and the damage extent of the bow. A simple model of a ship bow is a plated structure that can be divided into basic structural elements, for example, L-, T-, and X-shaped elements, or individual plating members as illustrated in Fig. 3.54. By analysing the crushing resistance of each element and summing them together, the total crushing resistance of the whole structure can be obtained. The formulae commonly used for the mean axial crushing force of a structure composed of L-, T-, and X-shaped elements or individual plates are summarized in the following:

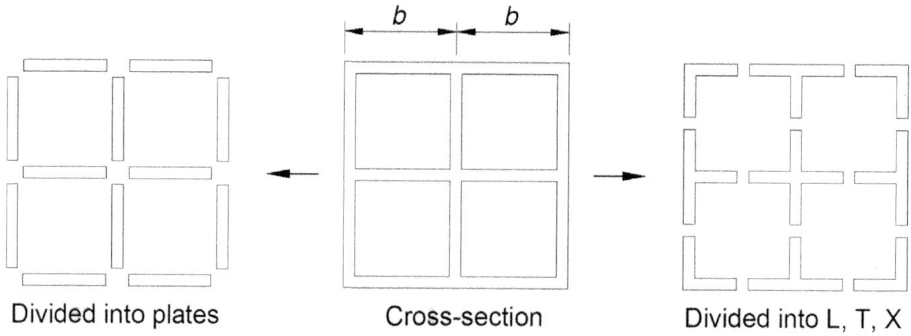

Fig. 3.54 A plated structure is divided into individual plates or into L-, T-, and X-shaped elements.

(1) Amdahl (1983)

$$F_m = 2.42\sigma_0 A\left(\frac{n_{AT}t^2}{A}\right)^{0.67} \cdot \left(0.87 + 1.27\frac{n_X + 0.31n_T}{n_{AT}}\left(\frac{A}{(n_X + 0.31n_T)t^2}\right)^{0.25}\right)^{0.67} \tag{3.112}$$

(2) Yang and Caldwell (1988)

$$F_m = \sigma_0\left(1.178/H\sum_{n_T} bt^2 + 0.215H\sum_{n_{AT}} t + 6.935\sum_{n_{AT}} t^2 + 0.265H\sum_{n_T} t + 0.589\sum_{n_T} t^2 + 0.75H\sum_{n_X}\sum_{4} t + 0.375\sum_{n_X}\sum_{4} t^2\right) \tag{3.113}$$

(3) Abramowicz (1994)

$$F_m = (n_L + 1.2n_T + 2.1n_X)\left(3.263\sigma_0 c^{0.33} t^{1.67}\right) \tag{3.114}$$

(4) Paik and Pedersen (1995)

$$F_m = \sigma_0\sum_i A_i\left(1.9514\left(\frac{t}{b}\right)^{0.5} + 0.3661\frac{t}{b}\right) \tag{3.115}$$

(5) Pedersen and Zhang (2000)

$$F_m = 3.5\sigma_0\sum_i A_i\left(\frac{t}{b}\right)^{0.67} \tag{3.116}$$

where

n_L = the number of L-shaped elements
n_T = the number of T-shaped elements
n_X = the number of X-shaped elements
t = the thickness of an individual plate
b = the width of an individual plate
A_i = the cross-sectional area of an individual plate
A = the summation of plate cross-sectional areas
L = the summation of plate cross-sectional widths
$c = L/n$
$n = n_L + n_T + n_X$
$n_{AT} = n_L + n_T$
$H = (1.178tL/(0.215n_L + 0.48n_T + 0.75n_X))^{0.5}$

In order to compare the predictions from the various formulae for crushing forces, different square tubes are considered as shown in Fig. 3.55. For cases 1–3, the width of the cross section of the tube is $a = 1.0$m, and the plating thickness is $t = 10$mm. A comparison of the mean crushing strength ($F_m/(\sigma_0 A)$) for the three cases is presented in Fig. 3.55. From the comparison, it is seen that reasonable agreement is achieved between the different methods, especially for the cases where the number n_T or n_X increases. Comparisons with the experimental results reported by Paik et al. (1999) are also presented in Fig. 3.55.

Case		Test result	Amdahl (1983)	Yang and Caldwel (1988)	Abramowicz (1984)	Paik and Pedersen (1995)	Pedersen and Zhang (2000)
Case 1 a = 1.0 m t = 10 mm	a	–	0.101	0.170	0.151	0.199	0.160
Case 2 a = 1.0 m t = 10 mm	a	–	0.235	0.264	0.240	0.283	0.255
Case 3 a = 1.0 m t = 10 mm	a	–	0.314	0.325	0.330	0.349	0.334
Paik et al. (1999) a = 100 mm	a t = mm	t = 2.2 0.216	0.171	0.302	0.253	0.297	0.271
		t = 2.8 0.268	0.201	0.363	0.297	0.337	0.319
		t = 3.0 0.272	0.210	0.382	0.311	0.349	0.334
		t = 4.2 0.395	0.264	0.498	0.390	0.415	0.418

Fig. 3.55 Comparisons of the mean crushing strength obtained by different formulae.

For longitudinally stiffened ship bows, the crushing strength of plate panels can be greatly increased since the plate panels of the crushing elements have longitudinal stiffeners, as discussed by Yang and Caldwell (1988). However, the final collapse mode of such plate panels can be regarded as bending failure of the plate panels. This means that the stiffeners behave as part of the plate in the crushing process. It therefore appears to be a valid approximation to include the bending collapse of a stiffened plate by forming plastic hinges into the bending of the plate flanges of the basic elements in the crushing analysis. The equivalent plate thickness for the bending collapse of a combined plate stiffener is given by $t_{eq} = \left(2\sqrt{3}M_1/\sigma_0 b\right)^{0.5}$ where b is the width of the plate attached to a stiffener and M_1 is the full plastic bending moment of the stiffener with attached plate. This approach for including the effect of longitudinal stiffeners in bow crushing analysis gives good predictions, as also demonstrated by Yamada (2006) in bow crushing experiments.

Pedersen et al. (1993) have further investigated the methods proposed by Amdahl (1983) and Yang and Caldwell (1988). To a large extent, both methods make use of the folding mechanisms proposed by Wierzbicki and Abramowicz (1983). Comparisons carried out by Pedersen et al. (1993) between the method of Amdahl (1983) and Yang and Caldwell (1988) with model tests (Amdahl, 1983) are shown in Fig. 3.56.

In Yamada and Pedersen (2008), the results of a benchmark test is presented where different simplified procedures have been compared with several experimental tests of relatively large-scale bow models; see Fig. 3.6.

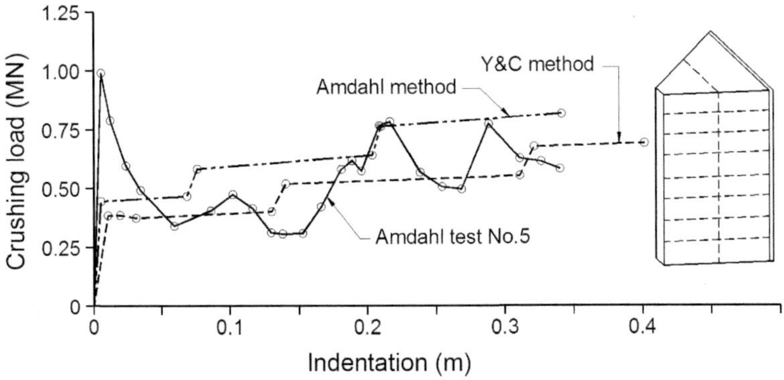

Fig. 3.56 Bow crushing load comparison between model test and analytical calculations. Based on Pedersen, P.T., Valsgaard, O., Olsen, D., Spangenberg, S., 1993. Ship impacts—bow collisions. Int. J. Impact Eng. 13 (2), 163–187.

3.8.3 Striking bow shape and ship main dimension

3.8.3.1 Striking bow shape

Ship bows vary in size and shape. It seems very difficult to characterize the scantlings and the geometry of striking bows by a simple model. Therefore, some assumptions must be made. Different ship types can be divided into the following groups: (1) oil tankers, (2) bulk carriers, (3) chemical tankers, (4) gas tankers, (5) general cargo ships, (6) container vessels, (7) passenger vessels, (8) Ro-Ro ships, (9) fishery vessels, (10) navy vessels, and (11) others.

Some years ago, statistics showed that about 40%–50% of all the above-mentioned ships have bulbous bows and the others have conventional bows (Nielsen, 1995). However, today, more and more bulbous bows are used in modern seagoing merchant vessels, since a ship with a bulbous bow most often requires less propulsive power and has better resistance characteristics than the same ship without bulbous bow (Schneekluth, 1987). Therefore, the descriptions of ship bows are here focused on bows with a bulb (**bulbous bow**).

The shape of a ship bow is influenced significantly by its hull form and the ship speed. More specifically, it is dictated by the Froude number that is defined as

$$F_n = \frac{V}{\sqrt{gL}} \tag{3.117}$$

where V is the ship design speed, g is the gravity ($=9.81\,\text{m/s}^2$), and L is the ship length. The block coefficient of a ship can also give good indication of a ship hull form and ship speed. The block coefficient of a ship at design draught can be approximated by

$$C_b = 1.07 - 1.68F_n \tag{3.118}$$

Typical values of Froude number and block coefficients for some commercial ships are the following:

$F_n = 0.14$; $C_b = 0.84$ for oil tankers and bulk carriers

$F_n = 0.19$; $C_b = 0.75$ for LNG ships

$F_n = 0.22$; $C_b = 0.70$ for container ships

The basic data for describing a bulbous bow can be separated into two parts: the first part is for the upper part of a bow, and the second part is the lower part for the bulb. Fig. 3.57 presents a detailed hull form of the bow of a container ship.

Through an investigation of actual designs, the upper deck profile can be represented by the following expression:

$$y = C_0 \cdot B^{0.5} x^{0.5} \tag{3.119}$$

where C_0 is a coefficient, B is the ship breadth, and x is the distance measured from the upper deck tip (see Fig. 3.58). The typical values for the coefficient C_0 and the stem angle φ are the following:

$C_0 = 0.57$; $\varphi = 75$ degrees for oil tankers
$C_0 = 0.47$; $\varphi = 65$ degrees for LNG ships
$C_0 = 0.40$; $\varphi = 56$ degrees for container ships

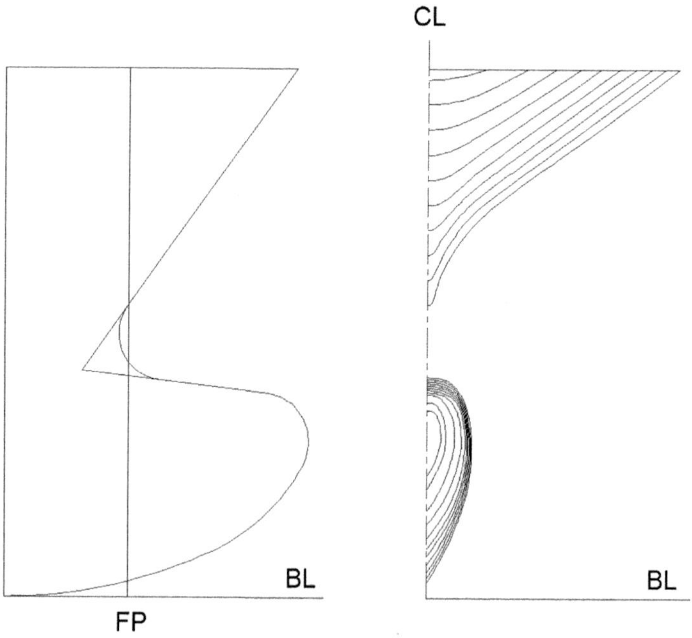

Fig. 3.57 Hull form of the bow of a container ship.

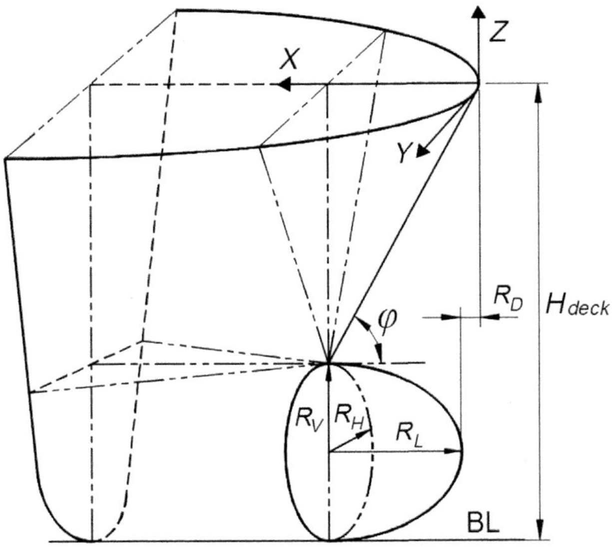

Fig. 3.58 Definitions of bulbous bow parameters.

The coefficient C_0 and the stem angle φ (in degrees) can be determined approximately from the block coefficient of a ship:

$$C_0 = 1.257C_b - 0.478$$

$$\varphi = 139.09C_b - 40.65$$

For the lower part of the bow, we assume that the bulb has three radii that are proportional to the height (H_{deck}) of the uppermost deck:

$$R_L = \lambda_L \cdot H_{deck}, \text{bulb length}$$

$$R_V = \lambda_V \cdot H_{deck}, \text{vertical radius}$$

$$R_H = \lambda_H \cdot H_{deck}, \text{horizontal radius}$$

Average values of the coefficients λ_L, λ_V, and λ_H, based on statistics of existing ships and the design guidelines for bulbs, are found as follows:

$$\lambda_L = 0.225; \;\; \lambda_V = 0.28; \;\; \lambda_H = 0.18 \;\; \text{for oil tankers}$$

$$\lambda_L = 0.29; \;\; \lambda_V - 0.20; \;\; \lambda_H = 0.10 \;\; \text{for LNG ships}$$

$$\lambda_L = 0.315; \;\; \lambda_V = 0.18; \;\; \lambda_H = 0.085 \;\; \text{for container ships}$$

The coefficients λ_L, λ_V, and λ_H may also be determined approximately from empirical expressions using the block coefficient C_b:

$$\lambda_L = -0.680C_b + 0.795$$

$$\lambda_V = 0.765C_b - 0.363$$

$$\lambda_H = 0.732C_b - 0.436$$

The distance R_D, between the bulb tip and the forecastle deck tip perpendicular, can be calculated from

$$R_D = \frac{H_{deck} - 2R_V}{\tan\varphi} - R_L$$

For example, for a container ship, if the stem angle is $\varphi = 56$ degrees, the distance R_D becomes

$$R_D = \frac{H_{deck} - 2 \cdot 0.18H_{deck}}{\tan 56\text{degrees}} - 0.315H_{deck} = 0.117H_{deck}$$

In order to describe the bulb mathematically, the origin of an $x_1y_1z_1$-coordinate system is positioned at the bulb tip, as shown in Fig. 3.59. The formulae for describing the bulb of a bow can be written as follows:

When $x \le R_L$,

$$\sin\left(\frac{x_1}{R_L}\frac{\pi}{2}\right) = \left(\frac{y_1}{R_H}\right)^{C_y} + \left(\frac{z_1}{R_V}\right)^{C_z} \tag{3.120}$$

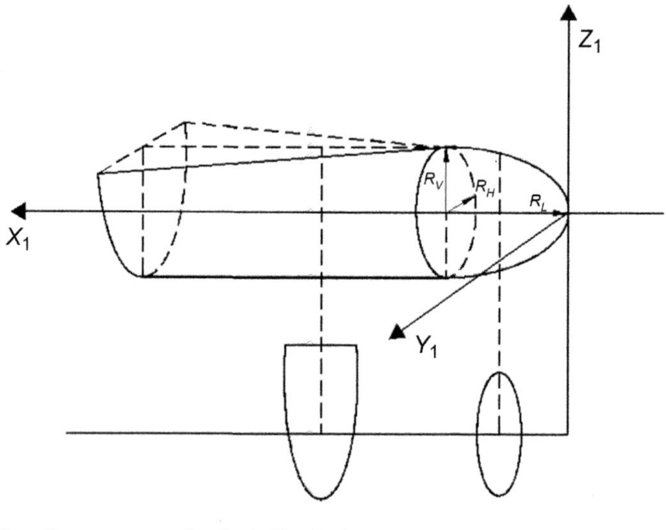

Fig. 3.59 Coordinate systems for the bulb of a bow.

where the coefficients C_y and C_z can be taken as follows (based on actual ship designs analysis):

$C_y = 2.5;\ \ C_z = 4.0$ for oil tankers

$C_y = 2.0;\ \ C_z = 2.5$ for LNG ships

$C_y = 1.7;\ \ C_z = 1.7$ for container ships

The coefficients C_y and C_z may be determined approximately from empirical expressions using the block coefficient C_b:

$$C_y = 5.952C_b - 2.469$$

$$C_z = 17.189C_b - 10.365$$

In the vertical x_1z_1-plane ($y_1 = 0$), the equation becomes

$$\sin\left(\frac{x_1}{R_L}\frac{\pi}{2}\right) = \left(\frac{z_1}{R_V}\right)^{C_z} \Rightarrow: z_1 = \left[\sin\left(\frac{x_1}{R_L}\frac{\pi}{2}\right)\right]^{1/C_z} \cdot R_V$$

In the horizontal x_1y_1-plane ($z_1 = 0$), the equation becomes

$$\sin\left(\frac{x_1}{R_L}\frac{\pi}{2}\right) = \left(\frac{y_1}{R_H}\right)^{C_y} \Rightarrow: y_1 = \left[\sin\left(\frac{x_1}{R_L}\frac{\pi}{2}\right)\right]^{1/C_y} \cdot R_H$$

When $x > R_L$,

In the vertical x_1z_1-plane ($y_1 = 0$), the approximation is

$$z = R_V$$

In the horizontal x_1y_1-plane ($z_1 = 0$), the equation becomes

$$\sin\left(\frac{x_1}{R_L}\frac{\pi}{2}\right) = \left(\frac{y_1}{R_H}\right)^{C_y} \Rightarrow: y_1 = \left[\sin\left(\frac{x_1}{R_L}\frac{\pi}{2}\right)\right]^{1/C_y} \cdot R_H, \text{but if } y_1 \geq B/2, \text{then } y_1 = B/2.$$

Fig. 3.60 presents a summary for a bulbous bow descriptions and equations for quick reference.

3.8.3.2 Ship main dimension

It is shown from the previous section that three parameters, ship breadth B, deck height H_{deck}, and block coefficient C_b, are important to determine a bow shape. For a given ship, these parameters may be known. But for probabilistic analyses, a large number of

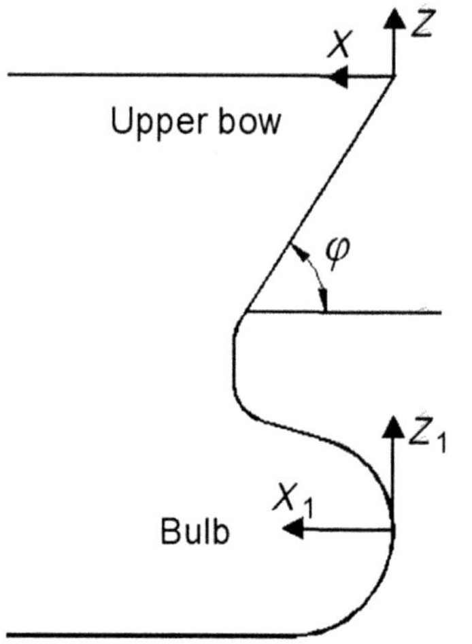

Upper bow

$$y = C_0 B^{0.5} x^{0.5}$$
$$C_0 = 1.257C_b - 0.478$$
$$\varphi = 139.09C_b - 40.65$$

Bulb

$$\sin\left[\frac{x_1}{R_L}\frac{\pi}{2}\right] = \left[\frac{y_1}{R_H}\right]^{C_y} + \left[\frac{z_1}{R_V}\right]^{C_z}$$

$$C_y = 5.952C_b - 2.469$$
$$C_z = 17.189C_b - 10.365$$
$$R_L = \lambda_L H_{deck}$$
$$R_V = \lambda_V H_{deck}$$
$$R_H = \lambda_H H_{deck}$$
$$\lambda_L = -0.680C_b + 0.795$$
$$\lambda_V = 0.765C_b - 0.363$$
$$\lambda_H = 0.732C_b - 0.436$$

Fig. 3.60 Summary of ship bow descriptions and equations.

ships are involved in the calculations. In this case, we do not have the exact dimensions of the striking ships. So, we would like to use one parameter, such as the ship length, to determine approximately the ship breadth and the forecastle deck height. Unfortunately, there are not such formulae. However, based on a large number of existing ships (Nielsen, 1995), we have found approximate data for the relations between the parameters, as shown in Table 3.13.

Based on the statistics of optimization results according to economic criteria, Schneekluth (1987) gave the following formula for determining the ship length:

$$L = C \cdot \Delta^{0.3} V^{0.3} \tag{3.121}$$

where

Table 3.13 Relationships for length, breadth, depth, and height of the forecastle deck.

	Tankers	**Bulk carriers**	**Chemical tankers**	**Gas tankers**	**Container ships**	**Passenger ships**
L/B	5.5–7.5	5.0–8.0	5.5–7.5	5.5–7.0	5.5–9.0	5.5–8.0
L/D	10.0–14.0	10.0–14.0	10.0–14.0	10.5–12.5	10.0–13.5	10.0–25.0
H_{deck}/B	0.55–0.65	0.65–0.80	0.60–0.75	0.60–0.85	0.70–0.80	0.70–0.80

L = ship length, B = ship breadth, D = ship depth, and H_{deck} = forecastle deck height.

L = length between perpendiculars in metres
Δ = displacement in tonnes
V = design speed in knots
$C = 3.2\frac{C_B + 0.5}{0.145/F_n + 0.5}$ where F_n is the Froude number.
Schneekluth (1987) formula can be also expressed as

$$L = 7.432(F_n \cdot \Delta)^{0.353} \tag{3.122}$$

The fleet database of oil tankers has been investigated by Zhang (2005) to study the relationship between ship length, breadth, depth, maximum draught, and displacement. The size of the oil tanker database is around 2000 ships, and they were built from 1950 to 2002. Figs 3.61–3.64 present the results of the relationships between the main dimensions of oil tankers. Empirical formulae for the relation with ship length L (m) can be expressed as

$$\Delta = 0.0048L^{3.12}, \quad \text{displacement (tonnes)}$$

$$B = 0.115L^{1.07}, \quad \text{ship breadth (m)}$$

$$D = 0.076L^{1.03}, \quad \text{ship depth (m)}$$

$$T = 0.048L^{1.05}, \quad \text{ship maximum draught(m)}$$

$$DWT = 0.0013L^{3.322}, \quad \text{ship deadweight (tonnes)}$$

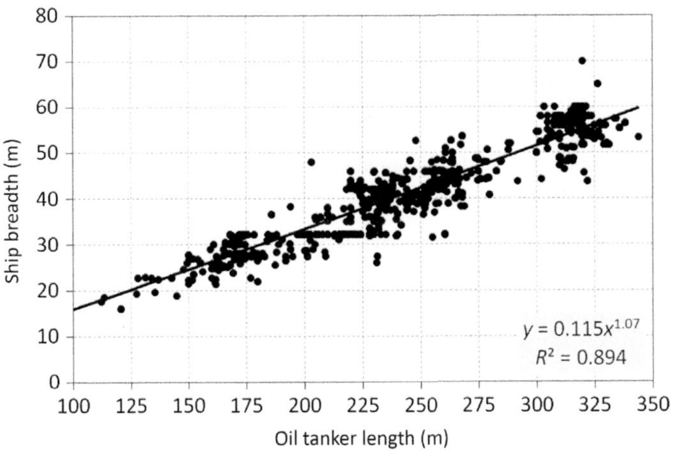

Fig. 3.61 Ship breadth and ship length relationship.

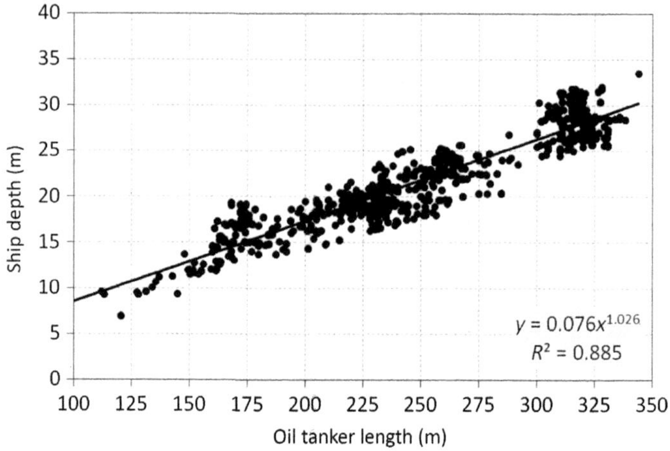

Fig. 3.62 Ship depth and ship length relationship.

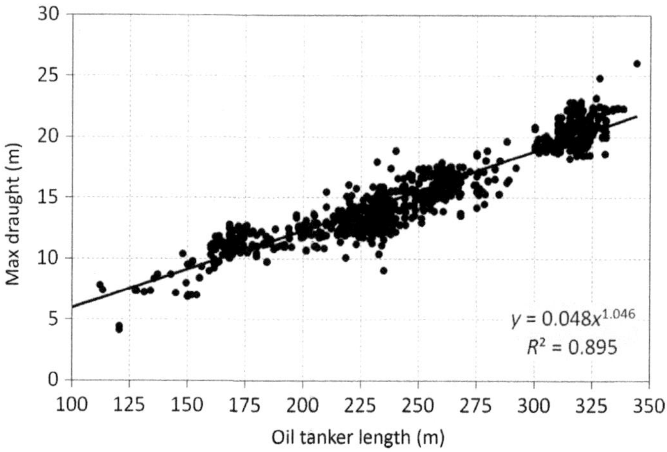

Fig. 3.63 Maximum draught and ship length relationship.

3.8.4 Bow crushing loads and bow damage in head-on collision

In this section, simple formulae for ship bow head-on crushing analyses are described based on the simplified theory presented in the previous sections (see Fig. 3.65). The derivation commences by defining the midship section since the geometric parameters amidships can be determined straightforward from classification rules and actual designs. By generalizing the shape of bows, the analysis is extended to the forward region.

In order to obtain a simple model, here, a number of assumptions are made with respect to the geometry of large double-hull tankers as described by Zhang (2002). The double-bottom height is taken as $0.1D$, and the ship breadth B is $2D$, where D

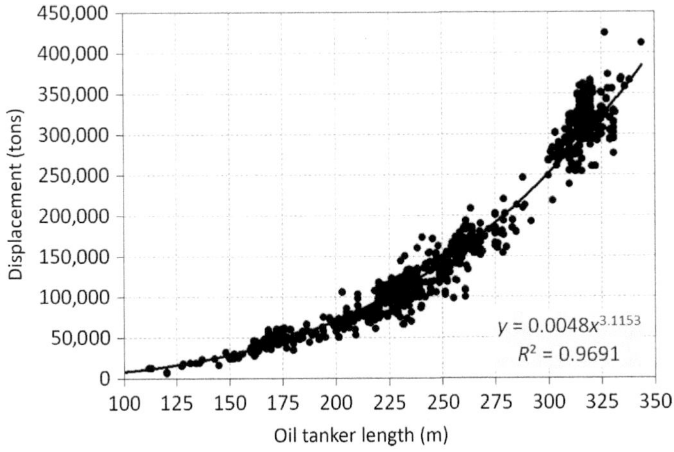

Fig. 3.64 Displacement and ship length relationship.

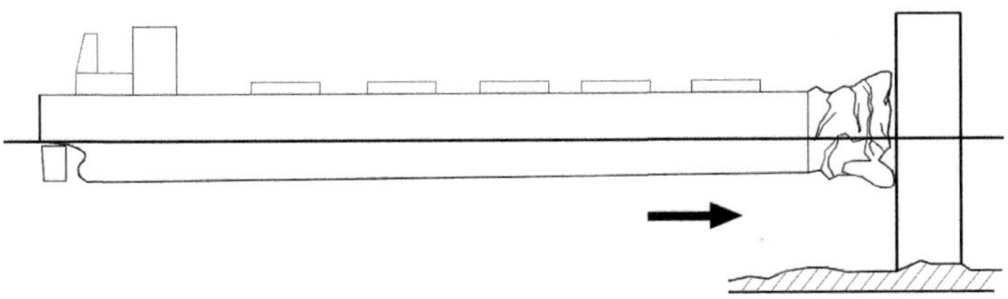

Fig. 3.65 Illustration of the damage to ship bow in head-on collisions.

is the ship depth. The equivalent thickness (plating plus longitudinal members) of the deck, side, longitudinal bulkheads, and bottom are identical. Under these assumptions, the midship section modulus at the deck is estimated as $Z=0.779B^2t_{eq}$, where t_{eq} is the equivalent thickness of the deck, side shell, etc. This formula gives a good estimate for the section modulus when compared with detailed calculations.

Using the requirement for the minimum midship section modulus $Z=k_LC_1L^2B(C_b+0.7)\times 10^{-6}$ (m^3), obtained from the Rules for the classification of ships (Lloyd's Register, 2018b), the equivalent thickness in millimetres can be written $t_{eq}=1.248k_LC_1L^2(C_b+0.7)/B\times 10^{-6}$, where k_L is the material factor, C_1 is a coefficient, C_b is the block coefficient, L is the ship length in metres, and B is the moulded ship breadth in metres. The cross-sectional area at the midship section can be expressed as $A_{mid}=k_LA_1C_1L^2(C_b+0.7)\times 10^{-6}$ (m^2), where A_1 is a coefficient depending on the midship configuration, for example, $A_1=6.0$ for a single-hull oil tanker and $A_1=7.7$ for a large double-hull oil tanker (VLCC). The coefficient C_1 generally increases with the ship length based on the Rules for ships. However, after analysing actual designs, it is found that the factor A_1C_1 can be taken as a constant value independently of the ship length and midship configuration, $A_1C_1=87$. By using

Eq. (3.116), the mean crushing resistance of the midship section can be calculated from

$$F_{mid} = \left(3.5\left(\frac{t}{b}\right)^{0.67} k_L \sigma_0\right)\left(87L^2(C_b + 0.7)\right) \tag{3.123}$$

where F_{mid} is the crushing resistance in N, σ_0 is the material flow stress in N/mm^2, t is the plating thickness in mm, and b is the plating width taken as the spacing of the longitudinal stiffeners in mm. Based on an analysis of modern designs of oil tankers and bulk carriers of length 150m and above, the following data are obtained to further simplify the formulations. The mean value of the ratio between plating width and plating thickness is $b/t = 48$. As these ships were constructed in high-tensile steel HT32, the flow stress of the material is $\sigma_0 = 0.5(315 + 440) = 377.5$ N/mm^2, and then, the material factor is $k_L = 0.78$. The block coefficient can be taken as $C_b = 0.84$ for tankers and bulk carriers. After simple substitutions, the expression for the midship crushing resistance becomes

$$F_{mid} = 10.39L^2 \text{ (kN)} \tag{3.124}$$

Eq. (3.124) is derived for longitudinally stiffened ships built from HT32 steel. The crushing resistance of transversely stiffened arrangement is much lower than longitudinally stiffened arrangement as demonstrated by Lützen (2001). The results for other steel grades differ slightly, and the effect of material on crushing resistance is small (Zhang et al., 2004b). By taking the average for the commonly used steel materials for ships, the expression for the midship crushing resistance becomes

$$F_{mid} = 10.57L^2 \text{ (kN)} \tag{3.125}$$

To obtain a relationship between the crushing force and crushing distance for the forward region of a ship, the bow shape as described in Section 3.8.3 is used, and Fig. 3.66 shows the idealized bow. In the horizontal plane, the upper deck of the bow is represented by the following parabolic curve:

$$y = \begin{cases} 0.57B^{0.5}x^{0.5}, & \text{for } x \le 0.4B \\ 0.48B^{0.7}x^{0.3}, & \text{for } 0.4B \le x \le 1.1B \end{cases}$$

Further investigations on scantlings of actual designs show that the bow cross-sectional area follows curves similar to the shape of the upper deck and can be described by

$$A_x = \begin{cases} \alpha_1 x^{0.5}, & \text{for } x \le 0.4B \\ \alpha_2 x^{0.3}, & \text{for } 0.4B \le x \le 1.1B \end{cases}$$

where $\alpha_2 = 0.833B^{0.2}\alpha_1$ and α_1 is a coefficient.

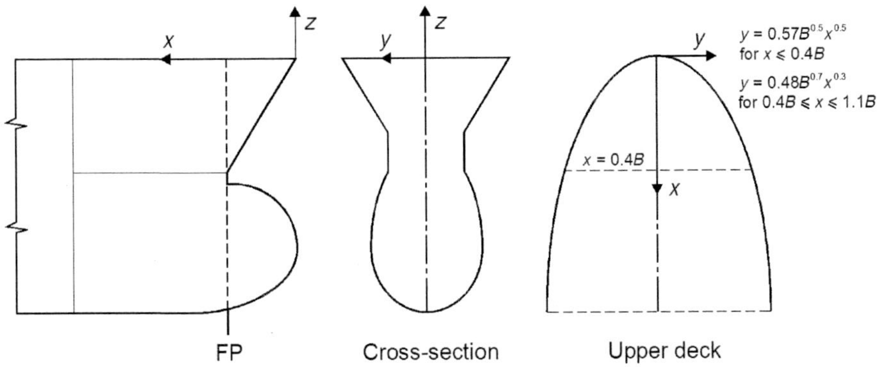

Fig. 3.66 Illustration of an assumed bow shape.

The relationship between the crushing force and the crushing distance d of a ship bow can be expressed as

$$F=\begin{cases} K_1 d^{0.5}, & \text{for}\, d \le 0.3B \\ K_2 d^{0.3}, & \text{for}\, d \ge 0.3B \end{cases}$$

where K_1 and K_2 are coefficients of the crushing force, d is the crushing distance measured from the start point of crushing to the front of the crushed zone, and x is the damage extent of the bow. After bow crushing, the damaged materials will pile up and will not be compressed further. The relationship between d and x can be taken as $d=0.75x$. It is seen that the bow force is nonlinear with the deformation. At a bow length of about $x=1.1B$, the deck width is approximately equal to the ship breadth B. At this damage extent, it can be assumed that the crushing resistance is equal to F_{mid}. Accordingly, the coefficient K_2 can be obtained as $K_2=11.20L^2/B^{0.3}$ (kN/m$^{0.5}$). At the damage extent $x=0.4$B, $K_1 d^{0.5}=K_2 d^{0.3}$, and the coefficient K_1 can be obtained as $K_1=14.25L^2/B^{0.5}$ (kN/m$^{0.3}$). Thus, the crushing force–crushing distance relationship is

$$F=\begin{cases} 14.25L^2B^{-0.5}d^{0.5}, & \text{for}\, d \le 0.3B \\ 11.20L^2B^{-0.3}d^{0.3}, & \text{for}\, d \ge 0.3B \end{cases} \quad (kN) \tag{3.126}$$

Eq. (3.126) can be further simplified by using the formula $B=0.1192L^{1.06}$. Thus, the following simple expression has been found:

$$F=\begin{cases} 41.26L^{1.47}d^{0.5}, & \text{for}\, d \le 0.3B \\ 21.20L^{1.68}d^{0.3}, & \text{for}\, d \ge 0.3B \end{cases} \quad (kN) \tag{3.127}$$

Integrating Eq. (3.127), the energy absorbed by the ship's bow is obtained as follows:

$$E_{bow} = \begin{cases} 27.51L^{1.47}d^{1.5} & \text{for } d \le 0.3B \\ 16.30L^{1.68}(d^{1.3} - 0.019L^{1.38}) & \text{for } d \ge 0.3B \end{cases} \quad (\text{kNm}) \tag{3.128}$$

where L, B, x, and d are in metres and E_{bow} is in kNm.

The collision scenario assumed here is that of a longitudinal stiffened tanker or bulk carrier collides head on with a rigid wall at speed V and that the initial kinetic energy of the ship is totally absorbed by crushing of the bow structure. The crushing distance of the ship bow can be determined from the energy balance equation $E_{bow}=0.5MV^2$, where M is the ship displacement. For ships with a bulbous bow, such as oil tankers and bulk carriers, the maximum damage extent is hardly beyond $0.4B$, which approximates to the collision bulkhead position. Therefore, for simplicity and practical use, the damage extent at the collision speed V can be determined by using the first formula of Eq. (3.128), as follows:

$$d_V = 0.069\left(\frac{MV^2}{L^{1.47}}\right)^{0.67} \quad (\text{m}) \tag{3.129}$$

Substituting Eq. (3.129) into the first formula of Eq. (3.127), the crushing resistance at the collision speed V is obtained as

$$F_V = 10.85\left(MV^2L^{2.94}\right)^{0.33} \quad (kN) \tag{3.130}$$

where L is in metres, M is in tonnes, and V is in m/s. The displacement of a bulk carrier in the full load condition can be approximated as $M=0.0015L^{3.3}$. Therefore, we have further simplified equations

$$d_V = 0.95\left(\frac{L}{300}\right)^{1.22} V^{1.33} \quad (\text{m}) \tag{3.131}$$

and

$$F_V = 1.27L^{2.08}V^{0.67} \quad (kN) \tag{3.132}$$

The damage extent to a ship bow is expressed as

$$x_V = \frac{d_V}{0.75} = 1.27\left(\frac{L}{300}\right)^{1.22} V^{1.33} \quad (\text{m}) \tag{3.133}$$

The crushing resistance can be also expressed as

$$F_V = 81.93V^{0.67}(DWT)^{0.63} \quad (kN) \tag{3.134}$$

where DWT is the deadweight tonnes of the ship. The maximum deceleration in the collision can be calculated from

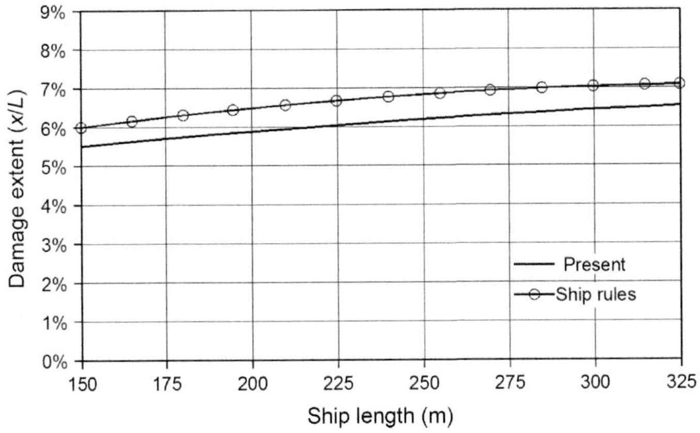

Fig. 3.67 Comparison of damage extent of ship bow between the analytical method (collision speed of 15 knots) and rule requirement on collision bulkhead position.

$$a_{\max} = \frac{F_V}{M} = 830.20 \frac{V^{0.67}}{L^{1.22}} \quad (\mathrm{m/s^2}) \tag{3.135}$$

For the ballast load condition, the displacement of a ship can be approximated as $M = 0.0006L^{3.3}$. A similar procedure to that above can be used to analyse the collision damage for the ballast load condition.

The impact duration can be approximated from (Pedersen et al., 1993)

$$t_\Delta = 1.67 \frac{d_V}{V} = 1.59 \left(\frac{L}{300} \right)^{1.22} V^{0.33} \quad (\mathrm{s}) \tag{3.136}$$

As an example, the present method is used to estimate the maximum damage extent of ship bows for comparison with the requirements in the Rules for ships (Lloyd's Register, 2018b) on collision bulkhead positions. Fig. 3.67 shows a comparison between the present calculations and the Rule requirements. The collision speed is taken as 15 knots. Good correlation is found. Fig. 3.68 shows the maximum decelerations induced in collisions for various ship sizes at a collision speed of 15 knots. It is seen that the decelerations are fairly low and are less than 1.0g for the relatively large ships. Therefore, the crew on board could generally avoid serious injury during such collision incidents.

3.8.5 Bridge pylon or pier design against ship collisions

One of the design standards for bridges against ship collisions is the AASHTO (1991) Load Resistance Factor Design (LRFD) bridge design specifications. It specifies that the head-on ship collision impact force on a pier shall be taken as

$$F_{bow} = 0.12V(DWT)^{0.5} \quad (MN) \tag{3.137}$$

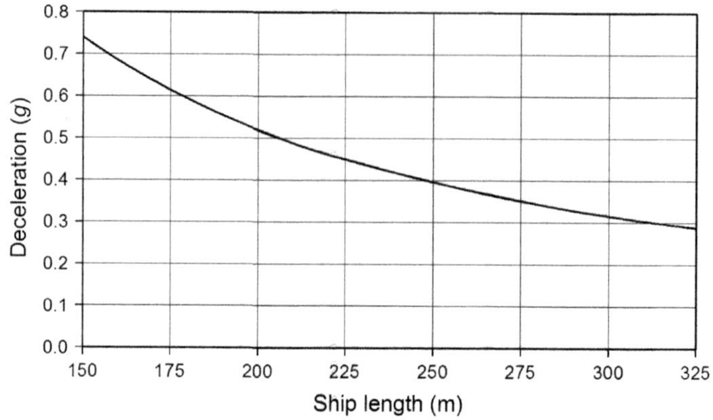

Fig. 3.68 Deceleration in collisions calculated by the present method at impact speed 15 knots.

where V is the collision speed in m/s and DWT is the ship deadweight in tonnes. This formula was developed based on the research conducted by Woisin (1979) in Germany to generate collision data for protecting the reactors of nuclear-powered ships from collisions with other ships. The model tests were at scales of 1:12 and 1:7.5. Due to the scatter of the test data, a 70% fractile force was used as an equivalent static impact force for bridge design.

Pedersen et al. (1993) studied the crushing load of ice-strengthened ship bows against rigid wall. In the theoretical analysis, the formulae proposed by Gerard (1957), Amdahl (1983), and Yang and Caldwell (1988) were applied and compared. Based on a series of calculations and analyses, an empirical expression was derived for the estimation of the maximum bow collision loads. The proposed expression for an ice-strengthened merchant vessel between 500 and 300,000 DWT has the following form:

$$F_{bow} = \begin{cases} 210L_1\left(E_1 + (5.0 - L_1)L_1{}^{1.6}\right)^{0.5} & \text{for } E_1 \geq L_1{}^{2.6} \\ 470.4(E_1L_1)^{0.5} & \text{for } E_1 < L_1{}^{2.6} \end{cases} \quad (MN) \tag{3.138}$$

where F_{bow} is the crushing force in MN, $L_1 = L/275$, $E_1 = 0.5MV^2/1425$, L is the ship length in metres, M is the displacement in 10^6 kg, and V is the collision speed in m/s. This design force has been adopted by Eurocode 1 (2010).

Another formula for bow crushing load, derived in Section 3.8.4, is represented here for easy comparison:

$$F_{bow} = 1.27L^{2.08}V^{0.67} \quad (kN) \tag{3.139}$$

where L is the ship length in metres and V is the collision speed in m/s.

Similarly, Pedersen (1998) derived an empirical expression for a broadside ship impact load, which can be used for ship collisions with bridge piers. The empirical formula is

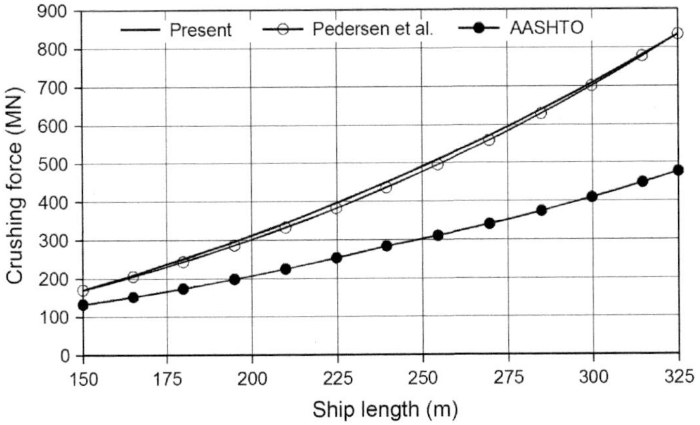

Fig. 3.69 Comparison of crushing forces obtained by different methods for tankers and bulk carriers with longitudinally strengthened ship bows at a collision speed of 15 knots.

$$P_{side} = 263c\left(1.0 + 0.88(b/D)^{1.06}\right) \cdot (L/300)^{2.20} \qquad \text{(MN)} \tag{3.140}$$

where P_{side} is the broadside collision load (MN), c is a factor accounting for the stiffening system of the ship, $c = 1.00$ for a longitudinally stiffened vessel, $c = 1.35$ for a transversely stiffened vessel, b is the width of the bridge pier in contact with the ship-side, D is the moulded depth of the vessel, and L is the ship length (m).

3.8.6 Calculation examples

Bow crushing forces are obtained by the three methods: the present method (Eq. 3.139), Pedersen et al. (1993) (Eq. 3.138), and AASHTO (1991) (Eq. 3.137), for various longitudinal strengthened ship sizes at a collision speed of 15 knots (7.72 m/s), are shown in Fig. 3.69. A comparison of crushing forces, calculated by the three methods, for a ship with length of 250 m at different collision speeds is shown in Fig. 3.70. It is seen from the result comparison that the present results agree reasonably well with Pedersen et al. (1993) method. It is also seen that the present results and Pedersen et al. results are about 30%–70% higher than AASHTO's results. This is to be expected since the forces derived by Pedersen et al. (1993) were for ships with high ice class.

3.9 Damage analysis for ship–ship collisions

3.9.1 Introduction and collision scenarios

A general scenario of a ship–ship collision is illustrated in Fig. 3.71. When a striking ship with a forward speed of V_1 impacts against a struck ship with a forward speed of V_2 at an angle β, the penetration angle of the striking bow into the struck ship is not

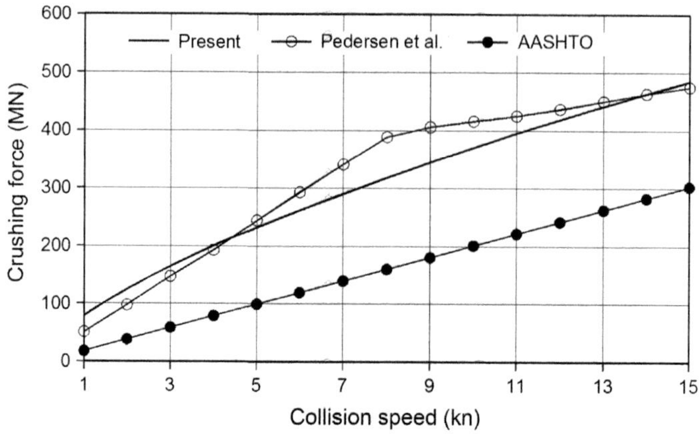

Fig. 3.70 Comparison of crushing forces for a tanker with a length of 250 m at different collision speeds.

equal to the collision angle β if the forward speed of the struck ship is not zero. It will follow the direction of the relative velocity: $V = V_1 - V_2$

In Chapter 2, we derived the expressions (2.16) and (2.17) for the relative accelerations of the striking ship and the struck ship at the contact point, which are components of the relative accelerations in the perpendicular direction ξ and in the parallel direction η to the struck ship:

$$\ddot{\xi} = -(D_\xi + \mu D_\eta) F_\xi$$

$$\ddot{\eta} = -\left(\frac{1}{\mu} K_\xi + K_\eta\right) F_\eta$$

It is assumed that the acceleration or the collision force has a sinusoidal function with time. Thus, the relative acceleration of the two ships at the impact point can be expressed as

$$\ddot{\xi} = A_1 \sin(\omega \cdot t) \text{ and } \ddot{\eta} = A_2 \sin(\omega \cdot t)$$

where A_1 and A_2 are constants to be determined by initial conditions, ω is the frequency, and t is time. Thus, the relative velocity and the relative penetration are expressed by

$$\dot{\xi} = -\frac{A_1}{\omega}\cos(\omega \cdot t) + B_1 \text{ and } \dot{\eta} = -\frac{A_2}{\omega}\cos(\omega \cdot t) + B_2$$

and

$$\xi = -\frac{A_1}{\omega^2}\sin(\omega \cdot t) + B_1 t + C_1 \text{ and } \eta = -\frac{A_2}{\omega^2}\sin(\omega \cdot t) + B_2 t + C_2$$

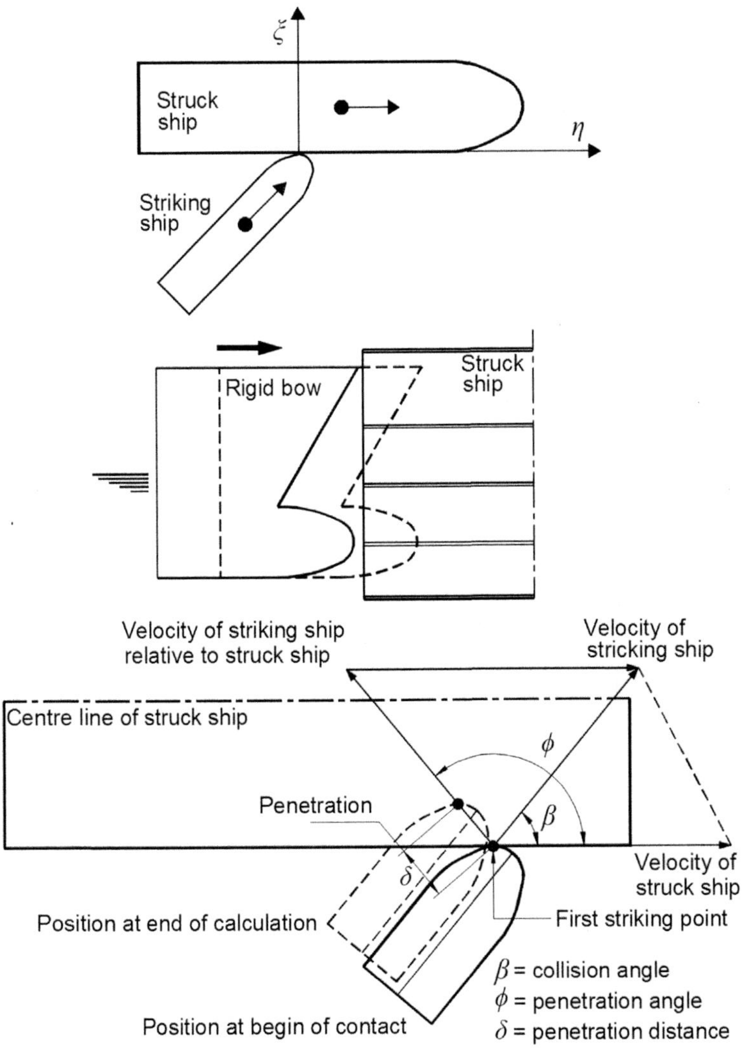

Fig. 3.71 General scenario of a ship–ship collision and the direction of the penetration.

At the beginning of a collision, $t=0$, the initial penetration is zero, and the relative velocity can be determined from the ship forward speeds V_1 and V_2. At the end of the collision, $t=t_\Delta$, the relative velocity is equal to zero. The conditions are expressed as follows:

When time $t=0$,

$$\dot{\xi}(0)=V_1\sin\beta \text{ and } \dot{\eta}(0)=V_1\cos\beta-V_2$$

$$\xi(0)=0 \text{ and } \eta(0)=0$$

When time $t = t_\Delta$,

$$\dot{\xi}(t_\Delta) = 0 \text{ and } \dot{\eta}(t_\Delta) = 0$$

where $t_\Delta = \frac{1}{4} \cdot \frac{2\pi}{\omega} = \frac{\pi}{2\omega}$ is the collision duration.

By the use of the above-mentioned conditions, the relative acceleration, the relative velocity, and the penetration are finally expressed as

$$\begin{aligned} \ddot{\xi} &= -\frac{\pi}{2t_\Delta}\dot{\xi}(0)\sin\left(\frac{\pi}{2t_\Delta}\cdot t\right) \\ \ddot{\eta} &= -\frac{\pi}{2t_\Delta}\dot{\eta}(0)\sin\left(\frac{\pi}{2t_\Delta}\cdot t\right) \end{aligned} \tag{3.141}$$

$$\begin{aligned} \dot{\xi} &= \dot{\xi}(0)\cos\left(\frac{\pi}{2t_\Delta}\cdot t\right) \\ \dot{\eta} &= \dot{\eta}(0)\cos\left(\frac{\pi}{2t_\Delta}\cdot t\right) \end{aligned} \tag{3.142}$$

and

$$\begin{aligned} \xi &= \frac{2t_\Delta}{\pi}\dot{\xi}(0)\sin\left(\frac{\pi}{2t_\Delta}\cdot t\right) \\ \eta &= \frac{2t_\Delta}{\pi}\dot{\eta}(0)\sin\left(\frac{\pi}{2t_\Delta}\cdot t\right) \end{aligned} \tag{3.143}$$

At the end of the collision, $t = t_\Delta$, the penetration in the perpendicular direction reaches the maximum, that is, $\xi_{max} = (2t_\Delta/\pi)\dot{\xi}(0)$. Thus, the collision duration is determined from

$$t_\Delta = \frac{\pi \xi_{max}}{2\dot{\xi}(0)} \tag{3.144}$$

The penetration angle ϕ, defined as the angle between the penetration direction and the forward direction of the struck ship, can be determined by

$$\tan\phi = \frac{\xi}{\eta} = \frac{\dot{\xi}(0)}{\dot{\eta}(0)} = \frac{V_1 \sin\beta}{V_1 \cos\beta - V_2} \tag{3.145}$$

This equation confirms that the penetration direction follows the direction of the relative velocity $\vec{V} = \vec{V}_1 - \vec{V}_2$. For special cases, the penetration angles are as follows:

If the speed of struck ship is zero, $V_2 = 0 \Rightarrow \phi = \beta$.

If the speeds of two ships are equal, $V_1 = V_2 \Rightarrow \pi = 90\,\text{degrees} + \beta/2$.

3.9.2 Analysis procedure for deformable striking ship and deformable struck ship

Lützen (2001) have proposed a simplified procedure to analyse how the collision damage can be shared between the striking and the struck vessels. The procedure is based on the concept of shared energy as it is described in Section 3.10.1, by Eq. (3.150) and Fig. 3.76.

The analysis of collision scenarios, where both the striking and the struck vessel can be damaged, is carried out in penetration steps. Only one of the involved vessels can be deformed in each step. By a comparison of the crushing forces for the bow and the shipside, it can be determined which vessel deforms during the considered step.

Before the calculation of the deformation of the two vessels, the following calculations are carried out:

(1) Force–penetration curve $F_{struck}(\delta_A)$ for the struck vessel, where the striking vessel is rigid
(2) Force–penetration curve $F_{striking}(\delta_B)$ for the striking vessel, where the struck vessel is rigid

If the striking vessel is fitted with a bulbous bow, the analysis of the crushing forces is separated into a bulb analysis and an analysis of the top of bow above the bulb.

The procedure for taking into account the deformation of both the bow and the struck structure indicated in Fig. 3.72 was to compare the two above-mentioned force–penetration curves, $F_{struck}(\delta_A)$ and $F_{striking}(\delta_B)$, at each step. This approach, however, only includes a very limited level of interaction. In reality, the force–penetration curve for the side of the struck vessel is a function of the deformation of the bow and vice versa. This stronger interaction is taken into account by comparing the forces F_A and F_B, which are determined as follows:

Struck vessel $F_A = F_{struck}(\delta_A)\frac{A'}{A''}$
Striking vessel $F_B = F_{Striking}(\delta_A + \delta_B)$

where

F_A is the force to crush the struck vessel;
F_B is the force to crush the striking vessel;

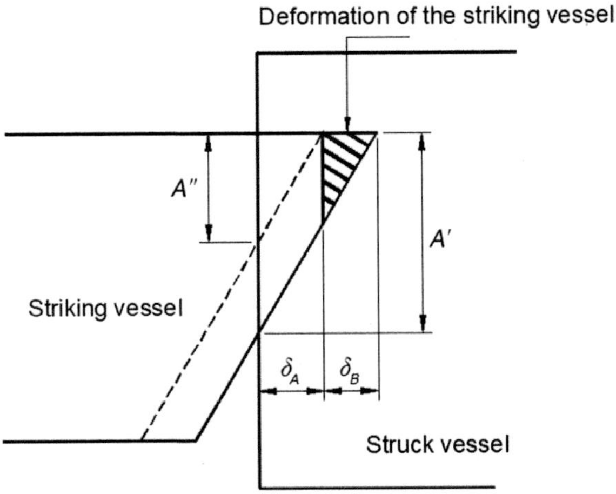

Fig. 3.72 Deformation of striking and struck vessels during collision. The *A*s relate to areas not lengths (Lützen, 2001).

F_{struck} is the force from the force–penetration curve for the struck vessel, where the striking vessel is rigid;

$F_{striking}$ is the force from the force–penetration curve for the striking vessel, where the struck vessel is rigid;

δ_A is the penetration into the struck vessel;

δ_B is the deformation of the striking vessel;

A' is the cross-sectional area of the striking vessel taken at a distance of $\delta_A + \delta_B$ from bow or bulb tip;

A'' is the cross-sectional area of the striking vessel taken at a distance of δ_A from bow or bulb tip.

The forces at the struck and the striking vessel F_A and F_B are compared:

- If $F_A > F_B$, the deformation of striking vessel, δ_B is increased.
- If $F_B > F_A$, the deformation of struck vessel, δ_A is increased.

The reason for correcting the resistance of the struck vessel is that when the bow is deformed, the resistance is approximately equal to the force at the side times the ratio between the areas. For a single-hull vessel, the correction will have nearly no influence, but for a double-hull vessel, there will be some corrections when the bow penetrates the inner side.

When the deformation patterns of the struck and the striking vessel are known, the total absorbed energy can be calculated and compared with the energy calculated by the external dynamics; see Chapter 2.

Lützen (2001) presented numerical results for collisions involving 11 different ships in order to investigate the amount of energy absorbed by the bow of the striking ship and the side structure of the struck vessel. Five different striking ice-strengthened vessels were considered (for the scantlings, see Pedersen et al., 1993):

- 150,000 DWT bulk carrier
- 40,000 DWT container vessel
- 3000 DWT general cargo vessel
- 2000 DWT tanker
- 500 DWT coaster

The struck vessels are three tankers and three Ro-Ro vessels. The respective length of the three tankers are 103, 198, and 317 m. The 103 m-long tanker is transversely stiffened, whereas the other two vessels are longitudinally stiffened.

The collisions were all right-angle collisions where the struck vessel has zero speed and the striking vessel has a speed of 4 m/s. The striking position was in all cases at the frame of midship of the struck vessel.

Table 3.14 shows the calculated actual penetrations into the struck vessels when the actual strength of the bow is considered. For comparison, the numerals in parentheses show the increased penetrations when the striking bows are assumed to be rigid. The results show a large reduction in the penetration for the three smaller striking vessels.

Table 3.15 shows the percentage of the energy absorbed in the striking bow as percentage of the total energy to be absorbed. These results show that even for these calculations where all striking vessels have high ice class, it is a common practice to assume that the bow of the striking ships in ship–ship collisions that do not deform during the impact do not always hold.

Table 3.14 Actual penetration into struck vessel, all numbers in metre.

Struck ship	Striking ship: Bulk carrier 150,000 DWT	Container vessel 40,000 DWT	General cargo 3000 DWT	Tanker 2000 DWT	Coaster 500 DWT
Tanker, $L=103\,m$	4.56 (4.56)	7.87 (7.87)	4.30 (4.30)	2.83 (3.03)	0.10 (1.63)
Tanker, $L=198\,m$	16.10 (16.10)	19.31 (19.31)	1.30 (2.30)	0.30 (1.50)	0.20 (1.63)
Tanker, $L=317\,m$	[a]	10.98 (10.98)	1.70 (3.50)	0.10 (2.10)	0.10 (0.80)
Ro-Ro, $L=58\,m$	3.32 (3.32)	3.00 (3.00)	2.40 (2.40)	0.10 (2.00)	0.20 (1.50)
Ro-Ro, $L=150\,m$	8.60 (8.60)	7.40 (7.40)	2.80 (3.60)	1.80 (2.10)	0.50 (1.00)
Ro-Ro, $L=180\,m$	9.00 (9.00)	8.20 (8.20)	3.60 (3.70)	1.90 (2.30)	0.40 (1.20)

Numbers in parentheses represent penetration where the striking vessel is considered rigid.
[a]The bulk carrier penetrates the whole breadth of the tanker.

Table 3.15 Energy absorbed in the striking bow as percentage of the total energy.

Struck ship	Striking ship: Bulk carrier 150,000 DWT	Container vessel 40,000 DWT	General cargo 3000 DWT	Tanker 2000 DWT	Coaster 500 DWT
Tanker, $L=103\,m$	0	0	0	24	97
Tanker, $L=198\,m$	0	0	36	95	89
Tanker, $L=317\,m$	0	0	52	99	94
Ro-Ro, $L=58\,m$	0	0	0	98	91
Ro-Ro, $L=150\,m$	0	0	20	48	82
Ro-Ro, $L=180\,m$	0	0	0.4	46	86

3.9.3 Hole created by a striking bow

When the penetration of a striking bow into the side of a struck ship is large enough, the strain in the side shell plating reaches its critical value, and the shell plating starts to rupture and crack. With increasing penetration, the crack extends/propagates, and finally, a hole is formed. Observations from accidents of full-scale ship collisions show that the exact shape of a damage hole in the shell plating is very complex. See Fig. 4.1 in Chapter 4 for reference, where a hole in the side shell of an oil tanker struck by a container ship is shown.

It is very difficult to predict the shape of the hole precisely. Here, we try to use a simple geometric analysis to calculate the size of a damage hole for right-angle collisions. Based on observations from experiments and collision accidents, it is appropriate to assume that the side shell plating conforms to the striking bow after the shell plating has ruptured.

Let us assume that the shell plating conforms to the striking bow after the shell plating has ruptured. Therefore, for a given penetration δ and knowing the shape of the striking bow, we can obtain the size of the hole by geometric analysis. Fig. 3.73 shows an illustration of holes created by a striking bow in a perpendicular penetration. Fig. 3.74 presents a number of examples of holes created by ship bulbs in ship collision incidents.

3.9.3.1 Upper hole

The length of the upper hole (along the ship length) can be determined from

$$D_X = 2C_0 \cdot B^{0.5}\left(\delta - \delta_f\right)^{0.5} \tag{3.146}$$

where δ is the penetration measured from the upper bow tip, δ_f is the critical penetration at which the shell plating starts to rupture, B is the breadth of the striking ship, and $C_0 = 1.257C_b - 0.478$ is the coefficient of the striking bow. If $D_X \geq B$, set $D_X = B$.

The height of the upper hole (in the vertical direction) can be calculated from

$$D_Z = (\delta - \delta_r)\tan\varphi \tag{3.147}$$

where φ is the stem angle of the striking bow. If $D_Z \geq H_{deck}$, set $D_Z = H_{deck}$. Here, H_{deck} is the height of the forecastle of the striking ship.

3.9.3.2 Lower hole

The length of the lower hole (along the ship length) can be determined from

$$D_X = 2\left[\sin\left(\frac{\delta_b - \delta_f}{R_L}\frac{\pi}{2}\right)\right]^{1/C_y} \cdot R_H \tag{3.148}$$

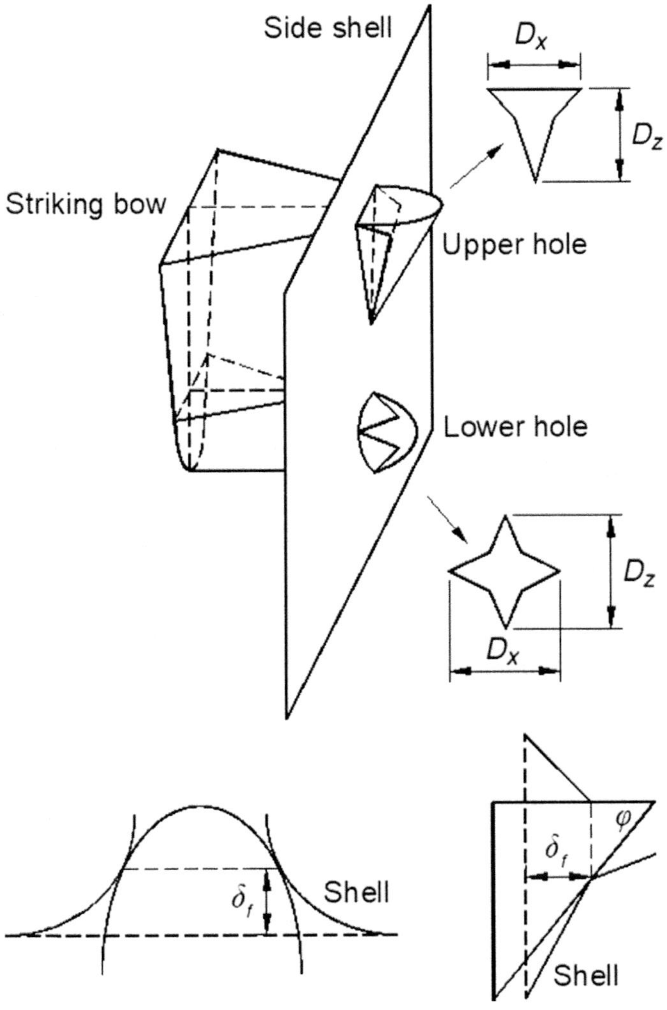

Fig. 3.73 Illustration of holes created by a striking bow.

where δ_b is the penetration measured from the bulb tip and δ_f is the critical penetration at which the shell plating starts to rupture. If $D_X \geq B$, set $D_X = B$, where B is the breadth of the striking ship.

The height of the lower hole (in the vertical direction) is calculated from

$$D_Z = 2\left[\sin\left(\frac{\delta_b - \delta_f}{R_L}\frac{\pi}{2}\right)\right]^{1/C_z} \cdot R_V \tag{3.149}$$

If $D_Z \geq 2R_V$, set $D_Z = 2R_V$. Coefficients C_y, C_z, R_L, R_V, and R_H for the geometry of a bulb are defined in Section 3.8.3.

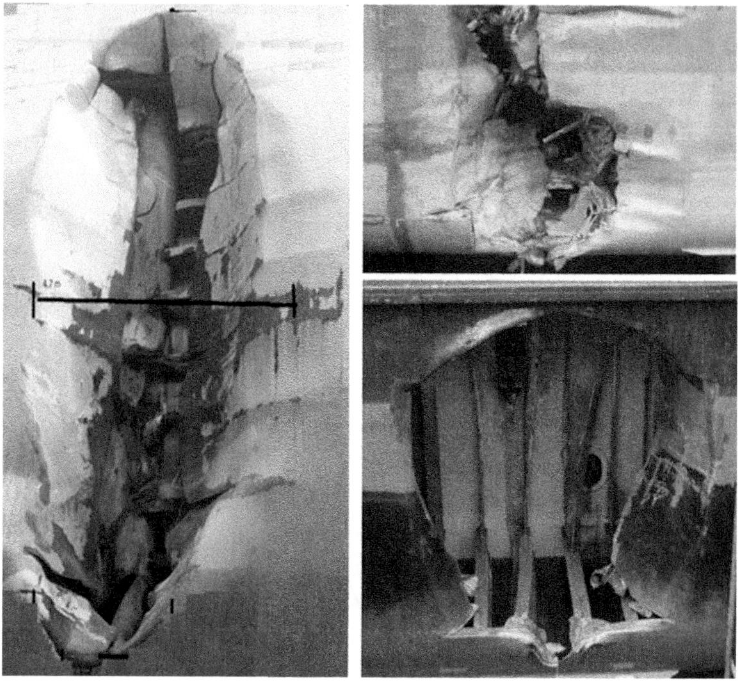

Fig. 3.74 Photo collections of holes created by striking ship bulbs in ship collision incidents.

3.10 Damage analysis of ship–offshore platform collisions

3.10.1 Introduction

The design principles of ship collision with platforms, as specified in NORSOK N-004 (2004), are categorized into three types:

(a) **Strength design** implies that the installation is strong enough to resist the collision force with minor deformation, so that the ship is forced to deform and dissipate the major part of the energy.
(b) **Ductility design** refers to a situation where the installation undergoes large, plastic deformations and dissipates the major part of the collision energy.
(c) **Shared energy design** implies that both the installation and ship contribute significantly to the energy dissipation.

An illustration of a ship collision with a platform and energy dissipations between installation and ship for the three categories is presented in Fig. 3.75. The total collision energy to be dissipated by ships and installations can be determined by the methods presented in Chapter 2, but the distribution of the energy absorbed by the ship and by the platform must be determined by their relative strength. Fig. 3.76 presents the energy dissipation in a ship and in a platform, and the structural responses of

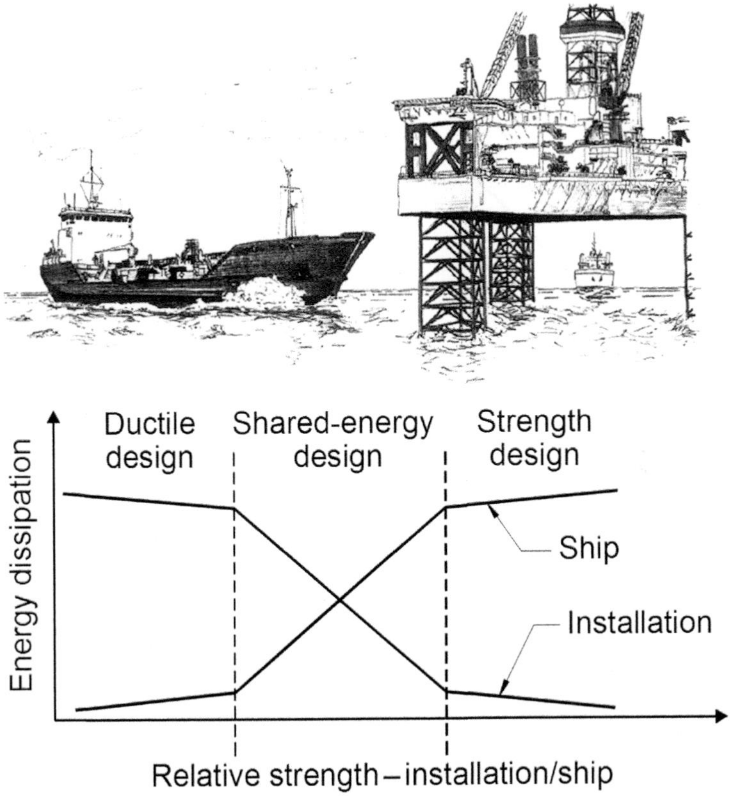

Fig. 3.75 Ship collisions with platform and energy dissipation between the installation and ship.
cf. Pedersen, P.T., 2010. Review and application of ship collision and grounding analysis procedures. Mar. Struct. 23 (3), 241–262; A concept based on NORSOK N-004, 2004. NORSOK Standard N-004, Rev. 2, October 2004. Standards Norway, Strandveien 18, N-1326 Lysaker, Norway.

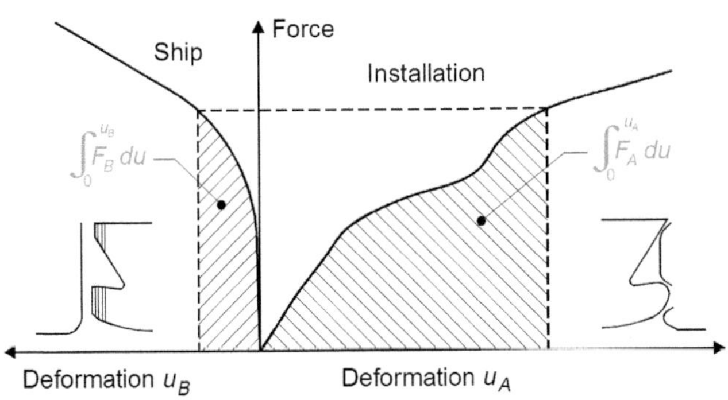

Fig. 3.76 Illustration of load–deformation curves of installations and ships and the energy dissipation determinations.

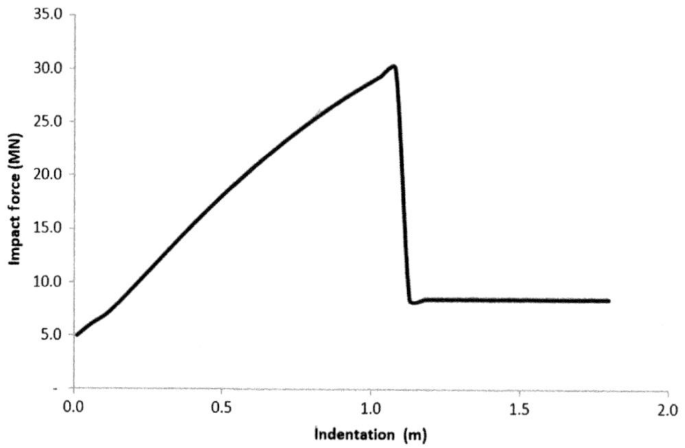

Fig. 3.77 A force–deformation curve for side impact of a supply vessel.

the ship and the installation as load–deformation curves assuming that the other object is infinitely rigid. The total energy dissipated by a ship and an installation equals the total area under the load–deformation curves:

$$E = E_{platform} + E_{ship} = \int_0^{u_A} F_A du + \int_0^{u_B} F_B du \tag{3.150}$$

Therefore, the load–deformation curves of ships and installations indented by rigid objects are to be determined or specified. This will be discussed in the next sections.

3.10.2 Rigid cylinder impact to side shell plating of a ship

An example of the force–deformation relationship for supply vessels with a displacement of 5000 tonnes is shown in Fig. 3.77 for a broadside impact. The curve for broadside impact is based upon penetration of an infinitely rigid vertical cylinder with a given diameter and may be used for impacts against jacket legs with columns of diameter $D = 1.5$ m. More curves of force–deformation relationships can be found in NORSOK N-004 (2004).

The force–deformation curves can also be determined analytically. The main deformation mode in the side shell is plastic tension and folding and crushing in deck and bottom. Approximations for the force–displacement relations for the side shell can be expressed as

$$F_p = \frac{4}{\sqrt{3}} \sigma_0 t \delta \cdot \left(\frac{D_H}{b} \cdot \frac{b - b_0}{b} \right), b_0 = \frac{R}{b} \delta, \text{and } b_0 \leq R \tag{3.151}$$

The force–displacement relations for the deck or for the bottom can be expressed as

$$F_d = \begin{cases} = 1.262\sigma_0 t^{1.83} b^{0.67} \dfrac{1}{\delta^{0.5}} + 1.290\sigma_0 t^{1.33} \delta \dfrac{1}{b^{0.33}}, & \text{(before plate rupture)} \\ = 6.77\sigma_0 t^{1.67} b^{0.33}, & \text{(concertina folding after rupture)} \end{cases} \tag{3.152}$$

where σ_0 is the flow stress of the materials, t is the thickness, δ is the displacement (or penetration), D_H is the contact height of the side shell, b is half of the frame spacing, and R is the radius of the rigid cylinder. The impact is assumed between transverse frames.

Fig. 3.78 illustrates a side collision scenario between a 5000 tonnes ice-strengthened supply vessel and a jack-up leg. The diameter of the chord of the leg is assumed as 1.5 m, and the cylindrical chord is assumed to be rigid that it penetrates into the shipside. The force–deformation curve is calculated using Eqs (3.151), (3.152), and it is presented in Fig. 3.78 (right figure). In the analytical calculation, it is assumed that the side shell plate ruptures at a deformation of 1.08 m. After the rupture of shell plates, the collision resistance from the shell plate disappears, and the collision force is a contribution from the deck and the bottom concertina folding. The force–deformation curve from NORSOK N-004 (2004) is also plotted in Fig. 3.78. It is seen that the collision force of the analytical method is considerably higher than that given by NORSOK N-004 (2004). This is due to the fact that the present example ship is an ice-strengthened and longitudinally stiffened ship. The same calculation is carried out for a nonice-classed supply vessel, and the calculated results are also plotted in Fig. 3.78. Eventually, it is seen that the analytical results are comparable with the NORSOK N-004 (2004) curve.

The energy–deformation curves for the three cases above are plotted in Fig. 3.79. The critical penetration point for the shipside is defined as the rupture of the side shell.

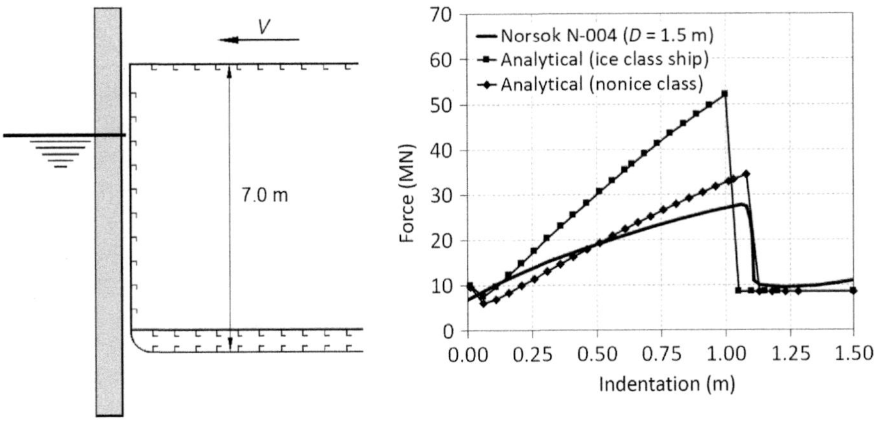

Fig. 3.78 Force–deformation curve comparisons for side impact between NORSOK N-004 (2004) and analytical calculations.

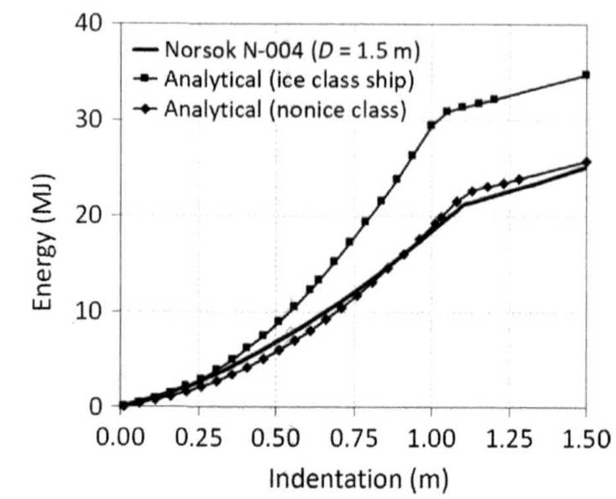

Fig. 3.79 Energy–deformation curve comparisons for side impact between NORSOK N-004 (2004) and analytical calculations.

At this point (deformation 1.08 m), the absorbed energies by the ship for the three cases are obtained as follows:

30.95 MJ for ice-strengthened vessel (analytical method)

19.97 MJ for nonice-strengthen vessel (analytical method)

20.55 MJ for nonice-strengthen vessel (NORSOK N-004)

This amount of absorbed energy (19.97–30.95 MJ) by the ship structures is larger than the collision energy of 14.5 MJ for the sideway collision of the 5000 tonnes supply vessel impact with the jack-up leg at 2.0 m/s (rigid case, in Section 2.3.5 'Calculation example for fixed platforms'). This means that the ship has sufficient energy absorption capability before the rupture of the side shell. So, in this example, the supply vessel is still intact and floating. Sinking of supply vessels after a collision has a high risk of damaging other structures under water such as risers and cables.

3.10.3 Tubular structures: Local plastic bending and tension

The methodology used in the following to determine the local capacity of the jack-up legs (tubular structures) against ship impact loads is a simplified approach based on stepwise determination of plastic hinges. Fig. 3.80 shows part of the leg of the considered jack-up rig for collision analysis. For a complete analysis, this portion of the leg has to be analysed for the five indicated load cases A–E.

For **load case A**, which is a bow collision on a horizontal bracing member (see Fig. 3.80), it is assumed that the bow of the colliding vessel is rigid, and the impact is against the midpoint of the brace. If we make the conservative assumption that the secondary bracing does not exist and that the diagonal bracings are ineffective, then we can determine the energy absorbed in the horizontal brace by bending (three-point plastic bending (Visser, 2004)):

$$E_b = P_u(L/2)\theta = 4M_p\theta = 4D^2 t\sigma_0\theta \tag{3.153}$$

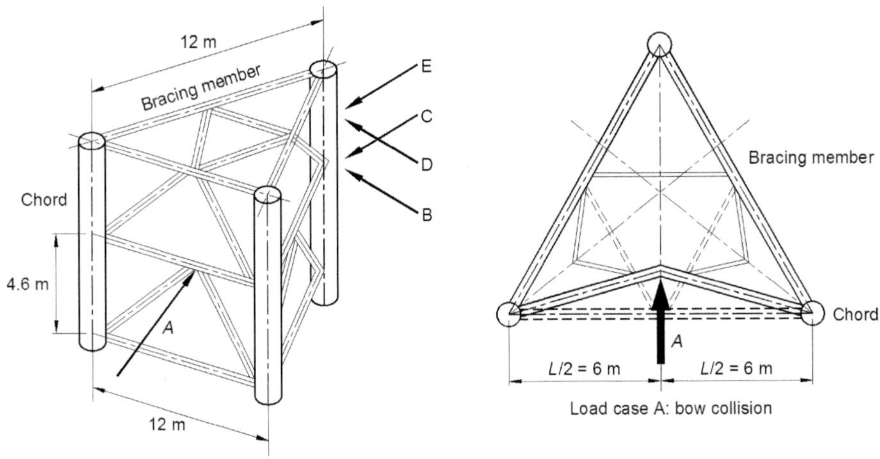

Fig. 3.80 Part of jack-up leg with five load cases and load case A: Bow collision.

and the energy absorbed in the horizontal brace by membrane tension:

$$E_D = \sigma_0 A \Delta L \tag{3.154}$$

where ΔL is the plastic elongation of the bracing:

$$\Delta L = \left(0.25L^2 + u_T^2\right)^{1/2} - 0.5L$$

where u_T is the plastic lateral displacement and D and t are the diameter and thickness of the brace, respectively. The cross-sectional area is $A = \pi Dt$. The lateral collision forces may be written as

$$P_c = \frac{8M_p}{L} + 2\sigma_0 A \sin u_T$$

or

$$P_c = \frac{8D^2 t\sigma_0}{L} + 2\sigma_0 A \frac{u_T}{\left(0.25L^2 + u_T{}^2\right)^{0.5}} \tag{3.155}$$

There is a practical limit to the rotation angle and the plastic deformation of a bracing before fracture occurs. The limit of rotation angle proposed by Visser (2004) is (based on the work by Marshall et al., 1977)

$$\theta_{\max} = 1250\left(\frac{t}{D}\right)^{2.75} \tag{3.156}$$

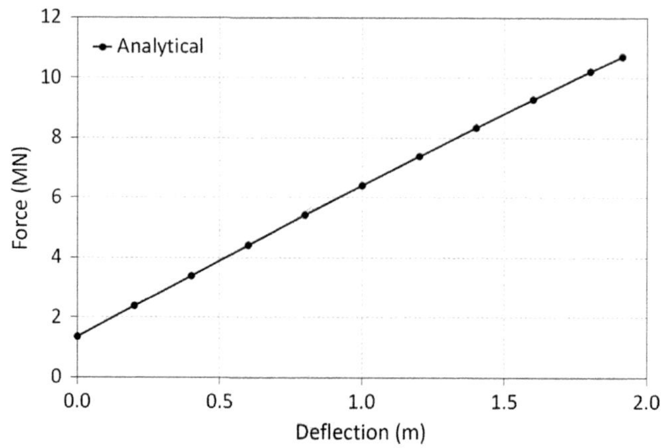

Fig. 3.81 Collision force–deflection curve of the bracing member in bending and membrane tension.

As an example, a horizontal brace has a diameter $D=410\,\text{mm}$, wall thickness $t=20\,\text{mm}$, cross-sectional area $0.0257\,\text{m}^2$, and length 12 m. The material flow stress of the bracing material is 600 MPa. Then, the limit of rotation angle is $\theta_{max}=1250\,(20/410)^{2.75}=0.309$. The maximum plastic elongation of the brace can be approximately determined from $\Delta L=L/\cos(\theta_{max})-L=12.596-12=0.596\,\text{m}$, that is, $\Delta L=5\%L$, and the maximum lateral deflection is $u_T=0.5L\tan(\theta_{max})=1.914\,\text{m}$. Therefore, the maximum energy absorption limit in load case A is (the collision force–deflection curve is also calculated and shown in Fig. 3.81):

$$\begin{aligned} E &= E_b + E_D \\ &= 4 \times 0.41^2 \times 0.02 \times 600 \times 0.309 + 600 \times 0.0257 \times 0.596 \\ &= 2.49 + 9.19 = 11.68\ \text{MJ} \end{aligned}$$

For load case A, the amount of absorbed energy (11.68 MJ) by the brace is larger than the collision energy for the bow collision of the 5000 tonnes supply vessel impacting with a rigid jack-up at 2 m/s ($0.5MV^2=10.5\,\text{MJ}$). Therefore, the bracing member has sufficient energy absorption capacity for the 5000 tonnes supply vessel.

With the large tension forces in the horizontal bracing, the adjacent chords are subjected to localized forces equal to the flow stress multiplied by the cross-sectional area of the bracings, that is, $P=\sigma_0 A=600 * 0.0257=15.45\,\text{MN}$; see Fig. 3.82. The effect of this load on the chords can be analysed by neglecting all other bracings and applying the load at the midpoint of the clamped-clamped chord of the double bay length using a simple three-hinge structural model. The chord cross section is also shown in Fig. 3.82. With the scantlings of the chord, diameter $D=1.0\,\text{m}$, wall thickness $t=0.053\,\text{m}$, total length $L=9.2\,\text{m}$, and flow stress of the material 700 MPa, it is found that the collapse load of the chord is

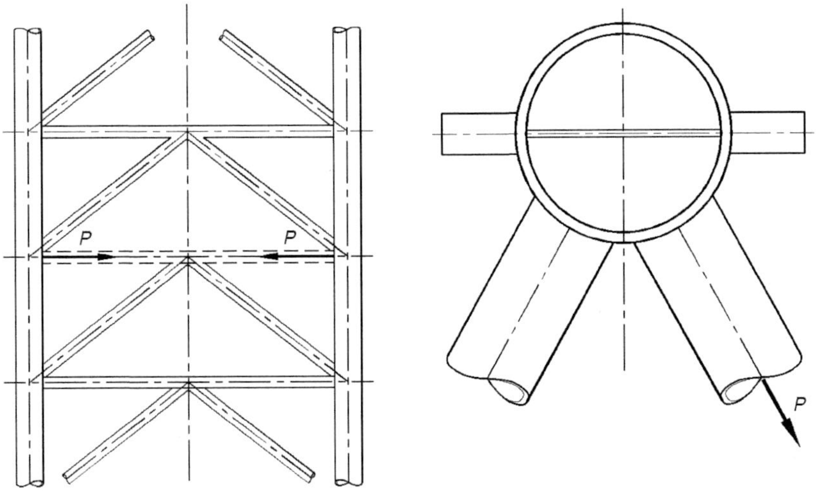

Fig. 3.82 Transverse loading on the chords in load case A.

$$P_c = \frac{8D^2 t\sigma_0}{L} = 32.26\,(\text{MN})$$

This load can be added without collapse ($P_c = 32.26\,\text{MN} > P = 15.45\,\text{MN}$).

3.10.4 Tubular structures: Local denting (ship drifting to chords)

For load cases B–E (ship drifting to chords), the local denting of the tubular chords is to be investigated. Local denting is the deformation concentrated around the contact point. Furnes and Amdahl (1980) and Amdahl (1983) were among the first to study the deformation behaviour of tubular members under lateral loading. Fig. 3.83 illustrates the assumed deformation pattern of a tubular chord impacted by a rigid shipside. In Amdahl (1983) and Amdahl et al. (2012), an expression for the denting resistance F_R that results in flattening of a circular tube with a diameter D and wall thickness t is given as

$$F_R = \sigma_0 \frac{t^2}{4}\sqrt{\frac{D}{t}} \cdot \left(22 + 1.2\frac{D_H}{D}\right) \cdot \left(\frac{w_d}{D}\right)^{\frac{1.925}{3.5 + \frac{D_H}{D}}} \cdot \left(\frac{4}{3}\left(1 - \frac{1}{4}\left(1 - \frac{N}{N_p}\right)^3\right)\right)^{1/2} \tag{3.157}$$

where w_d is the indentation and D_H is the contact height. The last term in the expression takes into account the effect of the axial force N in the tubular member (N_p is the plastic axial force), and this term was based on the work by Wierzbicki and Suh (1988). The expression is in accordance with the load curves given in NORSOK N-004 (2004).

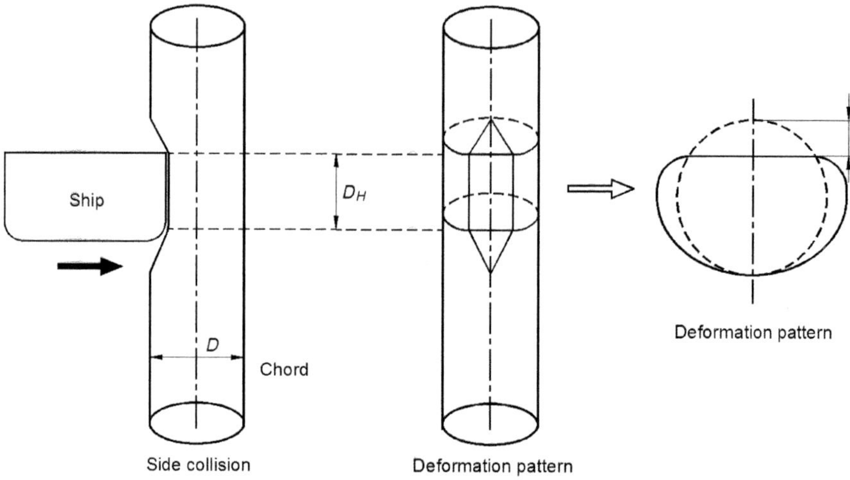

Fig. 3.83 Deformation of local denting of a tubular chord impacted by a drifting ship.

Yu and Amdahl (2018) followed the same analysis procedure as Wierzbicki and Suh (1988) on derivation of energy dissipation and extended a model to account for a rigid flat indenter with a contact height of D_H:

$$F_R = 16\sigma_0 \frac{t^2}{4}\sqrt{\frac{D}{t}} \cdot \left(\left(\frac{2\pi w_d}{3\ D} \right)^{1/2} \cdot \left(1 - \frac{1}{4}\left(1 - \frac{N}{N_p} \right)^3 \right)^{1/2} + \frac{D_H}{D}\sqrt{\frac{t}{D}} \right) \tag{3.158}$$

3.10.5 Deformation transition between local denting and local bending

For a tubular member, a local denting (a deformation around the contact area in the tube, as shown in Fig. 3.83) may occur first. With the increase of the local indentation, the plastic bending capacity (as a beam) of the tube decreases. At a certain indentation, a transition takes place where the brace/chord starts to deform with plastic bending at the end supports (three-point bending). With further deformation, the bending resistance remains constant, and axial membrane tension develops. Yu and Amdahl (2018) have reviewed the methods on deformation transition between local denting and local bending of tubular members. An illustration of the deformations of a chord hit by a ship bow is presented in Fig. 3.84.

With a local dent of w_d, the plastic bending capacity of the chord will be reduced, and it can be determined from (which is in accordance to NORSOK N-004, 2004)

$$M_{dent} = \left(\cos\frac{\theta}{2} - \frac{1}{2}\sin\theta \right) M_p$$

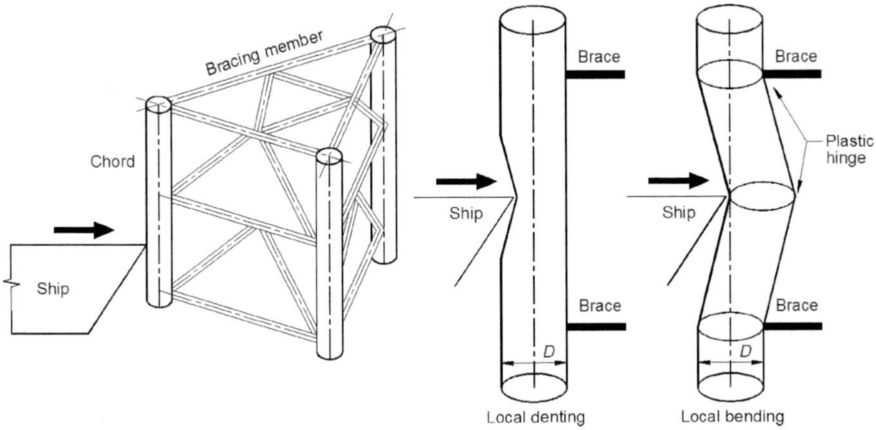

Fig. 3.84 Illustration of the deformation patterns of local denting and bending of a tubular chord hit by a ship bow.

where $\theta = \arccos(1 - 2w_d/D)$ and $M_p = D^2 t\sigma_0$ are the plastic bending moment without any dent. For a chord/brace with both clamped ends and impact at the midspan of the member length L, the residual bending resistance is determined from

$$R_d = \frac{4M_p}{L}\left(1 + \left(\cos\frac{\theta}{2} - \frac{1}{2}\sin\theta\right)\right) = \frac{4D^2 t\sigma_0}{L}\left(1 + \left(\cos\frac{\theta}{2} - \frac{1}{2}\sin\theta\right)\right) \quad (3.159)$$

At the transition point, the bending resistance is equal to the denting resistance, that is,

$$\frac{4D^2 t\sigma_0}{L}\left(1 + \left(\cos\frac{\theta}{2} - \frac{1}{2}\sin\theta\right)\right) = \sigma_0\frac{t^2}{4}\sqrt{\frac{D}{t}}\cdot\left(22 + 1.2\frac{D_H}{D}\right)\cdot\left(\frac{w_d}{D}\right)^{3.5 + \frac{1.925}{\frac{D_H}{D}}} \quad (3.160)$$

Solving this equation numerically, we can find the critical dent displacement w_{dc}. To get an accurate result, small increment steps for displacements may be needed in solving the equation. The final resistance and displacement curve can be determined from

$$F_R = \begin{cases} \sigma_0\frac{t^2}{4}\sqrt{\frac{D}{t}}\cdot\left(22 + 1.2\frac{D_H}{D}\right)\cdot\left(\frac{w_d}{D}\right)^{3.5 + \frac{1.925}{\frac{D_H}{D}}}, & \text{for } w_d \le w_{dc} \\ \frac{4D^2 t\sigma_0}{L}\left(1 + \left(\cos\frac{\theta_{dc}}{2} - \frac{1}{2}\sin\theta_{dc}\right)\right) + 2\sigma_0 A\frac{(w_d - w_{dc})}{\left(0.25L^2 + (w_d - w_{dc})^2\right)^{0.5}}, & \\ \text{for } w_d > w_{dc} & \end{cases} \quad (3.161)$$

where $\theta_{dc} = \arccos(1 - 2w_{dc}/D)$ and it is assumed that the bending resistance and denting resistance remain constant after reaching the critical transition w_{dc} and the axial membrane tension force develop.

3.10.6 Calculation example: Deformation transition between local denting and local bending of tubulars

The examples used here are taken from Yu and Amdahl (2018) where a tubular member is impacted by a rigid flat indenter. The first one is a brace of length $L = 20$ m, diameter $D = 1.5$ m, wall thicknesses $t = 30$ mm, and material yield stress 285 MPa. The height of the rigid flat indenter is $D_H = 0.6$ m.

The calculation results show that the critical transition indentation is found to be $w_{dc} = 0.425$ m or $w_{dc}/D = 0.283$. The resistance and indentation cures are calculated using the method in Section 3.10.5 using Eq. (3.161), and the results are presented in Fig. 3.85. The resistance is normalized to a characteristic resistance F_C, which is determined from $F_C = 0.25\sigma_0 t^2 (D/t)^{0.5}$. The finite element analysis results by Yu and Amdahl (2018) with a rigid flat indenter and a ship actual stern (both have a height of $D_H = 0.6$ m) are also presented in Fig. 3.85. It is seen that the analytical resistance–indentation curve agrees reasonably well with the FEA results.

The second example of a brace has the same dimensions as the first one but with a larger contact height of $D_H = 4.92$ m for the rigid indenter. Analysis results show that the critical transition indentation is found to be $w_{dc} = 0.165$ m or $w_{dc}/D = 0.11$. The resistance and indentation curves are calculated using the method in Section 3.10.5 using Eq. (3.161), and the results are presented in Fig. 3.86. FEA results by Yu and Amdahl (2018) with a rigid flat indenter are also presented in Fig. 3.86. It is observed that the analytical resistance–indentation curve has a reasonable agreement with the FEA results.

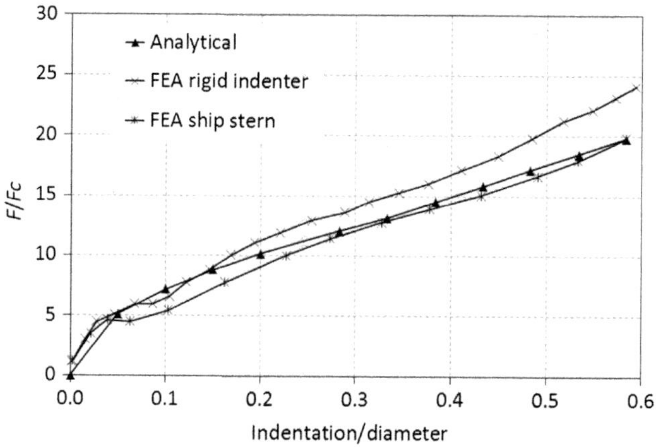

Fig. 3.85 The resistance–indentation curve comparison between FEA and analytical Eq. (3.161) (tube dimensions: $L = 20$ m, $D = 1.5$ m, and $t = 30$ mm; contact height: $D_H = 0.6$ m).

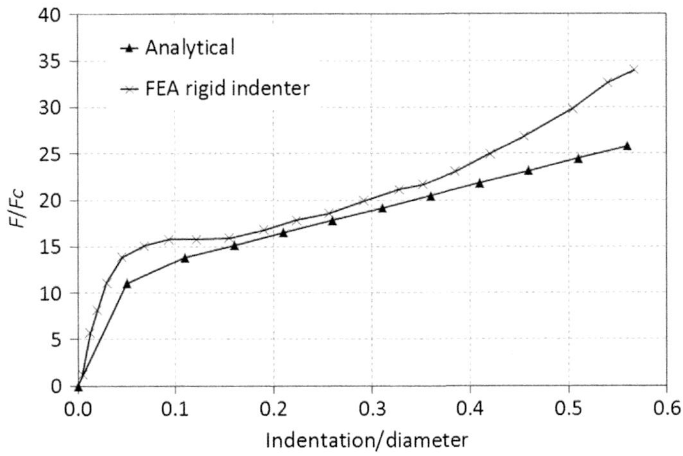

Fig. 3.86 The resistance–indentation curve comparison between FEA and analytical Eq. (3.161) (tube dimensions: $L=20\,\text{m}$, $D=1.5\,\text{m}$, and $t=30\,\text{mm}$; contact height: $D_H=4.92\,\text{m}$).

3.10.7 Calculation example: Collision between a deformable ship and a deformable tubular chord

This section provides an example of collision analyses between a ship and a platform where both structures are considered as deformable. The collision scenario can be represented by Fig. 3.78 where a 5000 tonnes supply vessel is drifting sideways with 2.0 m/s against a platform and collides with the chord of the platform. Thus, the maximum kinetic energy for collisions can be approximated as $E=0.5MV^2=14.5\,\text{MJ}$.

The task here is to determine how much energy is absorbed by the ship and by the platform chord and what are the damages (indentations) to the ship and the chord. The required data for the calculations are the following:

Supply vessel: the ship depth is 7.0 m. The effective thickness of the side shell taking into account the stiffeners is 12.9 mm. The thickness of the deck taking into account the stiffeners is 14.0 mm, and thicknesses of the inner bottom and outer bottom taking into account the stiffeners are 11.9 and 17.7 mm, respectively. The transverse frame spacing is 3.0 m, and the flow stress for all materials is taken as 422.5 MPa.

Platform chord: the diameter of the platform leg (the chord) is 1.0 m, and the wall thickness is 53.0 mm. The flow stress of the material is 700 MPa.

The impact point is at the midspan of two transverse frames of the ship. It is also assumed that the side shell plate of the ship will rupture at an indentation of 1.0 m.

The first calculation is to determine the force–displacement curve of the shipside indented by the chord (assume the chord is rigid in this case) using the method presented in Section 3.10.2 'Rigid cylinder impact to side shell plating of a ship' (Eqs 3.151, 3.152). The second calculation is to determine the force–displacement curve of the chord dented by the shipside (assume the ship is rigid in this case) using the method described in Section 3.10.4 'Tubular structures: local denting (ship drifting

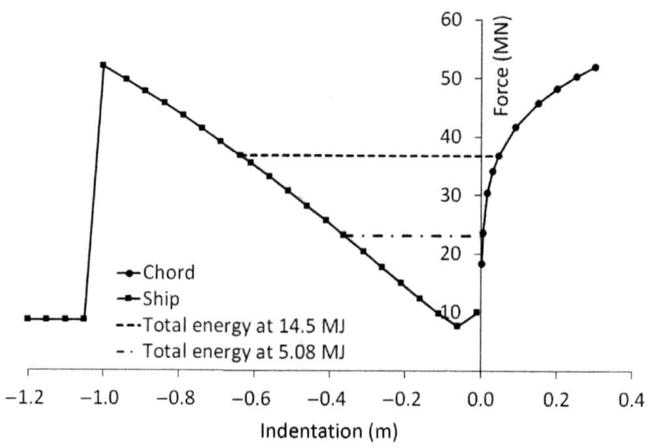

Fig. 3.87 Deformation of local denting of a tubular chord impacted by a drifting supply vessel (the ship is deformable).

to chords)' and ignoring the stress due to axial loads on the chord in Eq. (3.157). The calculated results of the two cases are presented in Fig. 3.87.

From the calculated result curves, it can be found that at the total absorbed energy of 14.5 MJ, the ship absorbed about 91% and the chord dissipated about 9% of the energy. The damage indentation to the shipside is 638 mm, and the dent depth into the chord is 45 mm. The collision force is 36.75 MN.

As shown in Chapter 2, Section 2.3.5 'Calculation example for fixed platforms', the available collision energy for crushing is only 5.08 MJ if the platform is considered flexible such that a large amount of energy is stored in an elastic/kinetic form in the platform and the ship. Therefore, the actual damages to the side of the 5000 tonnes supply vessel and to the chord of the jack-up rig will be much smaller. For this case, we obtain that the energy absorbed by the supply vessel is 5.01 MJ and the energy absorbed by the chord is only 0.06 MJ. The indentation to the shipside is 363 mm, and a dented depth into the chord is about 4 mm. The collision force is around 23.1 MN.

This example also shows that it is important and necessary to consider the flexibility of a platform when determining the collision energy for crushing since it has a significant effect on the collision damage analysis of the ship and platform.

3.11 Ship grounding on rocks

3.11.1 Introduction

In Chapter 2, Section 2.6, analyses of ship grounding on flat sea beds were presented. When a ship runs onto a rock, as illustrated in Fig. 3.88 (Zhu et al., 2002), the ship bottom can be deformed upwards and torn open. If the ship speed is high enough, the damage length could be up to the ship length. The deformation and damage

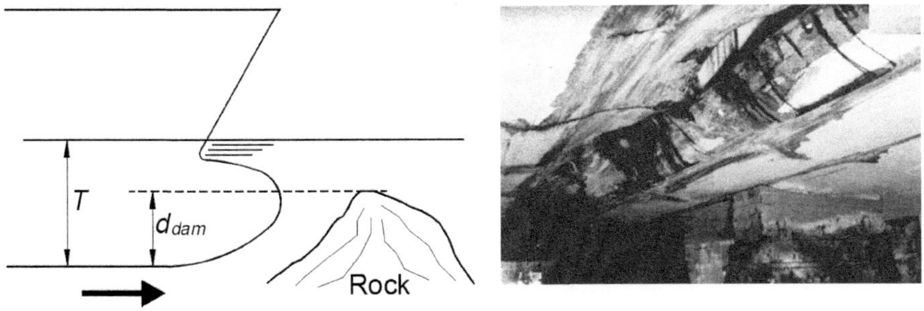

Fig. 3.88 Illustration of a ship grounding scenario and an example of grounding damage to a ship bottom.
Photo cf. Zhu, L., James, P., Zhang, S., 2002. Statistics and damage assessment of ship grounding, Mar. Struct. 15, 515–530.

patterns involve the bottom plating tearing and cutting, bottom floor membrane tension and rupture, and bottom longitudinal stiffener/girder twist and tension deformation. The actual damage could be even more complicated. However, idealizations of the grounding damage patterns must be made in order to establish theoretical analysis models.

There are two commonly used analytical analysis models. The first is that a rock is simplified as a rigid wedge, and the second is that a rock is modelled as a rigid cone. Due to the complexity of the problem, empirical methods are often proposed, as those discussed in the following sections.

3.11.2 Mechanics of plate cutting by a wedge

Due to the importance of the problem, many authors have investigated the process of plate cutting by experimental tests and theoretical analyses. Some studies have employed experimental approaches to develop empirical formulae, for example, Vaughan (1980), Jones and Jouri (1987), Lu and Calladine (1990), Paik (1994), and Lee and Hong (1997). These comprehensive experimental studies have helped to better understand the problem of plate cutting mechanics and to derive useful empirical formulae. Theoretical methods for understanding the energy absorption and failure mechanism are also important, but due to the complexity of the problem, these procedures are mainly based on constructed deformation patterns observed from experiments.

A pioneering theoretical analysis of the plate cutting problem was carried out by Wierzbicki and Thomas (1993). They carefully constructed a damage model and derived a closed-form solution to the problem. Crack extension at the front of the wedge tip was assumed in that model; for example, there is a gap between the wedge tip and the crack tip. Based on experimental observations, Ohtsubo and Wang (1995) argued that there is no evidence of crack extending into the plate ahead of the wedge tip. Then, using a similar model to that of Wierzbicki and Thomas (1993) for the curved surfaces and assuming a velocity field close to the wedge tip area, Ohtsubo

Fig. 3.89 Analysis model of a plate cutting by a rigid wedge.

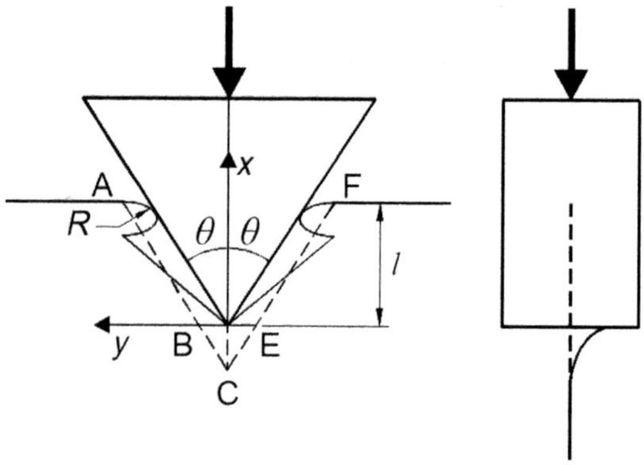

and Wang (1995) derived an alternative formula for analysing plate cutting. Later, Zheng and Wierzbicki (1996) and Simonsen and Wierzbicki (1997) derived formulae for a steady-state cutting of a plate.

In this section, the same assumption used in the method of Wierzbicki and Thomas (1993) for a separated curved surface is employed, in which the material rolls up into two curved surfaces behind the wedge tip. Also, the material in front of the wedge tip is stretched and ruptured due to the direct pushing of the wedge into the plate. By constructing a plastic deformation pattern in front of the wedge tip, analytical formulae for the plate cutting force and the absorbed energy are obtained where the critical rupture strain of the material enters in the solution.

In the present analysis model, we assume that the plate ruptures ahead of the wedge tip due to tension from the pushed wedge and that no crack extension exists. The analysis model is shown in Fig. 3.89. The hypothesis is that plastic deformation takes place inside of the line *ABCEF*, while outside this area, no plastic deformation occurs. The deformed area can be divided into two parts by line *BE*. The first part lies in front of the wedge tip, and it undergoes membrane stretching due to the wedge tip pushing. As the wedge advances, the material on line *BE* ruptures and separates to form two curved surfaces behind the wedge tip. These curved surfaces constitute the second part of the deformed plate that suffers bending deformation only. To allow for rupture failure of the tension zone in front of the wedge tip, the material property 'critical rupture strain' is introduced.

3.11.2.1 Bending energy rate

Wierzbicki and Thomas (1993) developed a method for calculating the rate of plastic bending energy dissipated in the two curved surfaces. It is assumed that all plastic bending is concentrated in the two moving hinge lines, *AB* and *EF*, and that no other plastic energy is dissipated in the continuous area. The length of the hinge line is $l_{AB} = l/\cos\theta$, where l is the cut length. The rate of plastic bending energy can be determined from

$$\dot{E}_b = 2M_0 \frac{V \sin\theta}{R} l_{AB} = \frac{1}{\sqrt{3}} \sigma_0 t^2 \frac{l}{R} V \tan\theta \tag{3.162}$$

where t is the plate thickness, V is the advancing speed of the wedge, and R is the radius of the curved surface that is to be determined.

3.11.2.2 Membrane energy rate

It is assumed that only membrane energy is dissipated in the area close to the wedge tip. It is postulated that, with direct pushing of the wedge tip, the plate stretches in the transverse direction y (see Fig. 3.89). The rate of membrane energy can be expressed as

$$\dot{E}_m = \iint_S \frac{2}{\sqrt{3}} \sigma_0 t \dot{\varepsilon}_{yy} dx dy \tag{3.163}$$

where S is the area of the tension field and $\dot{\varepsilon}_{yy}$ is the tension strain rate in the transverse direction y.

The transverse strain rate $\dot{\varepsilon}_{yy}$ can be further expressed as

$$\dot{\varepsilon}_{yy} = \frac{d\varepsilon_{yy}}{dx}\frac{dx}{dt} = V \frac{d\varepsilon_{yy}}{dx} \tag{3.164}$$

Substituting Eq. (3.164) into Eq. (3.163) and integrating with respect to x, it is obtained

$$\dot{E}_m = \frac{2}{\sqrt{3}} \sigma_0 t V \int_B^E \left[\varepsilon_{yy}(BE) - \varepsilon_{yy}(\infty)\right] dy = \frac{2}{\sqrt{3}} \sigma_0 t V \int_B^E \left[\varepsilon_{yy}(BE)\right] dy \tag{3.165}$$

Line BE is located in the transition area between the wedge tip near-field and the far-field curved surface behind of the wedge tip. The out-of-plane deflection of material fibres on line BE may be approximated as a linear function $w = R(1 - y/b)$, where b is a half-length of the line BE. The strain on the line BE is determined from $\varepsilon_{yy} = \frac{1}{2}\left(\frac{\partial w}{\partial y}\right)^2 = \frac{1}{2}\left(\frac{R}{b}\right)^2$. The material fibres on line BE is in the critical rupture state, and therefore, its strain reaches the material rupture strain ε_f. Therefore, $\varepsilon_{yy} = \frac{1}{2}\left(\frac{R}{b}\right)^2 = \varepsilon_f$, and $b = \frac{R}{\sqrt{2\varepsilon_f}}$. Thus, the membrane energy rate can be expressed as

$$\dot{E}_m = \frac{2}{\sqrt{3}} \sigma_0 t V \int_B^E \varepsilon_f dy = \frac{4}{\sqrt{6}} \sigma_0 t V {\varepsilon_f}^{0.5} R \tag{3.166}$$

3.11.2.3 Plastic deformation resistance

The equilibrium of the wedge-plate system can be expressed via the principle of virtual work:

$$F_p V = \dot{E}_b + \dot{E}_m$$

where F_p is the plastic deformation resistance. After substitutions, the resistance obtained as:

$$F_p = \frac{1}{\sqrt{3}} \sigma_0 t^2 \frac{l}{R} \tan\theta + \frac{4}{\sqrt{6}} \sigma_0 t \varepsilon_f^{0.5} R$$

Minimizing the plastic deformation resistance with respect to the rolling radius R, it is found that

$$\frac{\partial F_p}{\partial R} = 0 \Rightarrow : R = 0.5946 t^{0.5} l^{0.5} \varepsilon_f^{-0.25} (\tan\theta)^{0.5} \tag{3.167}$$

Thus, the expression for the minimized plastic deformation force is

$$\begin{aligned} F_p &= (0.971 + 0.971) \sigma_0 t^{1.5} l^{0.5} \varepsilon_f^{0.25} (\tan\theta)^{0.5} \\ &= 1.942 \sigma_0 t^{1.5} l^{0.5} \varepsilon_f^{0.25} (\tan\theta)^{0.5} \end{aligned} \tag{3.168}$$

This equation shows that the plastic deformation resistance depends on the properties of the material (σ_0 and ε_f) and on the geometric parameters (t, l, and θ). It is also seen that the contributions of the bending and the membrane to the plastic resistance force are equal.

3.11.2.4 Contribution of friction

Plate cutting experiments (Lu and Calladine, 1990) show that the contribution of friction between the wedge and the plate is important in the cutting process. For a mild steel plate, Lu and Calladine (1990) estimated the contribution of friction by measuring the ratio of the withdrawal force to the penetration force and found the mean value to be 0.4. Thick plate cutting experiments performed by Astrup (1994) showed that there were heavy scores on the wedge surface. Therefore, the friction force should be included in the plate cutting analysis.

Frictional resistance results from the relative motion on the contact surface between the wedge and the plate. The normal force on the surface of the plate is $F_n = F_p/(2\sin\theta)$. Using Coulomb's friction law and projecting the force in the direction of the advancing wedge, the friction contribution is obtained as

$$F_f = 2\mu F_n \cos\theta = F_p \frac{\mu}{\tan\theta} \tag{3.169}$$

3.11.2.5 Total cutting force

Summing the plastic deformation resistance, F_p, and the friction force, F_f, the final expression for the total cutting force is expressed as

$$F = 1.942\sigma_0 t^{1.5} l^{0.5} \varepsilon_f^{0.25} (\tan\theta)^{0.5} \cdot \left(1 + \frac{\mu}{\tan\theta}\right) \tag{3.170}$$

The total energy dissipated by plastic deformation and friction can thus be calculated from

$$E = \int_0^l F dl = 1.295\sigma_0 t^{1.5} l^{1.5} \varepsilon_f^{0.25} (\tan\theta)^{0.5} \cdot \left(1 + \frac{\mu}{\tan\theta}\right) \tag{3.171}$$

3.11.2.6 Summary of some formulations for plate cutting by a wedge

Some of the formulae developed by various investigators, both analytical and empirical, for the cutting force analyses are summarized below:

(1) Lu and Calladine (1990)—Empirical

Lu and Calladine (1990) carried out a series of quasistatic cutting experiments for mild steel plates. Based on experimental results, the following empirical formula was derived:

$$F = 1.3C_{1.3}\sigma_0 t^{1.7} l^{0.3} \tag{3.172}$$

$C_{1.3}$ is an empirical constant that for a mild steel material is $C_{1.3} = 2.3$.

(2) Wierzbicki and Thomas (1993)—Analytical

Using analytical approaches, Wierzbicki and Thomas (1993) developed the following expression:

$$F = 1.67\sigma_0 t^{1.6} l^{0.4} \left(\bar{\delta}_t\right)^{0.2} \frac{1}{(\cos\theta)^{0.8}} \cdot \left((\tan\theta)^{0.4} + \frac{\mu}{(\tan\theta)^{0.6}}\right) \tag{3.173}$$

$\bar{\delta}_t$ is the COD parameter and $\bar{\delta}_t$ is normally used as suggested by Wierzbicki and Thomas.

(3) Paik (1994)—Empirical

Paik (1994) carried out plate cutting experiments for unstiffened plates and stiffened plates. Through statistical analysis of the experimental results, an empirical formula was derived below:

$$F = 1.5C_{1.5}\sigma_0 t^{1.5} l^{0.5} \tag{3.174}$$

$C_{1.5} = 3.760\theta^2 - 1.156\theta + 1.112$ (angle θ is in rad).

(4) Ohtsubo and Wang (1995)—Analytical

Similarly, the analytical analysis expression derived by Ohtsubo and Wang (1995) is expressed as

$$F = 1.51\sigma_0 t^{1.5} l^{0.5} (\sin\theta)^{0.5} \cdot \left(1 + \frac{\mu}{\tan\theta}\right) \tag{3.175}$$

(5) Zhang (2002)—Analytical

The analytical formula developed by Zhang (2002), which takes into account material fracture strain, is expressed as

$$F = 1.942\sigma_0 t^{1.5} l^{0.5} \varepsilon_f^{0.25} (\tan\theta)^{0.5} \cdot \left(1 + \frac{\mu}{\tan\theta}\right) \tag{3.176}$$

3.11.2.7 Steady-state of plate cutting

The methods described above only apply for tearing and cutting of a plate for the initial cutting stage, and the cutting process does not reach a steady state. This initial stage is relatively short in a high-energy grounding incident, and thus, it is more relevant to obtain the steady-state cutting force.

It is possible to use the formulae for the initial cutting force to estimate the steady-state cutting force. When the wedge has a limited width and after the wedge shoulder comes into the plate, the cutting opening (or width) becomes constant. However, experimental results show that the cutting forces may still increase with the increase of cutting length at a certain distance. Simonsen (1997a) investigated the physical behaviours and found that when the cutting length is up to one to two times the wedge length (see Fig. 3.90), the cutting load remains constant. The wedge length can be determined from $l_w = 0.5B_d / \tan\theta$.

With these observations from experiments, we assume that when the cutting length is equal to two times the wedge length (cutting length $l = B_d / \tan\theta$), the steady-state

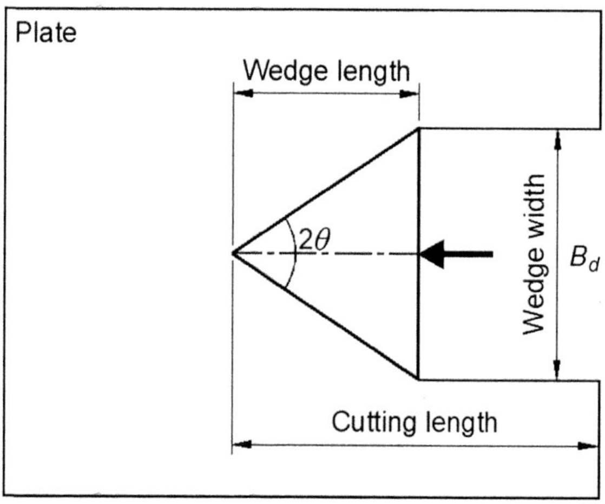

Fig. 3.90 Wedge length definitions in the plate cutting analysis.

cutting will be reached. Then, Eq. (3.176) can be further expressed as below for the steady-state cutting force estimation:

$$F = 1.942\sigma_0 t^{1.5} B_d^{0.5} \varepsilon_f^{0.25} \cdot \left(1 + \frac{\mu}{\tan\theta}\right) \tag{3.177}$$

Zheng and Wierzbicki (1996) developed an analytical method for the cutting force of a plate cut by a rigid wedge after the steady-state process has been reached. Their steady-state cutting force is determined from

$$F = 0.25\sigma_0 t^2 \left(2\frac{B_d + R}{R} + 1.27\frac{R}{t}\cos\theta + 1.28\theta^2 \frac{\cos(\theta/2)(B+R)^2}{\cos\theta \cdot Rt}\right) \cdot \left(1 + \frac{\mu}{\tan\theta}\right) \tag{3.178}$$

where the rolling radius R is calculated from

$$R = B_d \sqrt{\frac{2(t/B) + 1.28\theta^2 \cos(\theta/2)/\cos\theta}{1.27\cos\theta + 1.28\theta^2 \cos(\theta/2)/\cos\theta}}$$

3.11.3 Bottom plate tearing by a conical rock

Simonsen (1997b) studied the plate tearing induced by a conical rock and developed a set of analytical expressions for calculating the reaction force of a ship bottom deformed by a conical rock. Similarly, analytical solutions were also developed by Zeng et al. (2016) for ship grounding to a rock. The solutions derived by Zeng et al. (2016) need further numerical elaborations to find the minimum energy dissipation. The details of the methods can be found in Simonsen (1997b) and Zeng et al. (2016).

3.11.4 Estimation of the horizontal grounding force in ship grounding

When a ship runs into a rock with a high kinetic energy, the rock will indent the ship bottom vertically and eventually rupture the bottom plating. At the same time, the rock will also cut through the ship bottom in the horizontal (or longitudinal) direction. With the ship advancing forward, the damage to the bottom structure will extend along the ship until all the kinetic energy of the ship is dissipated. The horizontal force can be presented as a constant with periodic peaks at those instances where the rock touches the transverse members of the bottom. The damage to the bottom in a real ship grounding incident is much more complex than that described above for bare plate cutting and may consist of mixed damage modes of tearing, tension, crushing, and folding. The damage width (or rock width) may be several times the stiffener spacing. It is difficult to construct a single theoretical model to capture a complex damage of this nature, and so, a semiempirical method is an alternative approach. In a grounding accident, it is

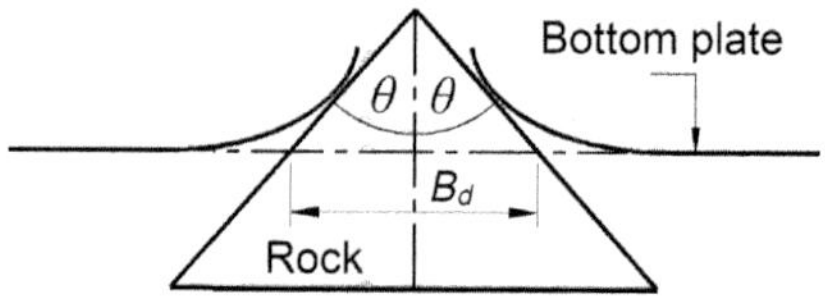

Fig. 3.91 Analysis model of a ship bottom plate tearing by a conical rock.

very difficult, even impossible, to know the exact shape of a rock. Through studies, it was found that the damage width (or rock width) is an important factor and that it has a significant effect on the steady-state cutting force. The effects from rock shape, rock angle, and friction can be embedded in an empirical factor. A semiempirical formula for estimating the average horizontal grounding force was proposed by Zhang (2002):

$$F_H = 3.58\sigma_0 t_{eq} t_e^{0.6} B_d^{0.4} \tag{3.179}$$

where σ_0 is the flow stress of the bottom material; t_e is the equivalent thickness of the bottom plating including longitudinal members (e.g. smearing of the longitudinal stiffeners to the shell plating $t_e = t_0 + A_L/s_L$, where $t_0 =$ plating thickness, $A_L =$ stiffener cross-sectional area, and $s_L =$ stiffener spacing); t_{eq} is the equivalent thickness of the bottom plating including longitudinal and transverse members ($t_{eq} = t_e + A_T/s_T$, $A_T =$cross-sectional area of the floor and $s_T =$floor spacing); and B_d is the damage width of the bottom, which is taken as the rock width at the level of the bottom plating (see Fig. 3.91). For double-bottom arrangement, this method applies to outer bottom and inner bottom separately.

Another empirical formulation for estimating the horizontal grounding force was developed by Simonsen et al. (2009) by fitting a series of FE simulation results of different ships that run a ground. The expression is

$$F_H = 0.77\sigma_0 t_{eq}^{1.17} B_d^{0.83} \varepsilon_f^{0.71} \tag{3.180}$$

where σ_0 is the flow stress of the bottom material and t_{eq} is the equivalent thickness of the bottom plating including longitudinal members and transverse members. The smearing of longitudinal and transverse stiffeners to the shell plating is here $t_{eq} = t_0 + A_L/s_L + 0.1Tt_t/s_T$, where t_0 is the plating thickness; A_L is the stiffener cross-sectional area; s_L is the stiffener spacing; $A_T = 0.1Tt_t$ is the assumed cross-sectional area of the floor, where T is the ship draught and s_T is floor spacing; and B_d is the damage width of the bottom. ε_f is the fracture strain of materials. It should be noted that $\varepsilon_f = 0.41$ was used in the FE numerical simulations for mild steels (Simonsen et al., 2009).

3.11.5 Horizontal grounding force for oil tankers of different sizes

Quantifying a ship's resistance to grounding impacts in terms of its principal particulars is a subject of much interest (Bockenhauer, 1996). Based on the foregoing studies, the following simple expressions for estimating damage resistance with a damage extent of oil tankers grounded on a rock are proposed. These formulae are considered valid for large oil tankers of 190 m and above in high-energy grounding scenario since the derivations are based on the analyses of this type of oil tankers.

The average horizontal grounding resistance for a **single-hull oil tanker** is expressed as

$$F_{SH} = 58\left(\frac{B_d}{B}\right)^{0.4} \cdot \left(\frac{L}{300}\right)^{2.04} \text{(MN)} \tag{3.181}$$

where F_{SH} is the average horizontal grounding force in MN, B_d is the rock width (or damage width) in m, B is the ship breadth in m, and L is the ship length in m.

For double-hull oil tankers, the double-bottom resistance in grounding damage is lower than the single-bottom arrangement if only the outer bottom is breached. A translation of Eq. (3.181) to double-hull structures is based on the following assumptions with respect to geometry: the double-bottom height is $0.1D$, and the ship breadth is $B=2D$, where D is the ship depth. The equivalent thickness (i.e. plating plus longitudinal members) of the deck, side, longitudinal bulkheads, and bottom are identical. The single-hull tanker and the double-hull tanker have the same principal dimensions and the same block coefficient. Under these assumptions, the midship section modulus at the deck for the single-hull tanker is $Z_{SH}=2.667D^2t_{SH}$, where t_{SH} is the equivalent thickness of the deck. Similarly, the midship section modulus at the deck for the double-hull tanker is $Z_{DH}=3.116D^2t_{DH}$, where t_{DH} is the equivalent thickness of the deck. Using the Rule requirement for the minimum midship section modulus and equating Z_{DH} to Z_{SH}, a relationship of the equivalent thickness for the single-hull tanker and the double-hull tanker is obtained as $t_{SH}=1.168t_{DH}$. The ratio of grounding forces for the two tankers is determined from $F_{SH}/F_{DH}=(t_{SH}/t_{DH})^{1.6}=1.283$ for the case where only the outer bottom is breached. This is based on VLCCs with two inner longitudinal bulkheads. For oil tankers with one central longitudinal bulkhead or without inner longitudinal bulkheads, the results obtained are similar. Thus, an approximation for the average horizontal grounding force of a **double-hull oil tanker**, when only the outer bottom is breached, is

$$F_{DH} = \frac{F_{SH}}{1.3} = 44.6\left(\frac{B_d}{B}\right)^{0.4} \cdot \left(\frac{L}{300}\right)^{2.04} \text{(MN)} \tag{3.182}$$

3.11.6 Calculation examples and comparisons with experimental results

3.11.6.1 Thin plate cutting experiment by Lu and Calladine (1990)

Lu and Calladine (1990) carried out a series of model tests of plate cutting. An example is selected here for comparison with the analysis of a mild steel plate cut by a rigid wedge with semiangle of 10 degrees and a width of 10 mm. The thickness of the plate

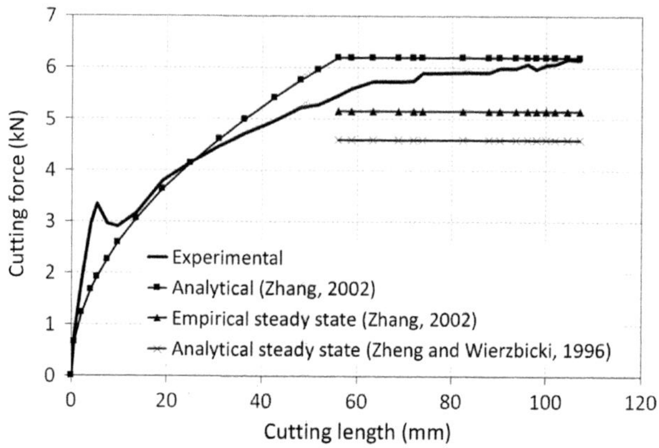

Fig. 3.92 Thin plate cutting force comparison between analytical and experimental (Lu and Calladine, 1990) results (mild steel plate thickness $t=1.6$ mm). (Notes on symbols: analytical (Zhang, 2002) using Eq. 3.176 for initial cut and Eq. 3.177 for steady-state cut, empirical steady state (Zhang, 2002) using Eq. 3.179, and analytical steady state (Zheng and Wierzbicki, 1996) using Eq. 3.178.)

is 1.6 mm, the flow stress of the material is 270 N/mm^2, the critical rupture strain of the material is taken as 0.35, and the coefficient of friction is taken as 0.25.

When the cutting length is equal to twice the wedge length (i.e. the cutting length is 10/tan 10 degrees = 56.8 mm), the steady-state cutting is reached. The cutting forces are calculated using Eq. (3.176) for initial cutting and Eq. (3.177) for steady cutting. The comparison between the analytical calculations and the experimental results for the cutting forces is presented in Fig. 3.92. It is seen from the comparison that the agreement is good.

The steady-state cutting force is also calculated using the analytical method by Eq. (3.178) and the empirical method given by Eq. (3.179), and the results are also presented in Fig. 3.92. These results are lower than the experimental result. It should be noted that the result using Eq. (3.180) gives only 1.16 kN, which seems to be too low when compared with the experimental results.

3.11.6.2 Thick plate cutting experiment by Astrup (1994)

Astrup (1994) carried out cutting tests of thick plates. The thickness of the tested plate is in the range of 15–25 mm, which are typical bottom thicknesses of large oil carriers. The experimental results for the 15 mm thickness plates, Tests P1-15 and P2-15 in Astrup (1994), are selected for comparison with the analytical analysis results. The semiangle of the rigid wedge is 30 degrees and the wedge width is 250 mm. The material properties obtained from uniaxial tensile tests are as follows: flow stress of the material (417+544)/2 = 480.5 MPa and the rupture strain 0.3. As Astrup pointed

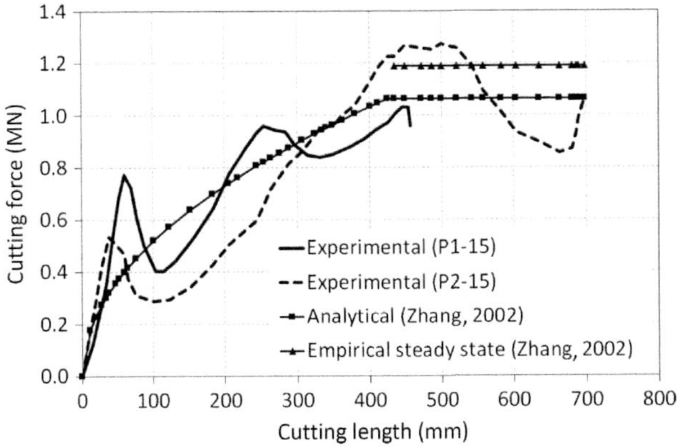

Fig. 3.93 Thick plate cutting force comparison between the analytical and the experimental (Astrup, 1994) results (steel plate thickness $t = 15$ mm). (Notes on symbols: analytical (Zhang, 2002) using Eq. 3.176 for initial cut and Eq. 3.177 for steady-state cut; empirical steady state (Zhang, 2002) using Eq. 3.179.)

out, there was a large friction force between the wedge and the thick plate in the cutting experiments, and Astrup believed that the friction in his experiment was larger than that in Lu and Calladine (1990) experiments. Therefore, here, it is assumed that the friction coefficient is 0.4.

As per the above example, when the cutting length is equal to twice the wedge length (cutting length is 250/tan 30 degrees = 433.3 mm), the steady-state cutting is reached. Therefore, the cutting forces are calculated using Eqs (3.176), (3.177). The comparisons with the experimental results are shown in Fig. 3.93. It is seen from the comparison that the agreement is also good. The steady-state cutting force calculated with the empirical method of Eq. (3.179) is also presented in Fig. 3.93. The resulting force is slightly higher than the experimental result. It should also be noted that by using Eq. (3.180), the force (only 0.37 MN) is too low when compared with the experimental result, while by using Eq. (3.178), the force (2.02 MN) is too high. It is interesting to note that for the two experimental tests, the cutting forces are not exactly the same, even though the plates have the same thickness.

3.11.6.3 Thick stiffened plate cutting experiment by Astrup (1994)

Astrup (1994) also carried out cutting tests of stiffened thick plates. The thickness of the selected plate is 20 mm (Tests S1-20 and S2-20 in Astrup, 1994), and it is stiffened by two longitudinal stiffeners (T $500 \times 12 + 200 \times 15$) spaced at 800 mm. The semi-angle of the rigid wedge is 30 degrees, the width is 250 mm, and it cuts the plate at the centre between the two stiffeners. The material properties obtained from uniaxial tensile tests are as follows: flow stress of the material$(393 + 526)/2 = 459.3$ N/mm^2 and rupture strain 0.317. The friction coefficient is set at 0.4, as the previous example.

It is also assumed that when the cutting length is equal to twice the wedge length (i.e. cutting length is 250/tan 30 degrees = 433.3 mm), the steady-state cutting is

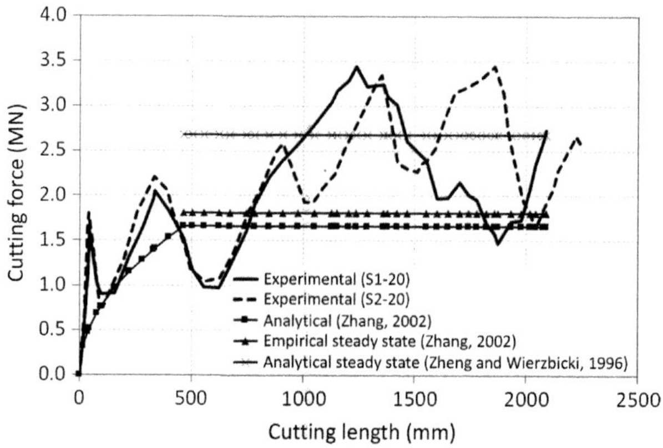

Fig. 3.94 Stiffened thick plate cutting force comparison between the analytical and the experimental (Astrup, 1994) results (steel plate thickness $t=20\,\text{mm}$). (Notes on symbols: analytical (Zhang, 2002) using Eq. 3.176 for initial cut and Eq. 3.177 for steady-state cut, empirical steady state (Zhang, 2002) using Eq. 3.179, and analytical steady state (Zheng and Wierzbicki, 1996) using Eq. 3.178.)

reached. The cutting forces calculated using Eqs (3.176), (3.177) and the comparison with the experimental results are shown in Fig. 3.94. It is seen from the comparison that the agreement is good for the initial cutting stage. However, the calculated steady-state cutting force is lower than the experimental results. The steady-state cutting force calculated using the empirical method of Eq. (3.179) is also presented in Fig. 3.94. Also these calculated results are lower than the experimental result. The reason for this can be explained by the fact that the stiffeners are ignored in the calculations. The stiffeners deform to a certain extent in the experimental tests; this will absorb energy. The stiffeners also restrict the plate curling as the wedge advances, and this will further increase the cutting resistance. Therefore, the experimental result is higher than the analytical analysis. However, the results using Eq. (3.178) give a good agreement with the experimental results (see Fig. 3.94).

Because the stiffeners get a certain level of deformations during the cutting, we may use the equivalent thickness concept to take into account the effect of the stiffeners. By smearing the stiffeners to the plate, we get the equivalent thickness of the stiffeners: $\Delta t=2\cdot(200\cdot 15+500\cdot 12)/2000=9$ (mm). Since the stiffeners are not fully deformed in the cutting process, here, only one-third of the equivalent thickness is taken as effective: $t_{eq}=20+\Delta t/3=23$ mm. The results using the equivalent thickness are compared with the experimental ones in Fig. 3.95. Generally, a better agreement is achieved, but the force given by Eq. (3.178) is a bit too high. Certainly, we should admit and be aware of the uncertainty in calculating the equivalent thickness and the uncertainty of the friction coefficients. It should be also noted here that for the two stiffened plate tests, the experimental cutting forces are not exactly same.

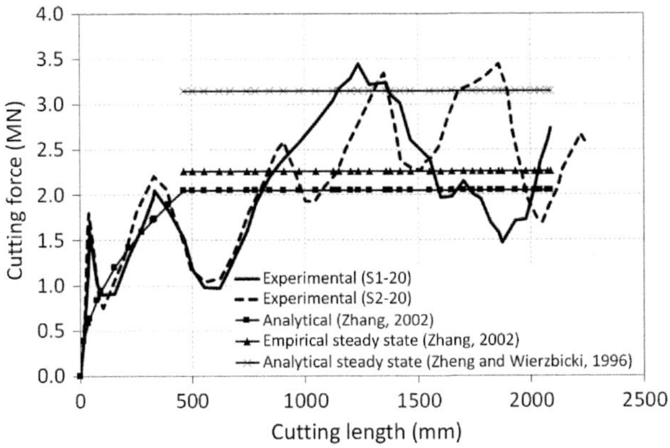

Fig. 3.95 Stiffened thick plate cutting force comparison between the analytical and the experimental results (steel plate thickness $t=20$ mm, plus 3 mm equivalent thickness of stiffeners used in the calculations). (Notes on symbols: analytical (Zhang, 2002) using Eq. 3.176 for initial cut and Eq. 3.177 for steady-state cut, empirical steady state (Zhang, 2002) using Eq. 3.179, and analytical steady state (Zheng and Wierzbicki, 1996) using Eq. 3.178.)

3.12 Concluding remarks

This chapter has provided a fundamental analysis foundation on simplified analytical methods for the internal mechanics of ship collisions and groundings.

Eleven sections have been presented to cover: (1) introduction on analysis methods, (2) material modelling, (3) theorems for structural collapse analysis, (4) lateral deformation of shell plating, (5) crushing of web frames and stringers, (6) crushing of web and stringer intersections, (7) axial crushing of structural elements, (8) crushing of ship bow against rigid walls, (9) damage analysis of ship–ship collisions, (10) damage analysis of ship–offshore platform collisions, and (11) ship grounding on rocks. Each section has provided the analysis methods, formulations, and experimental results wherever appropriate.

References

AASHTO, 1991. LRFD Bridge Design Specifications. American Association of State Highway and Transportation Officials.

Abramowicz, W., 1983. The effective crushing distance in axially compressed thin-walled metal columns. Int. J. Impact Eng. 1 (3), 309–317.

Abramowicz, W., 1994. Crushing resistance of T, Y and X sections. MIT-Industry Joint Program on Tanker Safety, Massachusetts Institute of Technology, USA, Report No. 24.

Abramowicz, W., Jones, N., 1984. Dynamic axial crushing of square tubes. Int. J. Impact Eng. 2 (2), 179–208.

Akita, Y., Ando, N., Fujita, Y., Kitamura, K., 1972. Studies on collision-protective structures in nuclear powered ships. Nucl. Eng. Des. 19, 365–401.

Alexander, J.M., 1960. An approximate analysis of the collapse of thin cylindrical shells under axial loading. Q. J. Mech. Appl. Math. 13, 10–15.

Amdahl, J., 1983. Energy Absorption in Ship-Platform Impacts (Doctoral thesis). Norwegian Institute of Technology.

Amdahl, J., 1995. Side Collision. 22nd WEGEMT Graduate School, Technical University of, Denmark.

Amdahl, J., Kavlie, D., 1992. Experimental and numerical simulation of double hull stranding. In: DNV-MIT Workshop on Mechanics of Ship Collision and Grounding, Høvik.

Amdahl, J., Watam, R., Hu, Z., Holmås, T., 2012. Broad side ship collision with jacket legs: examination of NORSOK N-004 analysis procedure. In: OMAE Proceedings, Rio de Janeiro. OMAE2012-84266.

Arita, K., Aoki, G., 1985. Strength of ship bottom grounding (1st report)—an investigation into the case of a ship stranded on a rock. (in Japanese). J. Soc. Nav. Archit. Jpn. 158, 359–367.

ASIS, 1998. Research on the Methodology for the Prediction of Accidental Damage to Tanker Structure in Case of Collision or Grounding and Development of New Hull Design With Improved Crashworthiness (1991–1997). The Association for Structural Improvement of the Shipbuilding Industry, Japan.

ASTM, 2004. ASTM E 8 Standard Test Methods of Tension Testing of Metallic Materials. American Society for Testing and Materials.

Astrup, O., 1994. Cutting of thick plates by a wedge—An experimental study. MIT-Industry Joint Program on Tanker Safety, Report No. 27, Dept. of Ocean Engineering, MIT, USA.

Bockenhauer, M., 1996. IMO's role in improving ship safety and mitigating environmental damage. In: Keynote Speech at the 1st International Conference on Design and Methodologies for Collision and Grounding Protection of Ships, San Francisco, USA.

Bodner, S.R., Symonds, P.S., 1962. Experimental and theoretical investigation of the plastic deformation of cantilever beams subjected to impulsive loading. J. Appl. Mech. 29, 719–728.

BS 7910: 2013 + A1: 2015, 2015. Guide to Methods for Assessing the Acceptability of Flaws in Metallic Structures. BSI Standards Publication.

BS EN 10002-1, 2001. Tensile Testing of Metallic Materials. Method of Testing at Ambient Temperature. European Standards.

BS EN ISO 2566-1, 1999. Steel—Conversion of Elongation Values—Part 1: Carbon and Low Alloy Steels. .

BS EN ISO 2566-2, 2001. Steel—Conversion of Elongation Values—Part 2: Austenitic Steels. .

Chang, P.Y., Seibold, F., Thasanatorn, C., 1980. A rational methodology for the prediction of structural response due to collisions of ships. In: SNAME Transactions.vol. 88. pp. 173–193.

Cowper, G.R., Symonds, P., 1957. Strain-hardening and strain-rate effects in the impact loading of cantilever beams. Technical Report No. 28, Division of Applied Mathematics, Brown University.

DiPaolo, B.P., Tom, J.G., 2006. A study on an axial crush configuration response of thin-wall, steel box components: the quasi-static experiments. Int. J. Solids Struct. 43, 7752–7775.

Ehlers, S., 2010. The influence of the material relation on the accuracy of collision simulations. Mar. Struct. 23, 462–474.

Ellinas, C.P., Valsgard, S., 1985. Collision and damage of offshore structures a state-of-the art. In: 4th Int. Symposium on Offshore Mechanics and Arctic Engineering, Dallas, Texas, pp. 17–22.

Eurocode 1, 2010. Actions on Structures—Part 1-7: General Actions—Accidental Actions. (Incorporated corrigendum February 2010). July 2006 CEN.

Forrestal, M.J., Sagartz, M.J., 1978. Elastic-plastic response of 304 stainless steel beams to impulse loads. J. Appl. Mech. 45, 685–687.

Furnes, O., Amdahl, J., 1980. Ship collisions with offshore platforms. In: Intermaritec'80, p. 310.

Gerard, G., 1957. The Crippling Strength of Compression Elements. American Institute of the Aeronautics and Astronautics.

Hollomon, J.H., 1945. Tensile deformation. Trans. Am. Inst. Min. Metall. Pet. Eng. 162, 268–290.

Hong, L., Amdahl, J., 2008. Crushing resistance of web girders in ship collision and grounding. Mar. Struct. 21, 374–401.

Ito, H., Kondo, K., Yoshimura, N., Kawashima, M., Yamamoto, S., 1984. A simplified method to analyze the strength of double hulled structures in collision. In: The Autumn Meeting of the Society of Naval Architects of Japan.

Jones, N., 1979. A literature survey on the collision and grounding protection of ships. Ship Structures Committee Report, SSC-283.

Jones, N., 2012. Structural Impact. Cambridge University Press.

Jones, N., Jouri, W.S., 1987. A study of plate tearing for ship collision and grounding damage. J. Ship Res. 31 (4), 253–268.

Jones, N., Wierzbicki, T. (Eds.), 1993. Structural Crashworthiness and Failure. Elsevier Applied Science.

Kierkegaard, H., 1993. Ship bow response in high energy collisions. Mar. Struct. 6, 359–376.

Kitamura, O., 1997. Comparative study on collision resistance of side structure. Mar. Technol. 34 (4), 293–308.

Ku, J.H., Seo, K.S., Park, S.W., Kim, Y.J., 2001. Axial crushing behavior of the intermittent tack-welded cylindrical tubes. Int. J. Mech. Sci. 43, 521–542.

Kuroiwa, T., 1996. Numerical simulation of actual collision & grounding accidents. In: International Conference on Design and Methodologies for Collision and Grounding Protection of Ships, San Francisco, USA.

Lee, J.W., Hong, S.J., 1997. A study on the dissipation energy of plate due to cutting. J. Ship Ocean Technol. 1, 48–56.

Lee, Y.W., Woertz, J.C., Wierzbicki, T., 2004. Fracture prediction of thin plates under hemi-spherical punch with calibration and experimental verification. Int. J. Mech. Sci. 46, 751–781.

Lehmann, E., Yu, X., 1998. Inner dynamics of bow collision to bridge piers. In: International Symposium Advances in Bridge Aerodynamics-Ship Collisions Analysis-Operation and Maintenance, Lyngby, Denmark.

Lehmann, E., Peschmann, J., 2002. Energy absorption by the steel structure of ships in the event of collisions. Mar. Struct. 15, 429–441.

Lehmann, E., Yu, X., 1995. Progressive folding of bulbous bows. In: PRADS'95, The Netherlands.

Lemmen, P.M., Vredeveldt, W., Pinkster, J.A., 1996. In: Design Analysis for Grounding Experiments.International Conference on Design and Methodologies for Collision and Grounding Protection of Ships, San Francisco, California, USA, pp. 22–23.

Liu, B., 2015. Energy Absorption of Ship Structural Components Under Impact Loading. (PhD thesis). Instituto Superior Técnico, University of Lisbon.

Liu, B., Villavicencio, R., Guedes Soares, C., 2014. On the failure criterion of aluminium and steel plates subjected to low-velocity impact by a spherical indenter. Int. J. Mech. Sci. 80, 1–15.

Liu, B., Villavicencio, R., Zhang, S., Guedes Soares, C., 2017. A simple criterion to evaluate the rupture of materials in ship collision simulations. Mar. Struct. 54, 92–111.

Liu, B., Pedersen, P.T., Zhu, L., Zhang, S., 2018. Review of experiments and calculation procedures for ship collision and grounding damage. Mar. Struct. 59, 105–121.

Lloyd's Register (LR), 2018a. Rules for the Manufacture Testing and Certification of Materials. Lloyd's Register, London.

Lloyd's Register (LR), 2018b. Rules and Regulations for the Classification of Ships. Lloyd's Register, London.

Lu, G., Calladine, C.R., 1990. On the cutting of a plate by a wedge. Int. J. Mech. Sci. 32 (4), 293–313.

Lützen, M., 2001. Ship Collision Damage (PhD thesis). Technical University of Denmark, Denmark.

Marinatos, J.N., Samuelides, M.S., 2015. Towards a unified methodology for the simulation of rupture in collision and grounding of ships. Mar. Struct. 42, 1–32.

Marshall, P.W., Gates, W.E., Anagnostopoulos, S., 1977. Inelastic dynamic analysis of tubular structures. OTC 2908.

Martens, I., 2014. Konstruktive Aspekte beim Entwurf von Bugwülsten zur Verbesserung des Energieaufnahmevermogens bei Schiffskollisionen. (in German). (PhD thesis), Schriftenreihe Schiffbau, Bericht 679, Technical University of Hamburg, Germany. ISBN 978-3-89220-679-8.

McDermott, J., Kline, R., Jones, E., Maniar, N., Chiang, W., 1974. Tanker structural analysis for minor collisions. In: SNAME Transactions.

Minorsky, V.U., 1959. An analysis of ship collision with reference to protection of nuclear power ships. J. Ship Res. 3 (2), 1–4.

Nielsen, L.P., 1995. Traffic and Route Data—Safety of Passenger RoRo Vessels. Technical University of Denmark, Dept. of Ocean Engineering.

NORSOK N-004, 2004. NORSOK Standard N-004. Rev. 2, October 2004, Standards Norway, Strandveien 18, N-1326 Lysaker, Norway.

Ohtsubo, H., Wang, G., 1995. An upper-bound solution to the problem of plate tearing. J. Mar. Sci. Technol. 1, 46–51.

Paik, J.K., 1994. Cutting of a longitudinally stiffened plate by a wedge. J. Ship Res. 38 (4), 340–348.

Paik, J.K., 2007. Practical techniques for finite element modeling to simulate structural crashworthiness in ship collision and grounding (Part I: Theory). Ship Offshore Struct. 2 (1), 69–80.

Paik, J.K., Pedersen, P.T., 1995. Ultimate and crushing strength of plated structures. J. Ship Res. 39 (3), 250–261.

Paik, J.K., Thayamballi, A.K., 2003. Ultimate limit state design of steel-plated structures. Wiley, Chichester, UK.

Paik, J.K., Chung, J.Y., Choe, I.H., Thayamballi, A.K., Pedersen, P.T., Wang, G., 1999. On rational design of double hull tanker structures against collision. In: The Society of Naval Architects and Marine Engineers. Transactions SNAME.vol. 107. paper number 14.

Pedersen, P.T., 1998. In: Gimsing, N. (Ed.), Ship Crushing Load Studies. The Storebælt Publications, East Bridge, pp. 44–57 (Chapter 3.6).

Pedersen, P.T., 2010. Review and application of ship collision and grounding analysis procedures. Mar. Struct. 23 (3), 241–262.

Pedersen, P.T., Zhang, S., 2000. Absorbed energy in ship collisions and grounding—revising Minorsky's empirical method. J. Ship Res. 44 (2), 140–154.

Pedersen, P.T., Valsgaard, O., Olsen, D., Spangenberg, S., 1993. Ship impacts—bow collisions. Int. J. Impact Eng. 13 (2), 163–187.

Qvist, S., Nielsen, K.B., Schmidt, M.H., Madsen, S.H., 1995. Ship collision—experimental and numerical analysis of double hull models. In: 9th DYMAT Technical Conference: Material and Structural Modelling in Collision Research, Munich Germany, October 10-11.

Reckling, K.A., 1983. Mechanics of minor ship collisions. Int. J. Impact Eng. 1 (3), 281–299.

Ringsberg, J., et al., 2018. MARSTRUCT benchmark study on nonlinear FE simulation of an experiment of an indenter impact with a ship side-shell structure. Mar. Struct. 59, 142–157.

Samuelides, E., Frieze, P.A., 1989. Fluid-structure interaction in ship collisions. Mar. Struct. 2, 65–88.

Sano, A., Muragishi, O., Yoshikawa, T., 1996. Strength analysis of a new double hull structure for VLCC in collision. In: International Conference on Design and Methodologies for Collision and Grounding Protection of Ships. San Francisco, California, USA.

Schneekluth, H., 1987. Ship Design for Efficiency and Economy. Butterworth & Co. (Publishers) Ltd.

Simonsen, B.C., 1997a. Mechanics of Ship Grounding (PhD thesis). Technical University of Denmark, Denmark. ISBN 87-89502-34-5.

Simonsen, B.C., 1997b. Ship grounding on rock—Part I theory & Part II validation and application. Mar. Struct. 10, 519–562.

Simonsen, B.C., Lauridsen, L.P., 2000. Energy absorption and ductile failure in metal sheets under lateral indentation by a sphere. Int. J. Impact Eng. 24, 1017–1039.

Simonsen, B.C., Ocakli, H., 1999. Experiments and Theory on Deck and Girder Crushing. Department of Naval Architecture and Offshore Engineering, Technical University of Denmark.

Simonsen, B.C., Wierzbicki, T., 1997. Plasticity, fracture and friction in steady-state plate cutting. Int. J. Impact Eng. 19 (8), 667–692.

Simonsen, B.C., Tornqvist, R., Lutzen, M., 2009. A simplified grounding damage prediction method and its application in modern damage stability requirements. Mar. Struct. 22, 62–83.

Storheim, M., Amdahl, J., Martens, I., 2015. On the accuracy of fracture estimation in collision analysis of ship and offshore structures. Mar. Struct. 44, 254–287.

Tarigopula, V., Langseth, M., Hopperstad, O.S., Clausen, A.H., 2006. Axial crushing of thin-walled high-strength steel sections. Int. J. Impact Eng. 32, 847–882.

Tautz, I., Schöttelndreyer, M., Lehmann, E., Fricke, W., 2013. Collision tests with rigid and deformable bulbous bows driven against double hull side structures. In: Amdahl, , Ehlers, , Leira, (Eds.), Proceedings of the 6th International Conference on Collision and Grounding of Ships and Offshore Structures. Taylor & Francis Group, London. ISBN 978-1-138-00059-9, pp. 93–100.

Vaughan, H., 1980. The tearing strength of mild steel plate. J. Ship Res. 24 (2), 96–100.

Visser, W., 2004. Ship collision and capacity of brace members of fixed steel offshore platforms. Research Report 220, UK Health and Safety Executive.

Wang, G., 1995. Structural Analysis of Ship Collision and Grounding. (PhD thesis)University of Tokyo.

Wang, G., Ohtsubo, H., 1997. Deformation of ship plate subjected to very large load. In: Proceedings of the 16th International Conference on Offshore Mechanics and Arctic Engineering (OMAE), Yokohamavol. 2, pp. 173–180.

Wang, G., Arita, K., Liu, D., 2000. Behavior of a double hull in a variety of stranding or collision scenarios. Mar. Struct. 13, 147–187.

Wierzbicki, T., 1994. Concertina tearing of metal plates-improved solution and comparison. Joint MIT-Industry Project on Tanker Safety, Report No. 22.

Wierzbicki, T., Abramowicz, W., 1983. On the crushing mechanics of thin-walled structures. J. Appl. Mech., ASME 50, 727–734.

Wierzbicki, T., Suh, M., 1988. Indentation of tubes under combined loading. Int. J. Mech. Sci. 30, 229–248.

Wierzbicki, T., Thomas, P., 1993. Closed-form solution for wedge cutting force through thin metal sheets. Int. J. Mech. Sci. 35 (3/4), 209–229.

Wierzbicki, T., Peer, D.B., Rady, E., 1993. The anatomy of tanker grounding. Mar. Technol. 30 (2), 71–78.

Woisin, G., 1979. Design against collision. In: Proceedings of Int. Symp. On Advances in Marine Technology, Trondheim, Norwayvol. 2.

Yahiaoui, M., Bracco, M., Little, P., Trauth, A.T., 1994. Experimental study on scale models for grounding. MIT-Industry Joint Program on Tanker Safety, Report No. 18.

Yamada, Y., 2006. Bulbous Buffer Bows: A Measure to Reduce Oil Spills in Tanker Collision. (PhD thesis)Technical University of Denmark, Denmark.

Yamada, Y., Endo, H., 2005. Collapse mechanism of the buffer bow structure on axial crushing. Int. J. Offshore Polar Eng. 15 (02), 133–141.

Yamada, Y., Pedersen, P.T., 2008. A benchmark study of procedures for analysis of axial crushing of bulbous bows. Mar. Struct. 21 (2–3), 257–293.

Yang, P.D.C., Caldwell, J.B., 1988. Collision energy absorption of ship bow structures. Int. J. Impact Eng. 7 (2), 181–196.

Yu, Z., Amdahl, J., 2018. Analysis and design of offshore tubular members against ship impacts. Mar. Struct. 58, 109–135.

Zeng, J., Hu, Z., Chen, G., 2016. A steady-state plate tearing model for ship grounding over a cone-shaped rock. Ships Offshore Struct. 11 (3), 245–257.

Zhang, S., 2002. Plate tearing and bottom damage in ship grounding. Mar. Struct. 15, 101–117.

Zhang, S., 2005. Cargo trading patterns and ship main dimensions. Report 01/05, Lloyd's Register, London, UK.

Zhang, L., Egge, E.D., Bruhns, H., 2004a. Approval procedure concept for alternative arrangements. In: Proc. 3rd International Conference on Collision and Grounding of Ships: Izu, Japan, pp. 87–96.

Zhang, S., Ocakli, H., Pedersen, P.T., 2004b. Crushing of ship bows in head-on collision. In: The Transactions of The Royal Institution of Naval Architects.vol. 146. Part A2.

Zhang, M., Liu, J., Hu, Z., Zhao, Y., 2018. On resistance of a rectangular thin plate under lateral indentation by a wedge indenter. Ship Offshore Struct. https://doi.org/10.1080/17445302.2018.1441618.

Zheng, Z., Wierzbicki, T., 1996. A theoretical study of a steady-state wedge cutting through metal plates. Int. J. Fract. 78, 45–66.

Zhu, L., James, P., Zhang, S., 2002. Statistics and damage assessment of ship grounding. Mar. Struct. 15, 515–530.

Damaged material volume method 4

4.1 Introduction

The most well-known empirical approach for collision analysis is probably that proposed by Minorsky (1959), who analysed 26 cases of full-scale ship collision accidents and developed the empirical formula $E=47.2R_t+32.7$, which relates the absorbed energy (E in MJ) to the destroyed material volume (R_t in m^3). Though the limitations of this approach are recognized, this empirical formula has been widely used by the industry for ship collision and grounding analyses because of its simplicity.

A simplified analytical method to estimate the damage extent and energy absorption during collision accidents was presented in Zhang (1999) and Pedersen and Zhang (2000). It was based on the relationship between the absorbed energy and the volume of the damaged material, but it takes into account the structural arrangements, the material properties, and the damage modes. Zhang and Pedersen (2016) compared a large number of experimental results with the energy estimated by the analytical method at a given penetration: usually at the final damage penetration for each collision experiment. However, it is also important to verify the energy response and the collision force behaviour during the entire impact event at any damage penetration. This was done by Zhang et al. (2018) to reflect the full picture of the collision assessment and to validate the robustness of the method. This chapter presents the details of the method and its application to full-scale ship collision and grounding accidents.

4.2 The damaged material volume method

4.2.1 Prediction methodology

The analysis of the structural damage caused by ship collisions and grounding is rather complex as the impact response is highly nonlinear, involving continuous changes in the geometry of the ship structures. The observed failure modes from ship collision and grounding accidents reveal that the primary energy-absorbing mechanisms are crushing of the decks, tearing of the bottom plating, folding of the web frames, and stretching of the shell plating.

Based on the characteristics of the structural arrangement and the observed failure modes in collision and grounding events, the ship structure can be divided into a number of plated components. A comparison of the derived expressions for the different structural units with different failure modes (folding, tearing, or crushing) shows that the mean resistance forces have some similarity. The mean resistance can be expressed as $F_m=D_1\sigma_0 t^\alpha c^{2-\alpha}$, where σ_0 is the flow stress of the material, t is the thickness of the plate, c is the cross-sectional length of a unit or the tearing length, D_1 is a

Probability and Mechanics of Ship Collision and Grounding. https://doi.org/10.1016/B978-0-12-815022-1.00004-9

coefficient related to the actual crushing or tearing configuration, and the exponent α is in the range $\alpha = 1.5 - 1.7$.

For a plate suffering lateral loading (such as those of the side shell), the structural behaviour is mainly driven by membrane tension. In this case, the relation of the resistance force F_p and the lateral indentation δ can be written as $F_p = D_2 \sigma_0 t \delta$, where D_2 is a coefficient related to the impact position and the plate size.

The actual damage in ship collisions and grounding is far more complicated than the damage of these basic elements. In fact, normally, we may not be able to predict the exact damage mode during the impact, as it may be a mode mixing crushing, folding, tearing, and stretching. However, it has been possible to establish simple expressions to describe collision and grounding damages, since the basic failure mechanisms have a certain common relationship.

4.2.2 The formulations

Three different energy absorption mechanisms have been identified: (1) plastic tension deformation, such as the indentation of the shell plating during a side collision; (2) folding and crushing damage mode, such as the crushing of the bow, and folding and crushing of the deck, web frames, and stringers; and (3) tearing damage mode, such as the raking damage to bottom structures during groundings. The proposed formulae for the relationship between the absorbed energy and the destroyed material volume are as follows:

4.2.2.1 Energy absorption by the plastic membrane tension damage mode

$$E_1 = 0.77 \varepsilon_f \sigma_0 R_{t1} \tag{4.1}$$

where E_1 is the absorbed energy, σ_0 is the flow stress of the material, ε_f is the rupture strain of the material, and R_{t1} is the material volume of the damaged/ruptured structural members in tension mode.

Before rupture of the shell plating, ε_f is to be replaced by $(\delta/\delta_f)\varepsilon_f$ where δ is the penetration and δ_f is the critical penetration at rupture that can be determined from experiments or by the expressions given in Chapter 3, Section 3.4.6, that is, $\delta_f = \varepsilon_f^{0.4} b^{0.4} R_b^{0.6}$.

4.2.2.2 Energy absorption by the crushing and folding damage mode

$$E_2 = 3.50 \left(\frac{t}{b}\right)^{0.67} \sigma_0 R_{t2} \tag{4.2}$$

where t is the thickness of the crushed plate, b is the width of the plate in the crushed cross section, and R_{t2} is the material volume of the crushed structural members.

4.2.2.3 Energy absorption by the tearing damage mode

$$E_3 = 3.21\left(\frac{t_{eq}}{\lambda}\right)^{0.6}\sigma_0 R_{t3} \tag{4.3}$$

where t_{eq} is the equivalent plate thickness, including longitudinal webs and stiffeners (in the tearing direction); λ is the critical tearing length where the steady tearing state has just been reached; and R_{t3} is the volume of torn material. In most practical cases, the critical tearing length can be taken as the width of the tearing object or the damage width. If the steady state is not reached, λ is equal to the tearing length.

The total absorbed energy can be obtained from the summation of the energy absorbed by each damaged structural member. For ship collisions and grounding,

$$E_{collision} = \sum(E_{1i} + E_{2i}) \tag{4.4}$$

$$E_{grounding} = \sum(E_{3i}) \tag{4.5}$$

4.2.2.4 Explanation of the parameters used in the calculations

The material properties are important parameters for energy absorption analyses. These parameters are normally obtained from standard uniaxial tensile tests that include yield stress σ_y, ultimate tensile stress σ_u, and elongation at rupture ε_f (referred here as rupture strain). In this book, the flow stress of the materials is taken as the average of the yield stress and the ultimate tensile stress, that is, $\sigma_0 = 0.5(\sigma_y + \sigma_u)$, as described in Chapter 3. The rupture strain to be used in the analytical calculations should be obtained from standard uniaxial tensile tests with a gauge length of $5.65\sqrt{S}$, if these data are available. On the other hand, material databases have to be consulted to identify a reliable rupture strain when information from the tensile tests is not available. Alternatively, the expression given in Chapter 3, Section 3.2.4, can be used, that is, $\varepsilon_f = 0.385(400/\sigma_u)$ for carbon steels.

The other important parameter is the volume of damaged materials. It is usually assumed that the striking bow is rigid in ship–ship collision analysis, and thus, all collision energy is absorbed by the struck ship. In such cases, the structural damage can be approximately determined by assuming that the damage to the struck ship is confined around the rigid striking bow. This means that the damaged space is equal to the volume of the striking bow penetrating into the ship side. This assumption is well supported by the observations from actual ship–ship collision damages. Fig. 4.1 illustrates structural members damaged in ship collision and grounding events.

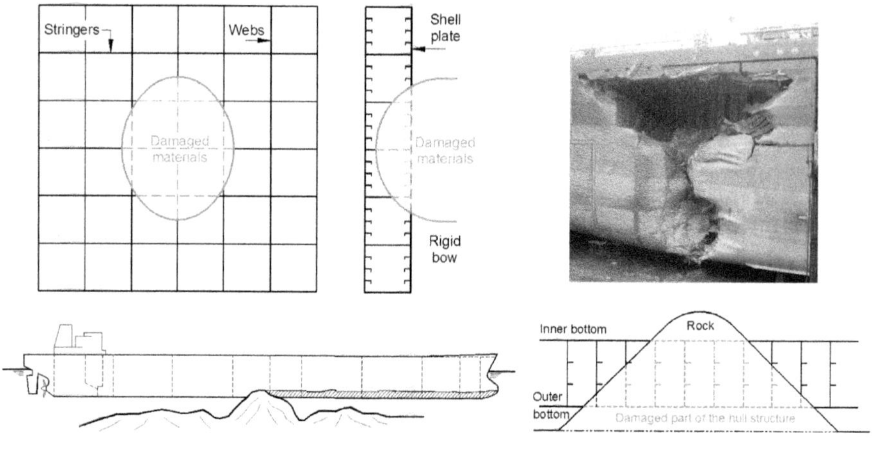

Fig. 4.1 Illustration of damaged materials in ship collision and grounding.

Friction contribution to the energy absorption in grounding analysis is important, and its effect has been considered and embedded in Eq. (4.3). For ship collision analysis, the friction between the rigid bow and the damaged structure can result in additional absorbed energy. However, friction effects have normally been ignored in the simplified analytical methods (Zhang, 1999; Pedersen and Zhang, 2000).

Although it is believed that the contribution from friction is not significant in ship collisions, further experiments are needed to confirm this. Here, finite element simulations of a double-hull model penetrated by a rigid cone are performed to investigate the contribution of the friction on the energy absorption; the double-hull model (S-II model) is selected from the experiments reported by Arita and Aoki (1985); see Fig. 4.3. The simulations use three friction coefficients μ: 0.23, 0.35, and 0.45. The results in Fig. 4.2 show that the effect of the friction is negligible before the rupture of the shell plate (about 0.07 m penetration) because the relative motion between the bow and the plate is small, while beyond that point, the effect increases with the penetration. It is also noticed that friction contributes only to about 10% of the total dissipated energy at relatively large penetrations and that such contribution is not very sensitive to the assumed friction coefficients. Similar results for other double-hull models were presented in Zhang et al. (2018). Considering this, the analytical method assumes that the contribution of friction is about 10% for collision analysis. This is implemented by first calculating the absorbed energy with Eqs. (4.1), (4.2) for ship collisions and then increasing the energy $E_{collision}$ by 11.5%. This method has been adopted for simplicity and to minimize the uncertainties of the input parameters.

The force–penetration response can be determined by the derivative of the energy–penetration curves obtained from Eq. (4.4) or (4.5). Numerically, the collision force at each penetration δ_i is obtained from

$$F_i = \frac{dE}{d\delta} = \frac{\Delta E}{\Delta \delta} = \frac{E_{i+1} - E_{i-1}}{\delta_{i+1} - \delta_{i-1}} \quad (4.6)$$

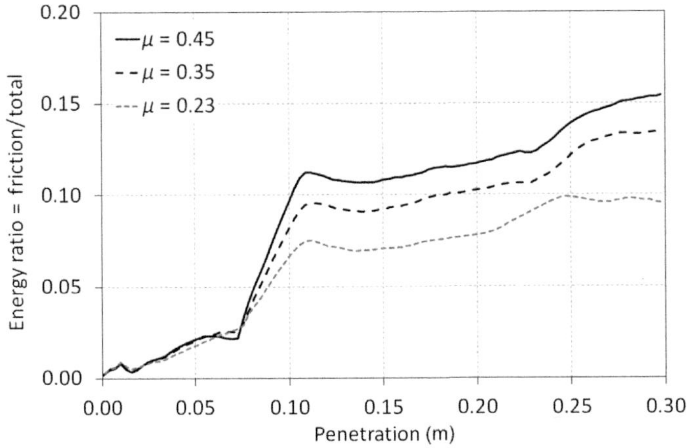

Fig. 4.2 Finite element results for the energy ratio between the energy absorbed by friction and the total absorbed energy.

By using this approach, the collision forces represent the average value within each analysed penetration step. However, the peak force caused by buckling, which can be observed in experiments, is not reflected in the calculation. Thus, the calculated force–penetration curves reflect only the overall trend of the collision resistance.

4.3 Analytical calculation results and experimental data

This section compares analytical calculation results with various experimental data found in the literature. In particular, this section deals with collision of ship side structures, bow crushing, and ship grounding scenarios.

4.3.1 Side collisions

Details of 18 collision experiments found in the literature will be used as the benchmark for the predictions of the analytical method. The analysis procedure described in Section 4.2 is consistently used for estimating the response of all selected collision experiments. Discussions of the results will be provided at the end of the present section.

4.3.1.1 Three model tests by Arita and Aoki (1985)

Arita and Aoki (1985) performed collision tests of three models of double-hull structures penetrated by a rigid conical indenter with a radius of 80 mm. The structural arrangement of the models is shown in Fig. 4.3, and the dimensions and material properties are given in Table 4.1. Models S-I and S-III have the same web spacing but different thicknesses, while Models S-II and S-III have the same thickness but different web spacing. The yield stress of the mild steel materials was given by Arita and Aoki

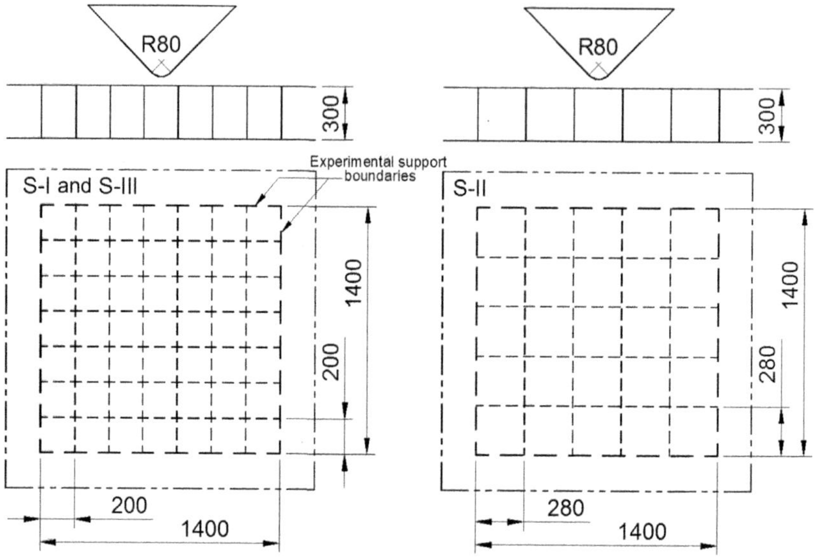

Fig. 4.3 Illustration of Arita and Aoki (1985) model tests.

Table 4.1 Dimensions and material properties of the three model tests by Arita and Aoki (1985).

ID	Model	Webs spacing (mm)	Depth (mm)	Plates and webs thickness (mm)	σ_y (MPa)	σ_u (MPa)	ε_f (−)
1	S-I	200	300	3.2	284.5	440.0	0.35
2	S-II	280	300	4.5	294.3	447.0	0.34
3	S-III	200	300	4.5	294.3	447.0	0.34

(1985); however, information of the ultimate tensile strength and the rupture strain was not provided. The magnitudes of these unknown properties are obtained from material data from the shipbuilding industry.

The absorbed energy–penetration and force–penetration curves are calculated, and the comparison between the analytical and experimental results is presented in Figs 4.4–4.6. It is seen that the analytical calculations agree very well with the experimental results.

4.3.1.2 Four model tests by Paik et al. (1999)

Paik et al. (1999) carried out collision tests of four double-hull models penetrated by a rigid conical indenter with a radius of 80 mm. The structural arrangement of a representative model is shown in Fig. 4.7, while the dimensions of all specimens are given

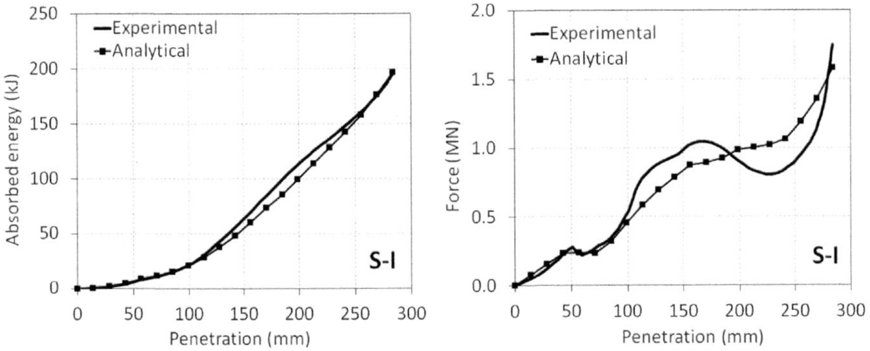

Fig. 4.4 Energy–penetration and force–penetration curves: Model S-I.

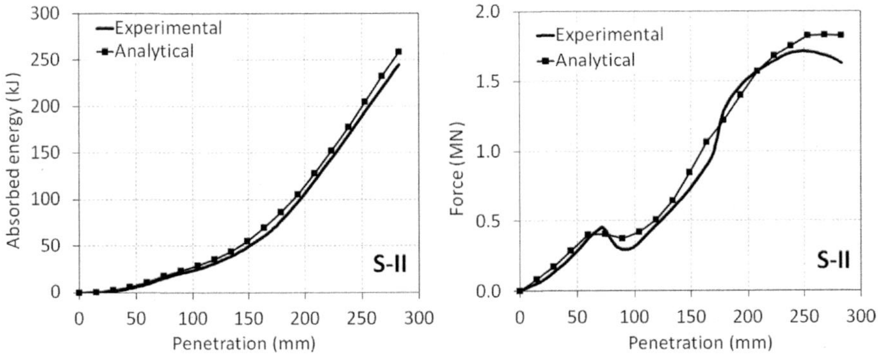

Fig. 4.5 Energy–penetration and force–penetration curves: Model S-II.

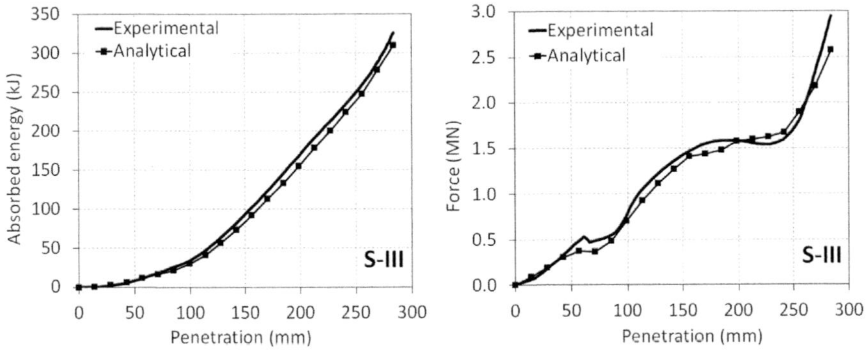

Fig. 4.6 Energy–penetration and force–penetration curves: Model S-III.

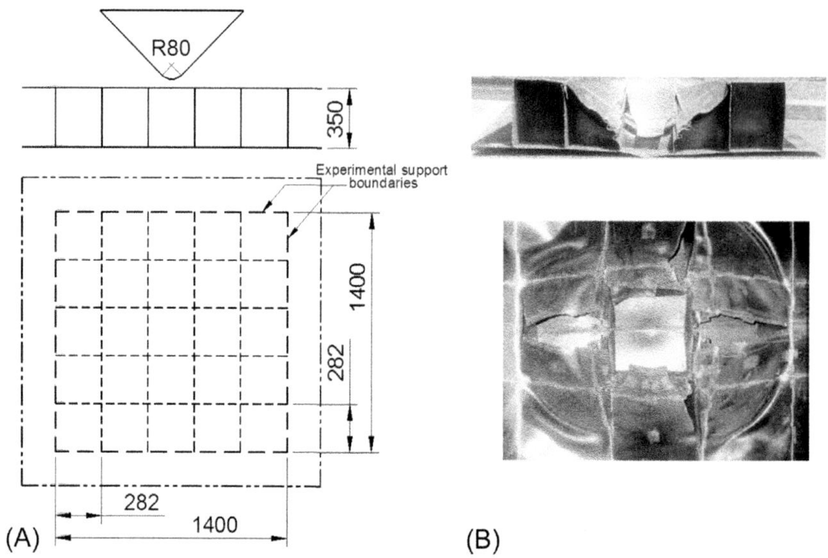

Fig. 4.7 (A) Illustration of Paik et al. (1999) model tests and (B) typical shape of the deformation.
Photos cf. Paik, J.K., Chung, J.Y., Choe, I.H., Thayamballi, A.K., Pedersen, P.T., Wang, G., 1999. On rational design of double hull tanker structures against collision. In: The Society of Naval Architects and Marine Engineers. Annual Meeting, SNAME, vol. 107.

in Table 4.2. The material properties presented in Table 4.2 were obtained from standard uniaxial tensile tests and can be found in Paik et al. (1999).

From the experiments, it is noticed that the inner hull suffered small deflections before the indenter gets into contact with it. Therefore, the penetrations are slightly adjusted, taking as reference the experimental data, so that to reflect this in the analytical calculations. It is also observed that the indenter did not provoke scratches on the plating of the models, or they were very small, and hence, the friction effect is negligible. Eventually, friction is ignored in the analytical calculations as this gives better agreement. The typical final damage pattern of the test models is shown in Fig. 4.7B. It is noticed that the webs folded with a regular pattern and that the shell plate is stretched.

The absorbed energy–penetration and force–penetration curves from the experiments and analytical analysis are compared in Figs 4.8–4.11. Similar to the previous comparison, the analytical calculations predict the experimental response with reasonable accuracy.

4.3.1.3 Nine model tests by Wang et al. (2000)

Collision tests of nine double-hull models were carried out by Wang et al. (2000). The radius of the rigid conical indenter varies from a relatively sharp radius of 10 mm to a relatively round radius of 300 mm. For all cases, the span between webs is 200 mm, the

Table 4.2 Dimensions and material properties of the four model tests of Paik et al. (1999).

ID	Model	Web spacing (mm)	Depth (mm)	Plates and webs thickness (mm)	Loading position	σ_y (MPa)	σ_u (MPa)	ε_f (−)
4	ST-3-BW	282	350	2.80	Between webs	245.3	337.8	0.46
5	ST-4-BW	282	350	3.95	Between webs	319.8	412.6	0.36
6	ST-3-OW	235	350	2.80	Web joint	245.3	337.8	0.46
7	ST-4-OW	235	350	3.95	Web joint	319.8	412.6	0.36

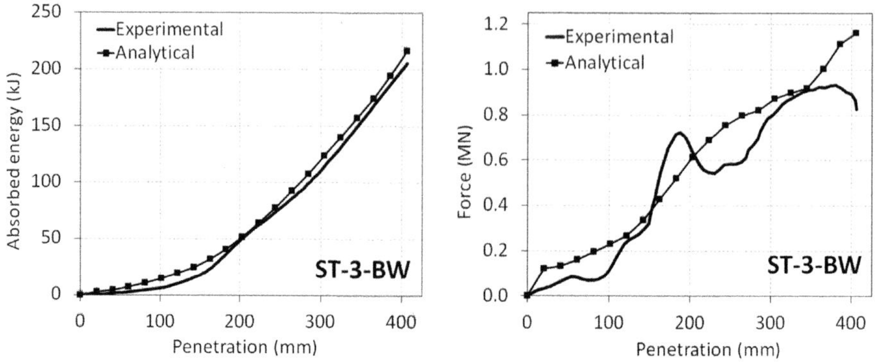

Fig. 4.8 Energy–penetration and force–penetration curves: Model ST-3-BW.

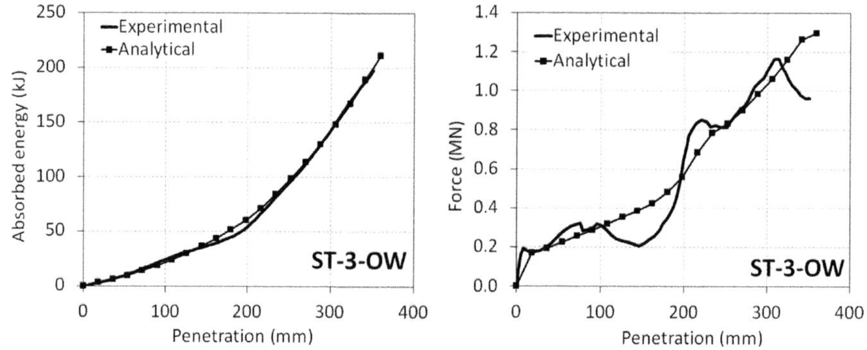

Fig. 4.9 Energy–penetration and force–penetration curves: Model ST-3-OW.

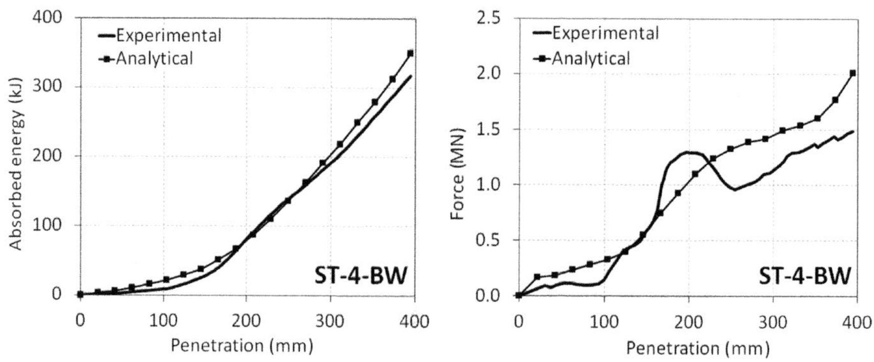

Fig. 4.10 Energy–penetration and force–penetration curves: Model ST-4-BW.

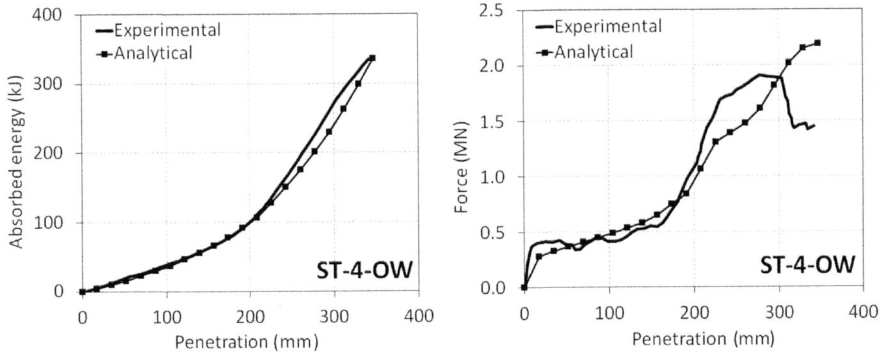

Fig. 4.11 Energy–penetration and force–penetration curves: Model ST-4-OW.

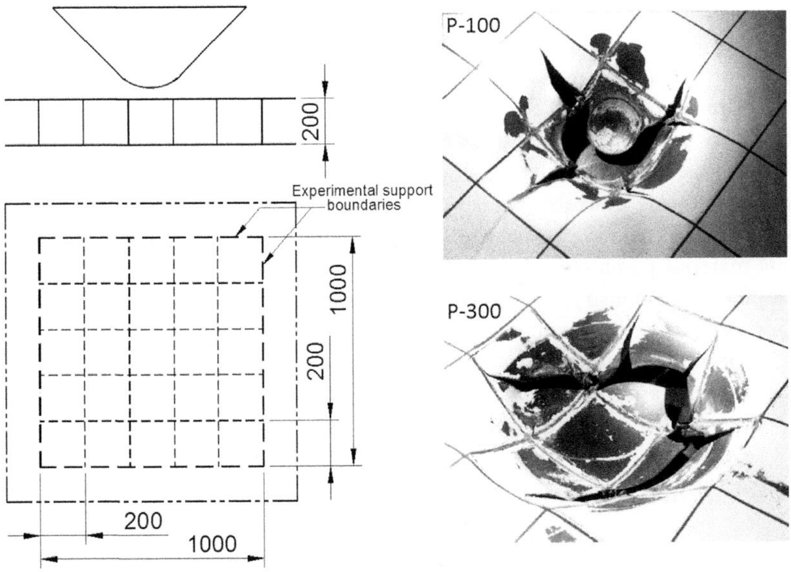

Fig. 4.12 Illustration of Wang et al. (2000) model tests and shape of deformation of specimens P-100 and P-300.
Photos cf. Wang, G., Arita, K., Liu, D., 2000. Behavior of a double hull in a variety of stranding or collision scenarios. Mar. Struct. 13, 147–187.

depth of the specimens is 200 mm, and the thickness of the outer and inner plating and the webs is 2.3 mm. The penetration was enforced until reaching the inner hull, and the collision positions were at three different locations: (1) between the webs, (2) on the webs, and (3) at the webs' intersection. The geometry of the specimens is illustrated in Fig. 4.12 together with the damage patterns of specimens P-100 and P-300 after the collision tests. Details for each experiment and the material properties are summarized in Table 4.3.

Table 4.3 Details and material properties of Wang et al. (2000) experiments.

ID	Model	Cone radius (mm)	Contact location	Plate (mm)	σ_y (MPa)	σ_u (MPa)	$\varepsilon_f(-)$
8	P-10	10	Between webs	2.3	282	438	0.35
9	P-50	50	Between webs				
10	P-100	100	Between webs				
11	P-200	200	Between webs				
12	P-300	300	Between webs				
13	W-50	50	On web				
14	W-200	200	On web				
15	C-50	50	Intersection of webs				
16	C-200	200	Intersection of webs				

The material is mild steel, but the exact values of the material properties are not available in Wang et al. (2000). Zhang and Pedersen (2016) reported analytical calculations of these experiments where the yield stress was taken as 282 MPa and the ultimate stress as 400 MPa. The rupture strain was determined as 0.45 by calibrating the analytical results of model P-100 with its corresponding experimental results; friction effects were omitted. Although this magnitude of the rupture strain of 0.45 is not unusual for Japanese steels, as also demonstrated by the uniaxial tensile tests performed by Ito et al. (1984), the value of 0.45 can be considered as an upper limit for this material parameter. Therefore, after analysing material property data from the shipbuilding industry, the properties given in Table 4.3 are found to be more appropriate and used for calculations.

The comparison of the experimental and analytical absorbed energy–penetration and force–penetration curves is presented in Figs 4.13–4.21. Overall, the analytical calculations follow the experimental response sufficiently.

4.3.1.4 Model test by Tautz et al. (2013)

Tautz et al. (2013) performed a laboratory collision experiment with a rigid bulbous bow driven against a model of a double-hull side structure.

The thickness was 4.0 mm for both the top and the bottom plates stiffened by bulb profiles HP 140 × 7 spaced at 280 mm. The web frames of thickness 5.0 mm were spaced at 800 mm. The total height from the top to the bottom plate was 900 mm. In the analytical calculations, the cross-sectional area of the stiffeners is smeared into both the top and bottom plates. A sketch of the model test and the rigid bulb is illustrated in Fig. 4.22 (Tautz et al., 2014) together with the deformed shape of the specimen CE-1 after collision test. The material properties obtained from standard tensile tests are given in Table 4.4. Both the top and bottom plates were fully penetrated (see Fig. 4.22), and thus, the experiment represents a good opportunity to validate theoretical methods for severe collisions.

The absorbed energy–penetration and force–penetration curves are presented in Fig. 4.23. While the analytical calculations overestimate the energy response before

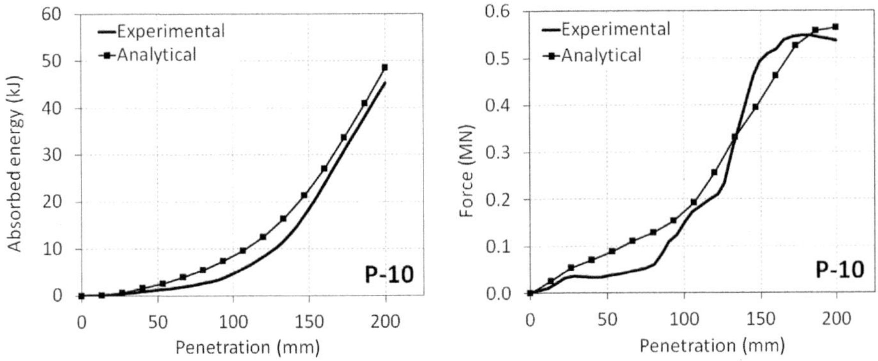

Fig. 4.13 Energy–penetration and force–penetration curves: Model P-10.

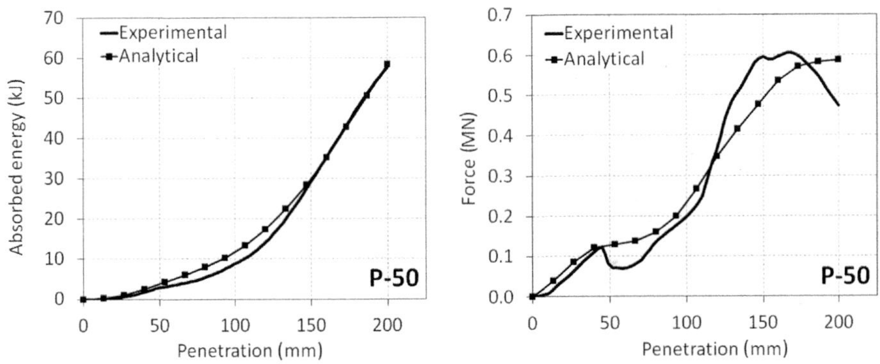

Fig. 4.14 Energy–penetration and force–penetration curves: Model P-50.

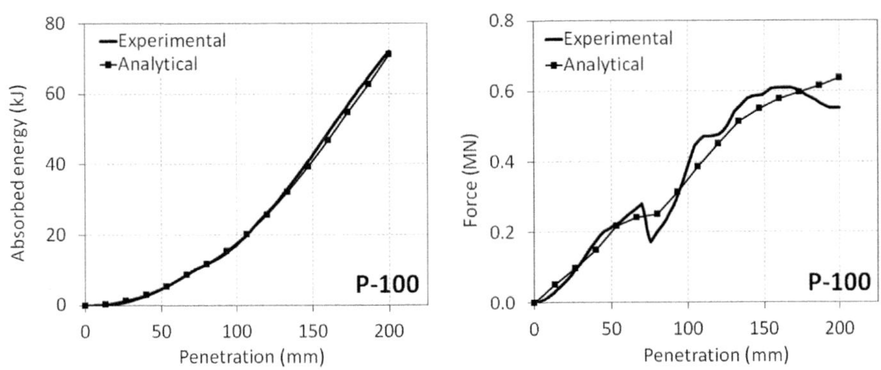

Fig. 4.15 Energy–penetration and force–penetration curves: Model P-100.

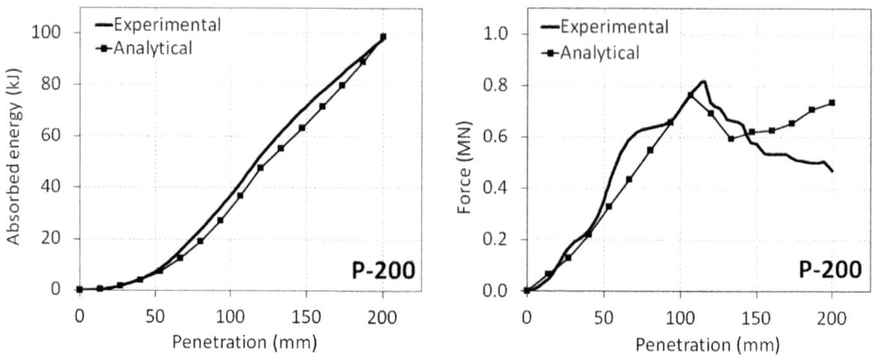

Fig. 4.16 Energy–penetration and force–penetration curves: Model P-200.

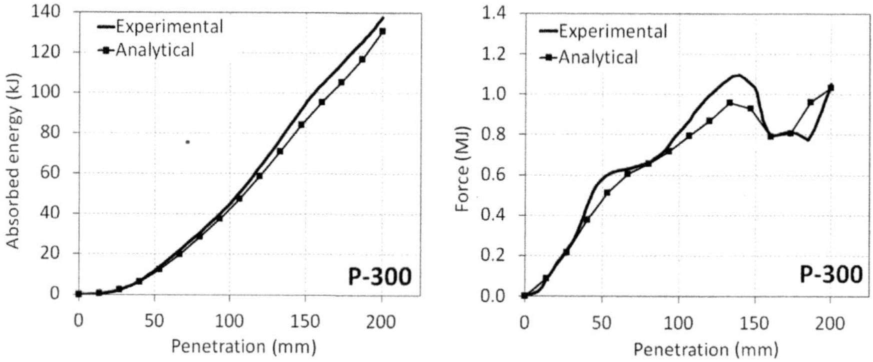

Fig. 4.17 Energy–penetration and force–penetration curves: Model P-300.

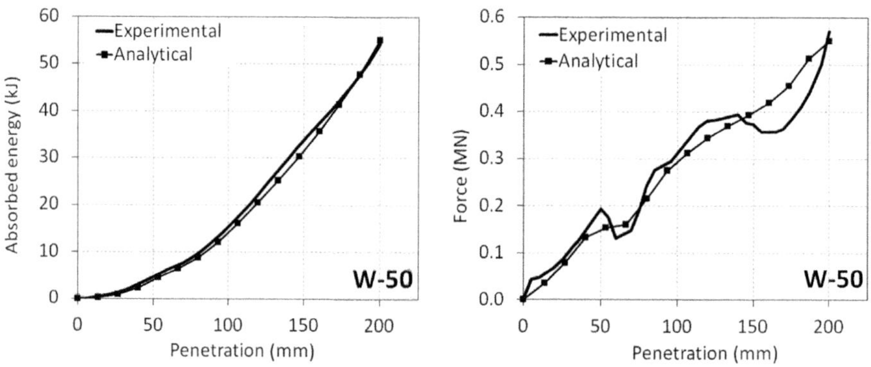

Fig. 4.18 Energy–penetration and force–penetration curves: Model W-50.

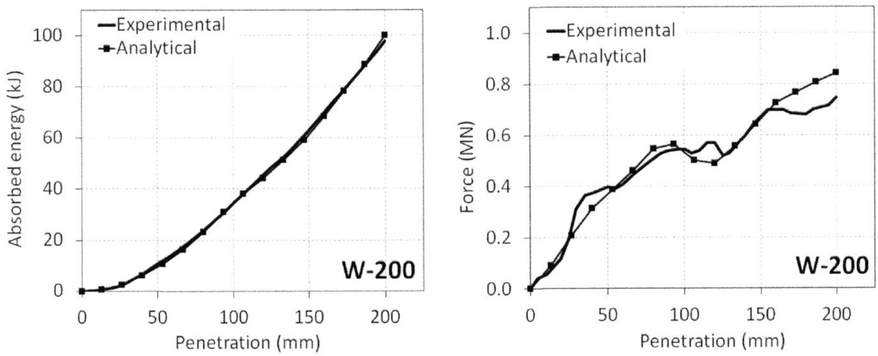

Fig. 4.19 Energy–penetration and force–penetration curves: Model W-200.

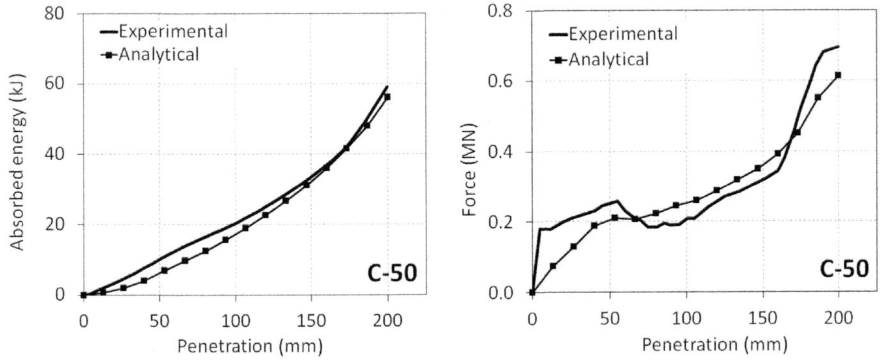

Fig. 4.20 Energy–penetration and force–penetration curves: Model C-50.

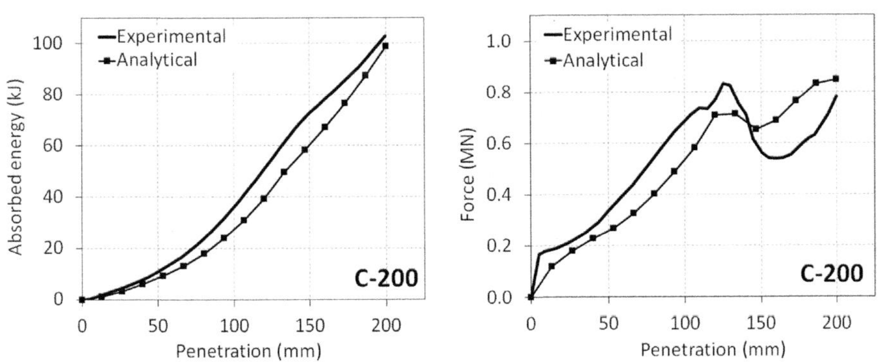

Fig. 4.21 Energy–penetration and force–penetration curves: Model C-200.

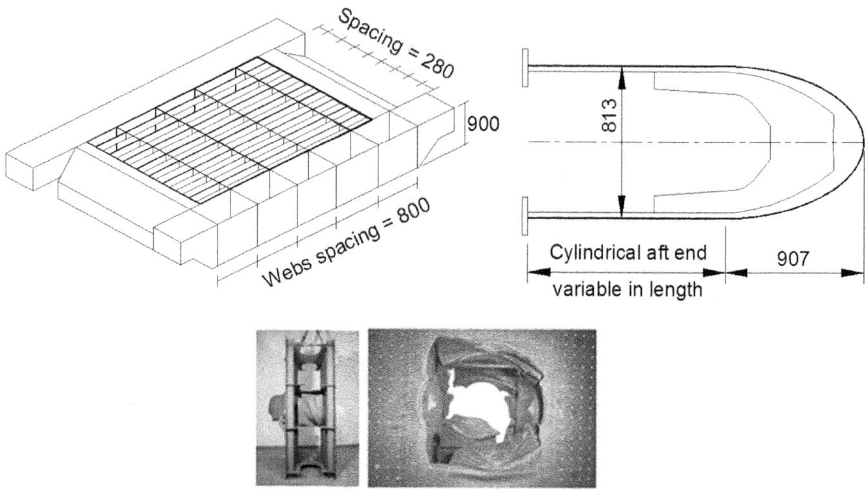

Fig. 4.22 Illustration of Tautz et al. (2013) collision experiment CE-1.
Photos cf. Tautz, I., Fricke, W., Schöttelndreyer, M., 2014. Validierung von Kollisionsberechnungen durch Großversuche an Konstruktionsvarianten von Seitenhüllen (in German). Technical Report, Technische Universität Hamburg. ISBN 978-3-89220-676-7.

Table 4.4 Material properties of the top and bottom plate: Tautz et al. (2013).

ID	Model	Plate thickness (mm)	σ_y (MPa)	σ_u (MPa)	ε_f (−)
17	CE-1	4.0	340	460	0.32

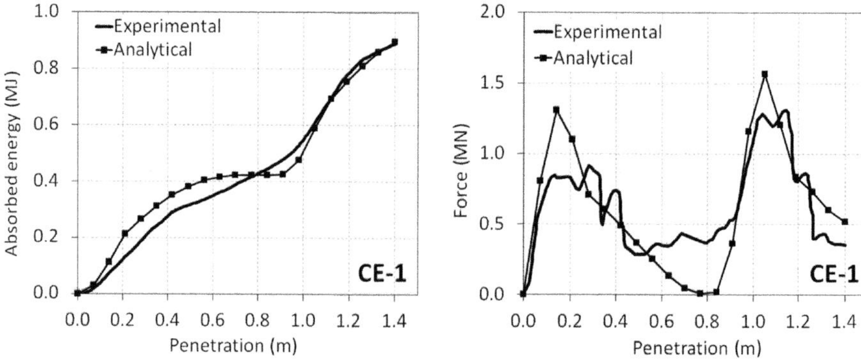

Fig. 4.23 Energy–penetration and force–penetration curves: Model CE-1.

the penetration of about 800 mm, the agreement beyond that point is very good. The initial overestimation in the energy response causes the deviation in the force–displacement response before 800 mm.

4.3.1.5 Large scale model test by Lehmann and Peschmann (2002)

Lehmann and Peschmann (2002) reported the results of a large-scale collision test carried out in association with the Netherlands Organization for Applied Scientific Research (TNO) (one of these experiments is presented in Fig. 3.3 of Chapter 3). This collision test can be also found in Peschmann (2001).

Two inland waterway vessels of about 80 m length were used in the collision tests. A double-hull structure (referred here as test model) was inserted at the midship region of the struck ship, representing 1/3 of a medium-size double-hull tanker of approximately 30,000 DWT (see Fig. 4.24). The dimensions of the test model were length 7.5 m, double side width 0.7 m, and total height 4.2 m. The thickness of the outer and inner shell plating was 5.0 mm stiffened by flat bars 100 × 5.0 mm equally spaced at 225 mm. The vertical webs were 6.0 mm thickness spaced at 1125 mm, while the horizontal stringers were 5.0 mm thickness spaced at 1130 mm. The striking ship was fitted with a rigid bulbous bow, and the collision speed was 2.55 m/s.

As ship motions occurred during the collision, this test is considered a rather complicated experiment. Since a large penetration was achieved during the test, both the outer and inner hulls suffered severe damage, and thus this test provides valuable information for the validation of theoretical models.

The material properties were obtained from the standard uniaxial tensile test performed by Peschmann (2001): yield stress $\sigma_y = 319$ MPa, ultimate tensile stress

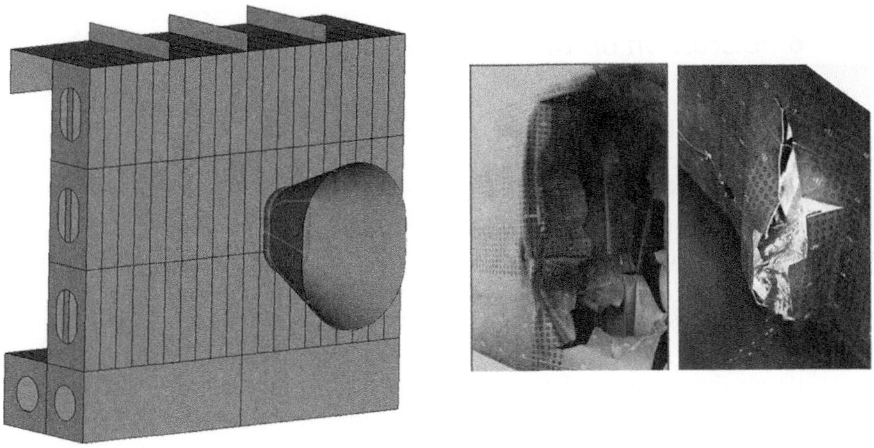

Fig. 4.24 Large-scale collision experiment conducted in the Netherlands.
Photos cf. Peschmann, J., 2001. Berechnung der Energieabsorption der Stahlstruktur von Schien bei Kollisionen und Grundberuhrungen (in German). (PhD thesis), Technical University of Hamburg, Germany.

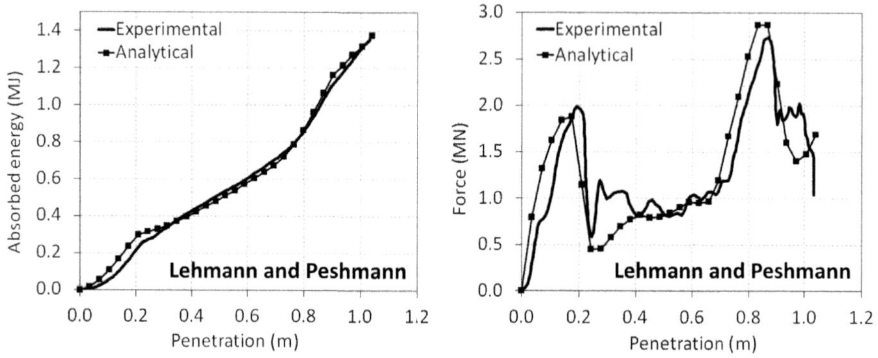

Fig. 4.25 Energy–penetration and force–penetration curves: Model by Lehmann and Peschmann.

$\sigma_u = 423$ MPa, and elongation at break $\varepsilon_f = 0.3$. Peschmann (2001) reported that the scratches on the model after the collision were very small. Therefore, he used a relatively small friction coefficient ($\mu = 0.1$) in his finite element analysis, as it gave the best agreement with the experimental results. Considering that the contribution of friction to the total absorbed energy is negligible, the friction effect was ignored in the analytical calculation.

The absorbed energy–penetration and force–penetration curves from experiments and analytical analysis are presented in Fig. 4.25. It is seen from the comparison that the analytical method gives an excellent agreement with the experimental results for such a complicated collision test.

4.3.1.6 Discussion on the validation of the method with side collision experiments

The predictions of the analytical method have been compared with 18 experiments. In general, the analytical calculations follow the experimental response quite accurately.

From the comparison with the experiments conducted by Arita and Aoki (1985), it can be observed that the energy- and force–displacement curves are well predicted by the analytical method for specimens S-II and S-III. For specimen S-I, the analytical calculation reaches a reasonable agreement with the experimental results.

The comparisons with specimens ST-3-BW and ST-4-BW also show that the analytical prediction of the energy responses is close to the experimental results. It should be noted here, that the material parameters are obtained from uniaxial tensile tests, and consequently, the material uncertainty has been diminished. In spite of some deviations with the force–displacement curves, all analytical calculations predict accurately the energy absorbed during the entire impact event. It should be recalled that the calculated collision force only represents the mean value of each analysed step and it does not reflect the peak force caused by buckling in the experiments.

The predictions of the results from the experiments of Wang et al. (2000) lead to more challenges as the radius of the indenters varies considerably and the impact occurs at different locations. For the collisions with the centre of the indenter positioned between webs, the analytical calculations follow the 'entire' energy–displacement response quite well. For the impacts on a web (W-50 and W-200) and at intersection of webs (C-50 and C-200), the analytical calculations follow the energy response equally well.

For the model of Tautz et al. (2013), the analytical calculation follows the experimental curves with a relatively good accuracy. Although it overestimates the energy absorption until penetration of 800mm, the predictions are quite accurate beyond that point. The special feature of this test model, which is different from the models discussed above, is that the outer and inner bottoms were made from stiffened panels. Here, the smearing of the stiffeners to the equivalent plate seems to work well.

The large-scale collision test by Lehmann and Peschmann (2002) is definitely the more complex scenario presented here. However, the analytical approach predicts the experimental response very well.

Table 4.5 and Fig. 4.26 show the summary of the calculated and experimental absorbed energy at the end of each experiment. It can be concluded that the standard deviation of the analytical method is 0.04 and all differences are within the range ±10%.

The analysis of all results presented here indicates that the analytical method can be used for collision assessments with a good degree of confidence when it is required to estimate the extent of structural damage.

It should be mentioned that one practical application of this method has already been established. This analytical procedure is recommended in the published *Guidance Notes for Collision Assessment for the Location of Low-flashpoint Fuel Tanks* (Lloyd's Register, 2016).

4.3.2 Bow crushing

This section provides analysis and comparison with experimental results on ship bow crushing.

4.3.2.1 Comparison with Woisin's (1979) bow crushing experiments

Woisin (1979) conducted a series of model collision tests for protection of nuclear powered ships against collisions. An example of the collision situation and the damaged bow after a collision test is illustrated in Fig. 4.27.

The collision tests showed that the bow suffered significant damage while the damage to the side structure was small, mainly because the tests were designed to protect the side structure. The average flow stress of the materials was (320+470)/2=395MPa. The average thickness of the bow structure was 1.375mm, and the stiffener spacing of the horizontal deck was 135mm. Thus, the absorbed energy related to the destroyed material volume can be expressed as (Eq. 4.2)

$$E = 3.50 \cdot \left(\frac{1.375}{135}\right)^{0.67} \cdot 395 \cdot R_t = 64 \cdot R_t$$

Table 4.5 Summary of absorbed energy at the end of the collision tests.

		Penetration	Absorbed energy (kJ)		
ID	**Model**	**(mm)**	**Calculated**	**Experimental**	**Ratio = Cal./ Exp.**
1	S-I	284	196.4	195.4	1.01
2	S-II	283	258.8	244.7	1.06
3	S-III	284	309.9	325.4	0.95
4	ST-3-BW	406	216.8	205.6	1.05
5	ST-4-BW	394	349.1	317.4	1.10
6	ST-3-OW	352	200.0	197.3	1.01
7	ST-4-OW	342	333.0	334.4	1.00
8	P-10	200	48.4	45.4	1.07
9	P-50	200	58.3	57.8	1.01
10	P-100	200	71.2	72.1	0.99
11	P-200	200	99.1	98.2	1.01
12	P-300	200	131.0	137.8	0.95
13	W-50	200	55.2	54.3	1.02
14	W-200	200	100.0	97.8	1.02
15	C-50	200	56.1	59.0	0.95
16	C-200	200	98.8	103.0	0.96
17	CE-1	1400	895.4	888.9	1.01
18	Lehmann and Peschmann	1032	1370.0	1353.4	1.01
18 models		Mean value			1.01
		Standard deviation			0.04

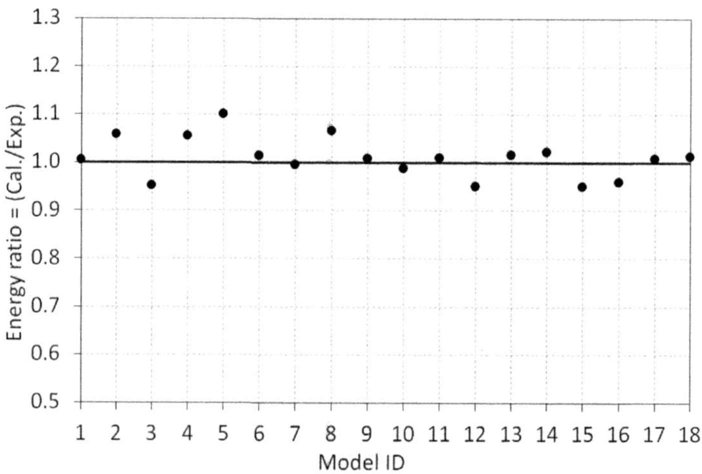

Fig. 4.26 Collision energy ratio of the calculated and experimental result at the end of collision tests for the 18 models.

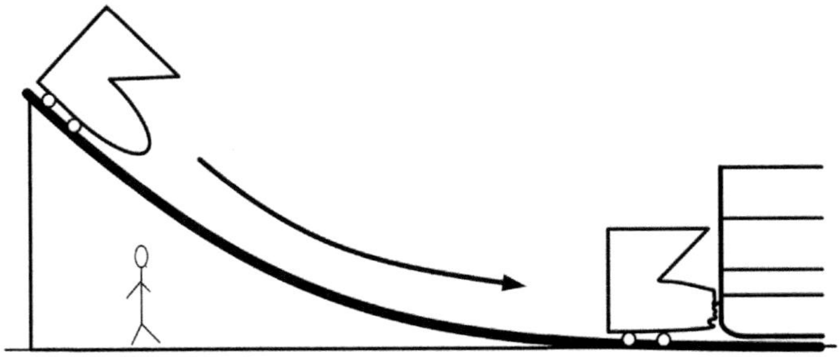

Fig. 4.27 Illustration of bow damage collision tests by Woisin (1979).

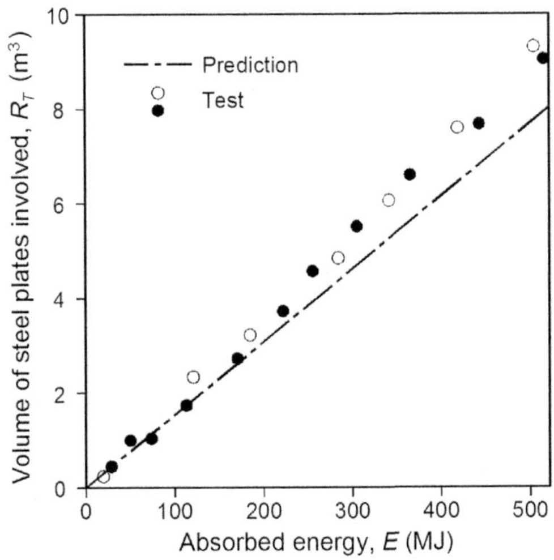

Fig. 4.28 Absorbed energy and the destroyed material volume obtained by the present calculations and Woisin's test results of bow crushing.

The analytical results using Eq. (4.2) and the test results of Woisin are compared in Fig. 4.28 for different collision tests; it is seen that the correlation is good.

4.3.2.2 Comparison with Amdahl's (1983) bow crushing tests

Amdahl (1983) performed a series of model tests on bow crushing. Four test specimens are chosen here for the comparison. The main dimensions and structural arrangements are shown in Fig. 4.29.

The models were made of 2.0mm-thick mild steel plates, while frames and stiffeners were made of 3.0mm flat bars. The average flow stress of the material is

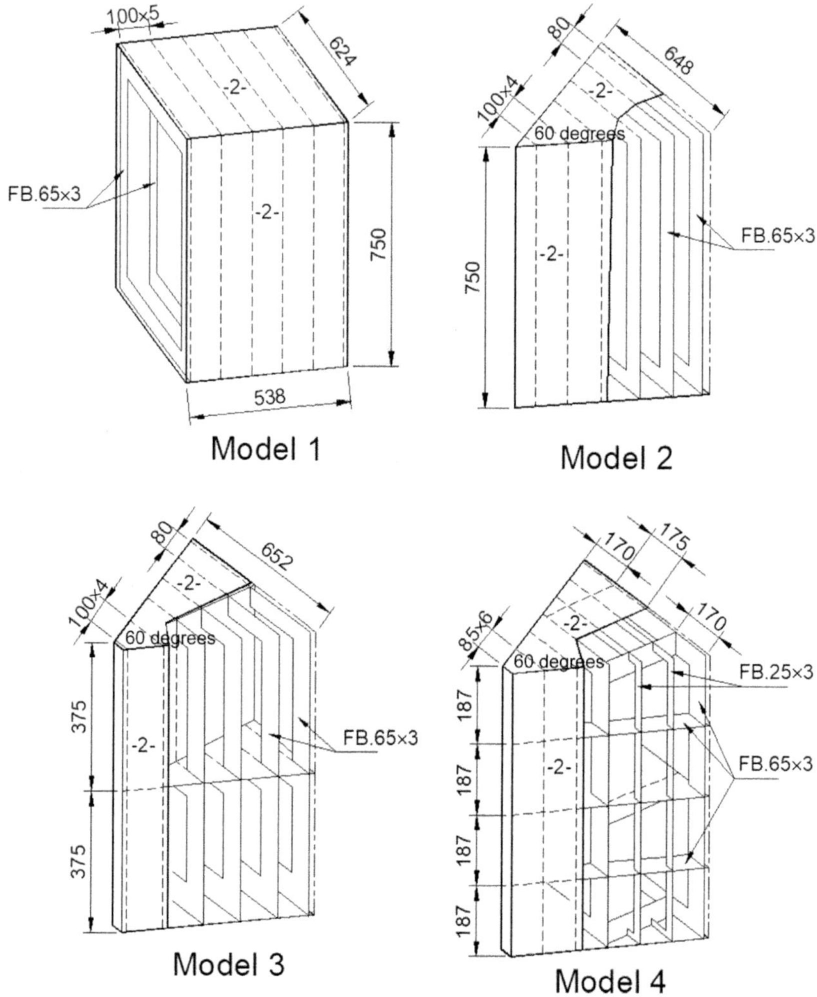

Fig. 4.29 Bow models for crushing test performed by Amdahl (1983).
Based on Amdahl, J., 1983. Energy Absorption in Ship-Platform Impact. (PhD thesis), Norwegian Institute of Technology, Report No. UR-83-34.

$(220+310)/2=265\,\text{MPa}$. The bow models were crushed progressively during the tests until a damage length of 0.2 m. The material volume and dissipated energy for these four models at a damage length of 0.2 m are given next:

Model 1—Rectangular box with transverse frames

Destroyed material volume: $R_t = 2 \times (0.75 + 0.624) \times 0.2 \times \left(2 + \frac{65 \times 3}{100}\right)/1000 = 2.17 \cdot 10^{-3}\,(\text{m}^3)$

Average width of the plates in the crushed cross section: $b = (750 + 624)/2 = 687$ (mm)

Energy dissipated: $E_{model1} = 3.50 \times \left(\frac{2}{687}\right)^{0.67} \times 265 \times R_t = 18.55 \cdot R_t = 40.3$ (kJ)

Model 2—Wedge-shaped bow

Destroyed material volume: $R_t = 1.37 \times 10^{-3}$ (m^3)
Average width: $b = 750/2 + 100 \times \tan 30$ degrees $= 432.7$ (mm)

Here, the width b is a linear function of the crushing distance, and thus, it is reasonable to use the mean value at half of the considered damage length.

Energy dissipated: $E_{model2} = 3.50 \times \left(\frac{2}{432.7}\right)^{0.67} \times 265 \times R_t = 25.28 \cdot R_t = 34.63$ (kJ)

Model 3—Wedge-shaped bow with an additional deck and a longitudinal bulkhead

Destroyed material volume: $R_t = 1.81 \times 10^{-3}$ (m^3)
Average width: $b = (375 + 100 \times \tan 30$ degrees$)/2 = 216.4$ (mm)

Energy dissipated: $E_{model3} = 3.50 \times \left(\frac{2}{216.4}\right)^{0.67} \times 265 \times R_t = 40.22 \cdot R_t = 72.79$(kJ)

Model 4—Wedge-shaped bow with shell longitudinal stiffeners

Destroyed material volume: $R_t = 1.07 \times 10^{-3}$ (m^3)
Average width: $b = 94$ (mm)
This is because the spacing of the longitudinal stiffeners is 94 mm.

Energy dissipated: $E_{model4} = 3.50 \times \left(\frac{2}{94}\right)^{0.67} \times 265 \times R_t = 70.31 \cdot R_t = 75.23$ (kJ)

Results summary of the four model crushing test

Table 4.6 presents the comparison of the absorbed energy obtained by the analytical method and the test results for the four models.

In general, the agreement between the present method and the test results varies. While the agreement for Model 3 is excellent, the difference for Model 2 is relatively large (+28%). It should be mentioned that a better agreement can be achieved if smaller calculation steps are used. From the analytical calculations, it is interesting to observe that the energy absorption efficiency (E/R_t) is quite different for the four

Table 4.6 Absorbed energy of the bow models at the crushing damage length of 0.2 m.

	Model 1	Model 2	Model 3	Model 4
Test result	48.0 kJ	27.0 kJ	70.0 kJ	90.0 kJ
Calculated	40.3 kJ	34.6 kJ	72.8 kJ	75.2 kJ
Error	−16.0%	+28.1%	+4.0%	−16.4%
E/R_t (MJ/m^3)	18.55	25.28	40.22	70.31

bow models (see Table 4.6); this energy represents the absorbed energy per unit volume of destroyed material. It is noticed that the Model 3 and Model 4 have a relatively high efficiency of energy dissipation.

4.3.3 Ship grounding

This section provides analysis and comparison with experimental results on ship grounding.

4.3.3.1 Comparison with Rodd (1997) grounding tests

A series of grounding model tests were performed in the Naval Surface Warfare Center in the United States (Rodd, 1997). The test facility was used to simulate the grounding of small-scale bottoms of oil tankers of 30,000–40,000 DWT on a pinnacle rock. The scale of the test model was 1/5, and the test setup is shown in Fig. 4.30. The experimental characteristics of four of these model tests are first summarized, and then, the experimental results are compared with the present analytical method.

Model 1 is a conventional double-bottom design with transverse webs (floors) and a central longitudinal girder (2.3 mm) between the outer (3.0 mm) and the inner hull (3.0 mm); from the 14 floors spaced at 460 mm, 7 were placed between the aft and mid bulkhead and the other 7 – between the mid and fore bulkhead. The outer and inner plates were stiffened by longitudinally oriented stiffeners spaced at 130 mm (respective profiles: L 69.9 × 3/31.8 × 3 and L 35 × 3/35 × 3).

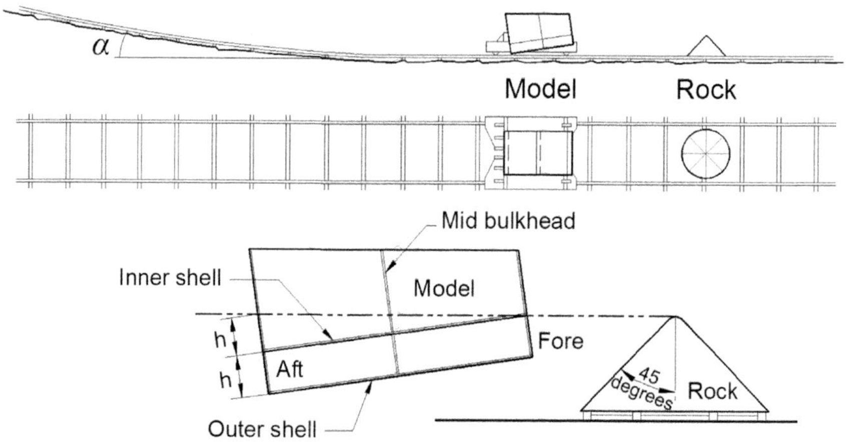

Fig. 4.30 Experimental setup of grounding tests performed in the United States.
Based on Rodd, J.L., 1997. Frame design effects in the rupture of oil tanks during grounding accidents. In: Int. Conf. on Advances in Marine Structures III, DERA, ROSYTH, UK and Pedersen, P.T., Zhang, S., 2000. Absorbed energy in ship collisions and grounding—revising Minorsky's empirical method. J. Ship Res. 44(2), 140–154.

Model 2 is a nonconventional double-hull construction, as it does not have floors between the inner and the outer hull. Instead, the model was reinforced with 10 longitudinal webs of 3.0mm equally spaced at 230mm and stiffened by two flat bar profiles 25.4 × 3mm. The outer and inner shells and the mid bulkhead had thickness of 3.0mm.

Model 3 is also a nonconventional double-hull construction similar to Model 2. However, Model 3 was designed with 14 longitudinal web girders spaced at 170mm, and the mid bulkhead was constructed with a double skin of width 170mm.

Model 4 is similar to Model 3, but the longitudinal girders are 3.4mm thick, as they were not stiffened. Details of all four models can be found in Rodd (1997) or Pedersen and Zhang (2000).

For all four models, the yield stress of the material is 283MPa, and the ultimate stress is 345MPa. In order to calculate the energy absorbed by the damaged bottom structures using Eq. (4.3), we must determine the critical tearing length and the volume of the damaged material. As discussed in Chapter 3, Section 3.11.2, the steady tearing state is reached when the tearing length is approximately equal to twice the wedge length. Since the cone-shaped rock has a circular section in the horizontal plane, the length of the wedge can be assumed to be the radius of the rock at the outer bottom plate. This implies that the critical tearing length is equal to the diameter of the cone rock at the outer bottom level.

Fig. 4.31 shows the damaged section of Model 3 after the grounding experiment (Rodd, 1997). It is seen that the damaged material of the double bottom is just around the outer contour of the rock. It can be also noticed that the width of the damage in the outer and the inner bottom is similar. Therefore, it has been assumed here, that the inner bottom and the outer bottom suffer the same damage width equal to the cone diameter at the outer bottom level.

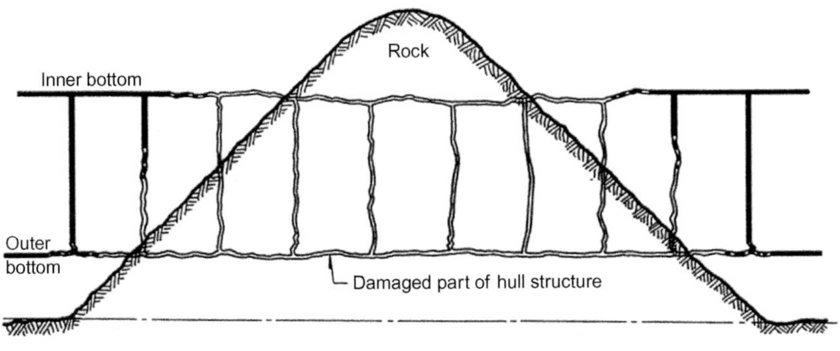

Fig. 4.31 Illustration of damage section of Model 3 (ADH/PD328) in the grounding test. Based on Rodd, J.L., 1997. Frame design effects in the rupture of oil tanks during grounding accidents. In: Int. Conf. on Advances in Marine Structures III, DERA, ROSYTH, UK and Pedersen, P.T., Zhang, S., 2000. Absorbed energy in ship collisions and grounding—revising Minorsky's empirical method. J. Ship Res. 44(2), 140–154.

Details of the calculation procedure for each model at a penetration length of 5.46 m are given next:

Model 1

The thickness of the outer bottom plating is 3.0 mm. Considering the longitudinal stiffeners on the outer hull, the equivalent thickness of the outer bottom is calculated from

$$t = 3.0 + \frac{(69.9 + 31.8) \times 3.0}{130} = 5.35\,(\text{mm})$$

The semiangle of the rock is 45°. Therefore, the damage width at the penetration length of 5.46 m is obtained as 2(0.37 + 0.37(5.46/7.32)) = 1.292 m, where 7.32 m is the total length of Model 1. Thus, the average damage width is (0.74 + 1.292)/2 = 1.016 m. The damaged material volume at the damage length of 5.46 m is approximately $R_t = 0.0687\,\text{m}^3$.

Then, the energy dissipated by Model 1 at the damage length of 5.46 m is calculated from

$$E_{model1} = 3.21 \times \left(\frac{5.35}{1016}\right)^{0.6} \times 314 \times 0.0687 = 43.28 \cdot R_t = 2.97\,(MJ)$$

Model 2

The calculation procedure is similar to that of Model 1. The outer hull thickness is 3.0 mm. Considering the longitudinal girders on the outer hull, the equivalent thickness of the outer bottom is $t = 5.74$ mm.

The damage width at the penetration length of 5.46 m is obtained as 2(0.37 + 0.37 (5.46/6.7)) = 1.342 m, where 6.7 m is the total length of Model 2. Thus, the average damage width is (0.74 + 1.342)/2 = 1.041 m. The destroyed material volume at the damage length of 5.46 m is obtained approximately as $R_t = 0.0675\,\text{m}^3$. Therefore, the energy dissipated by Model 2 at the damage length of 5.46 m is determined as

$$E_{model2} = 3.21 \times \left(\frac{5.74}{1041}\right)^{0.6} \times 314 \times 0.0675 = 44.50 \cdot R_t = 3.0\,(MJ)$$

Model 3

The outer hull thickness is again 3.0 mm. Considering the longitudinal girders on the outer hull, the equivalent thickness of the outer bottom is calculated as $t = 6.98$ mm.

The damage width at a penetration length of 5.46 m is obtained as 2(0.4 + 0.4 (5.46/5.48)) = 1.6 m, where 5.48 m is the total length of Model 3. Thus, the average damage width is (0.8 + 1.6)/2 = 1.2 m. The destroyed material volume at the damage length of 5.46 m is calculated as $R_t = 0.0967\,\text{m}^3$. Therefore, the energy dissipated by Model 3 at the damage length of 5.46 m is found to be

$$E_{model3} = 3.21 \times \left(\frac{6.98}{1200}\right)^{0.6} \times 314 \times 0.0967 = 45.95 \cdot R_t = 4.45\,(MJ)$$

Model 4

The outer hull thickness is also 3.0 mm. Considering the longitudinal girders on the outer hull, the equivalent thickness of the outer bottom is calculated as $t = 70\,\text{mm}$.

The damage width at the penetration length of 5.46 m is $2(0.4 + 0.4(5.46/5.48)) = 1.6\,\text{m}$, where 5.48 m is the total length of Model 4. Thus, the average damage width is $(0.8 + 1.6)/2 = 1.2\,\text{m}$. The destroyed material volume at the damage length of 5.46 m is found to be $R_t = 0.0969\,\text{m}^3$. Then, the energy absorbed by Model 4 at the damage length of 5.46 m is calculated from

$$E_{model4} = 3.21 \times \left(\frac{7}{1200}\right)^{0.6} \times 314 \times 0.0969 = 46.03 \cdot R_t = 4.46\,(MJ)$$

Summary of the results on grounding tests

The summary of the comparison between the analytical and experimental absorbed energies at the damage length of 5.64 m is given in Table 4.7. Although the actual damage of the grounding tests is very complex, it can be seen that an acceptable or even good agreement can be achieved by the present simplified method as it manages to catch the main damage mechanisms of a grounding event. The energy absorption efficiency (E/R_t) of the four models is similar.

4.3.3.2 Comparison with ASIS grounding tests

During 1994–95, a series of grounding experiments sponsored by the Japanese ASIS (1998) were carried out. Some of these test results were presented by Wang et al. (1997). The experiments were used to simulate the grounding of a 1/4 scaled very

Table 4.7 Comparison of the absorbed energy obtained by the present method and that from the grounding tests when the damage length is 5.46 m (Rodd, 1997).

	Model 1	Model 2	Model 3	Model 4
Test result	3.25 MJ	2.65 MJ	5.34 MJ	6.03 MJ
Calculation	2.97 MJ	3.00 MJ	4.45 MJ	4.46 MJ
Error	−8.6%	+13.2%	−16.7%	−26.0%
E/R_t (MJ/m^3)	43.28	44.50	45.95	46.03

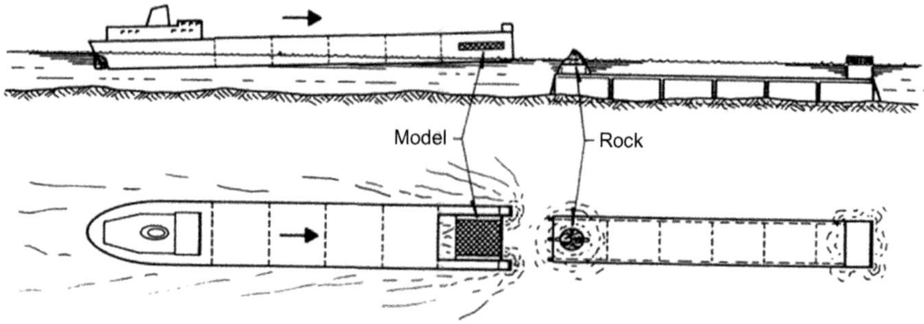

Fig. 4.32 Illustration of test setup of the grounding tests.

large crude carrier (VLCC). The general test setup is shown in Fig. 4.32. The test model is fixed on a ship that runs towards an artificial pinnacle rock.

The test models were double-bottom constructions of height 750 mm where the thickness of the outer and inner hull was 5.0 mm, the spacing of the 5.0 mm transverse floors was 1250 mm, and the spacing of the longitudinal stiffeners of the hull plates was 250 mm (flat bars 150 × 5 mm). The flow stress of the material was 340 MPa. Two tests have been selected for comparison with the simplified method.

Test No. 1 was designed to simulate a minor grounding situation. Only the bottom outer hull was damaged, while the inner hull remained intact. The initial position of the top of the artificial rock was 180 mm below the inner hull. The test result revealed that the damage of the outer hull was mainly confined within three times the spacing of the longitudinal stiffeners (3 × 250 mm = 750 mm). The equivalent thickness of the outer hull plating, including longitudinal stiffeners, is calculated as $t = 5.0 + (150 \times 5.0)/250 = 8.0$ mm. Thus, the destroyed material volume at the penetration length of 5.0 m is $R_t = 0.0356\,\text{m}^3$.

Similarly to the US grounding tests (Rodd, 1997), the critical tearing length is assumed to be equal to the damage width (750 mm in this case). Thus, the absorbed energy at a penetration length of 5.0 m is

$$E = 3.21 \times \left(\frac{8}{750}\right)^{0.6} \times 340 \times 0.0356 = 71.58 \cdot R_t = 2.55\,(MJ)$$

The test result of the absorbed energy at the penetration length of 5.0 m is 2.20 MJ. For reference, the calculation result given by Wang et al. (1997) is 1.75 MJ. The comparison of the absorbed energy penetration determined by the present method and that of the test result is shown in Fig. 4.33; the agreement is reasonable.

Test No. 2 was designed to simulate a severe grounding accident, and thus, both the outer and inner hull were damaged. The initial position of the top of the artificial rock was 540 mm above the inner hull. The test results revealed that the damage mode of the outer hull was very complex and the inner hull suffered mainly tearing damage. The damage width in the transverse direction varied from four to seven times the

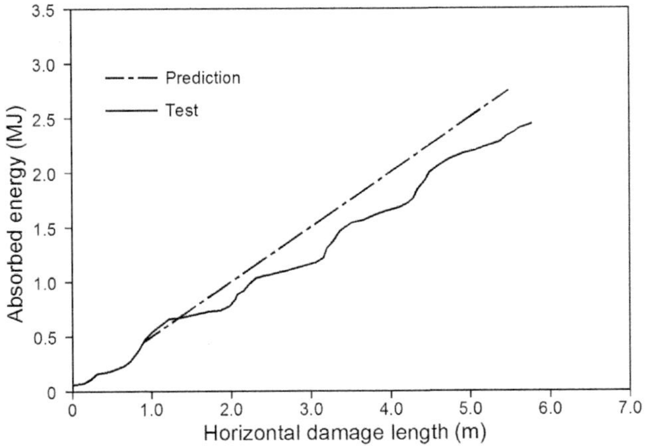

Fig. 4.33 Comparison of absorbed energy obtained by the present method and test results (Test No. 1 by Wang et al., 1997).

stiffener spacing at the outer hull and three to five times the stiffener spacing at the inner hull.

Here, it is assumed that the average damage width in the outer and the inner hull is four times the spacing of the longitudinal stiffeners ($4 \times 250\,\text{mm} = 1000\,\text{mm}$). Thus, the destroyed material volume at the penetration length of 3.0 m is calculated as $R_t = 0.057\,\text{m}^3$.

As per Test No. 1, the critical tearing length is assumed to be equal to the damage width in the outer hull (1000 mm). Then, the absorbed energy is

$$E = 3.21 \cdot \left(\frac{8}{1000}\right)^{0.6} \cdot 340 \cdot R_T = 60.23 \cdot R_t = 3.43\,(MJ)$$

The test result for the absorbed energy at the penetration of 3.0 m is 3.1 MJ. For reference, the result given by Wang et al. (1997) is 2.7 MJ. The comparison of the absorbed energy penetration determined by the present method and that of the test result is shown in Fig. 4.34; again, the agreement is good.

The absorbed energy for Test No. 1 and Test No. 2 obtained by the present method and the grounding tests is presented in Table 4.8.

4.4 Application

Applications of the damaged material volume method to three actual cases of ship collision and grounding are presented and discussed in this section. An estimate of the absorbed energies of oil tankers at initial rupture of the side shell in ship collisions is also presented.

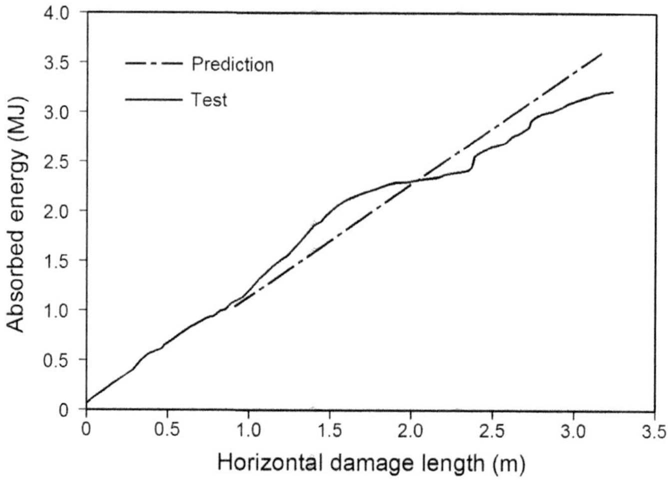

Fig. 4.34 Comparison of absorbed energy obtained by the present method and the test results (Test No. 2 by Wang et al., 1997).

Table 4.8 Comparison of the dissipated energy obtained by the analysis and the experiments (ASIS grounding tests).

	Test No. 1 at 5.0 m penetration	Test No. 2 at 3.0 m penetration
Test result	2.20 MJ	3.10 MJ
Present calculation	2.55 MJ	3.43 MJ
Error	+15.9%	+10.6%
E/R_t (MJ/m^3)	71.58	60.23

4.4.1 Application to a full-scale collision accident

Martens (2014) reported a full-scale collision accident that occurred in the German harbour Travemünde in 2012: the collision happened between the roll on/roll off passenger (RoPax) vessels URD (struck ship) and the NILS HOLGERSSON (striking ship). Martens (2014) also performed a detailed finite element analysis of the collision accident; the outcomes of the FEA analysis can also be found in Storheim et al. (2015).

The struck vessel was moored alongside the quay when the striking vessel lost control during manoeuvring and impacted the struck vessel at 6.5 knots (3.34 m/s); the impact direction was almost perpendicular. The struck ship suffered considerable damages both above and below the waterline, whereas the striking ship received only minor plate dents, as the NILS HOLGERSSON is an ice-strengthened ship (see Fig. 4.35). The maximum indentation into the struck vessel was 7.6 m at the upper deck level (measured from deck edge) and about 4.6 m at the waterline level (measured from the ship side shell).

Fig. 4.35 Damage in the RoPax URD and the NILS HOLGERSSON.
cf. Martens, I., 2014. Konstruktive Aspekte beim Entwurf von Bugwülsten zur Verbesserung des Energieaufnahmevermogens bei Schiffskollisionen. (PhD thesis), Schriftenreihe Schiffbau, Bericht 679. Technical University of Hamburg, Germany. ISBN 978-3-89220-679-8 (in German).

The mass of the NILS HOLGERSSON was 20,500 tons at the draught she had at the impact. The added mass of the vessel may have been increased as a result of shallow water effects as the depth of the harbour was 10 m. Under the assumption of a 10% added mass, the total available kinetic energy was estimated at 126.5 MJ.

Martens (2014) estimated that 12.1 MJ of the dissipated energy was consumed by the motions of the striking and struck vessels and the deflection of the fender system along the quayside. This leaves 114.4 MJ to be dissipated through strain energy and friction.

The yield stress of the materials used in the present analytical calculation is the same as the yield stress used by Martens (2014) and Storheim et al. (2015) in the FE simulations. On the other hand, the ultimate tensile stress and the rupture strain are estimated based on material data from the shipbuilding industry (see Table 4.9). The thickness of the upper part of the side shell is 14 mm, and the lower part (region hit by the bulb) is 19 mm.

The energy predicted by the analytical method, 111.4 MJ, at the damage penetration of 4.6 m from ship side is very close to the actual value of 114.4 MJ. Fig. 4.36 shows the comparison between the analytical calculation and the FE simulation results (Martens, 2014); excellent agreement is observed.

4.4.2 Application to a full-scale grounding accident

In 1975, a 1-year-old single-hull tanker ran aground at a speed of 11.5 knots (5.92 m/s) in the coast of Singapore. The bottom of the ship was torn about 180 m from forward to the midship, and more than 10,000 tons of oil were spilled. The main particulars of the tanker are shown in Table 4.10, and the grounding situation is illustrated in Fig. 4.37.

Kuroiwa (1996), Wang et al. (1997) and Simonsen (1997a,b) have analysed this grounding accident. Good correlation between their predictions and the accidental

Table 4.9 Material properties used in the analytical calculation of the full-scale collision accident.

Model	σ_y (MPa)	σ_u (MPa)	ε_f (−)
RoPax URD incident	314	462	0.33

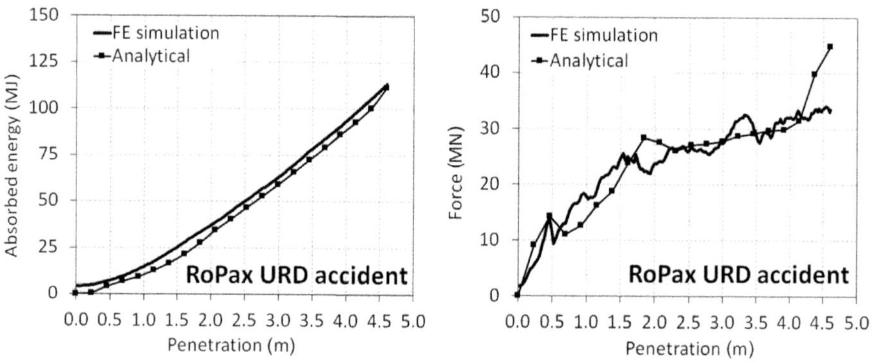

Fig. 4.36 Analytical results of energy–penetration and force–penetration curves of the full-scale collision accident (FE results were taken from Martens, 2014). The penetration is measured from the shipside at the waterline level.

Table 4.10 Main particulars of the grounded oil tanker.

Length between perpendiculars	304.0 m
Breadth moulded	52.4 m
Depth moulded	25.7 m
Draught	19.8 m
Displacement	237,000 tons
Grounding speed	11.5 knots

damage was found, though under certain assumptions. Here, the present damage material volume method is used to analyse this grounding event.

The thickness of the single bottom plate was 28.5 mm. After considering the longitudinal stiffeners, among other structural components, the equivalent thickness of the bottom is estimated as 57.94 mm. The flow stress of the material is 320 MPa. The damage width in the transverse direction was approximately 7.0–8.0 m; here, it is assumed as 7.75 m, which is the same width as used by Kuroiwa (1996) in his FE calculation. Eventually, the destroyed material volume at the damage length L_{dam} is determined as

$$R_t = L_{dam} \times 7.75 \times 57.94/1000 = 0.449 L_{dam} \left(\mathrm{m}^3\right)$$

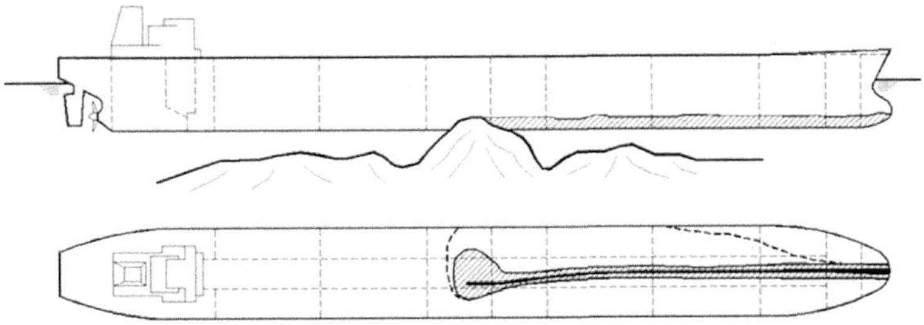

Fig. 4.37 The grounding accident and the resulting damage to the 237,000 tons single-hull tanker.

The initial kinetic energy of the tanker including the 5.0% of added mass effect is

$$E_0 = \frac{1}{2} M(1 + 0.05)V^2 = 4354.2\,(\text{MJ})$$

It is assumed that the entire initial kinetic energy of the tanker is dissipated by destroying the bottom structure. The critical tearing length is again taken to be the damage width of 7.75 m, and thus, the damage length for the grounding accident is calculated by Eq. (4.3):

$$L_{dam} = \frac{4354.2}{3.21 \cdot \left(\frac{57.94}{7750}\right)^{0.6} \cdot 320 \cdot 0.449} = 183.3\,(\text{m})$$

The actual damage length of the accident was about 180 m. Considering the complexity of the phenomenon, the agreement is very satisfactory.

4.4.3 Collision management and risk control for LNG carriers

It is a concern in the industry that LNG carriers during the loading and/or unloading at ports could be struck by ships entering the ports. As the characteristic bow of some ships is relatively sharp, the consequence of such collisions would be significant with potential gas leakage into the port and surrounding areas. In order to avoid such situations, it is of interest to determine the maximum critical collision speed that a typical striking ship should have, before reaching the LNG cargo tank boundary.

An LNG carrier of length 270 m is selected for this application example. The main dimensions of the LNG carrier are given in Table 4.11 together with those of a striking container ship. The collision scenario assumes that the LNG carrier is fully loaded and it is hit by the fully loaded container ship while waiting to be unloaded. It is also assumed that the LNG carrier is stationary and floating 'freely' before the perpendicular collision occurs at any location along its length, as illustrated in Fig. 4.38.

Table 4.11 Main particulars of the striking (container ship) and struck (LNG carrier) ships.

Ship	Length (m)	Breadth (m)	Depth (m)	Displacement (t)
LNG carrier (struck)	270	44.0	26.0	105,000
Container ship (striking)	230	32.2	21.1	55,800

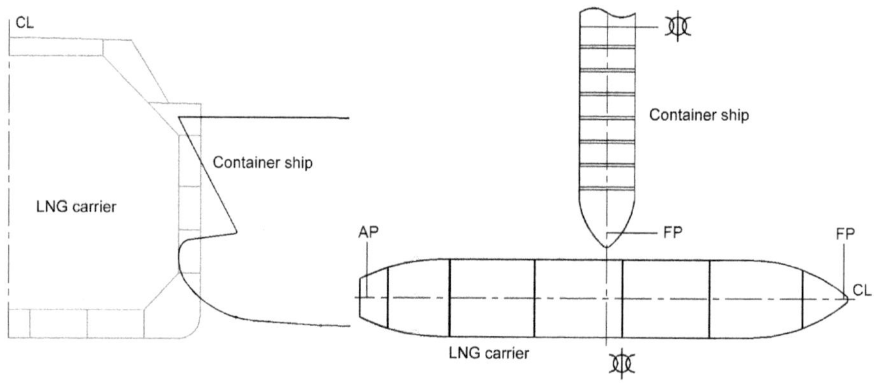

Fig. 4.38 The LNG carrier (struck ship) and the container ship (striking ship) collision scenario.

The task here is to determine the critical speed of the striking container ship so that the inner hull of the LNG carrier remains intact.

The width of the double-hull side structure is 2.5 m, and thus, the critical penetration was set at 2.5 m to avoid potential leakage of the cargo containment system. The properties of the mild steels are yield stress 282 MPa, tensile strength 438 MPa, and rupture strain 0.35.

The first step of the analysis is to determine the energy–penetration curves using the present analytical approach; the results are presented in Fig. 4.39 (left side). For reference, the energy–penetration and force–penetration curves created by the upper part of the striking bow and the lower part (the bulb) of the striking bow are presented in Fig. 4.40. The calculation results show that the structure of the LNG carrier can absorb 105 MJ at the critical penetration of 2.5 m.

The next step is to determine the critical collision speed resulting in the collision energy of 105 MJ. Here, the analytical method presented in Chapter 2 is used for estimating the collision speed, and the calculated results are shown in Fig. 4.39 (right side). In the calculations, the added mass coefficients were taken as 0.05 for surge motion, 0.85 for sway motion, and 0.21 for yaw motion. The radius of gyration was taken as a quarter of the ship length.

It is seen from the analysis results that the critical striking speed of the container ship is 4.2 knots for the collision amidship of the LNG carrier. When the collision location is at 20% of the ship length (from the aft perpendicular, AP), the critical

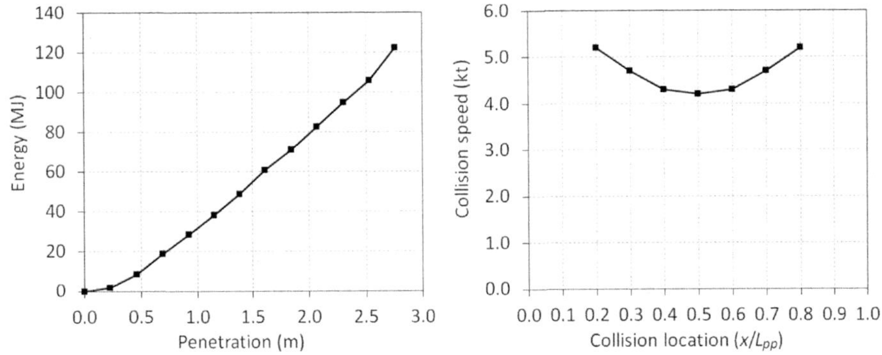

Fig. 4.39 Energy–penetration curve and critical collision speeds along the LNG carrier length.

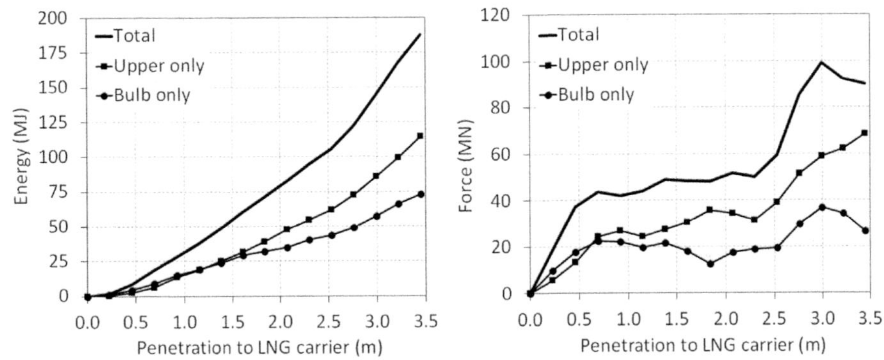

Fig. 4.40 Energy–penetration and force–penetration curves by the upper part of the striking bow and the lower part (the bulb) of the striking bow; the total energy and force are also included.

striking speed is 5.2knots (2.7m/s). It may be concluded that if the container ship is managed and controlled within a speed limit of 4.2 knots (2.2m/s) in the port and surrounding areas, the risk of causing LNG cargo leakage in the event of a collision would be very small.

A collision accident (February 2017) of a large LNG carrier of length 303m struck by a large oil tanker of length 263m while stationary at port has further enhanced the value of this application example. In the accident, the bulb of the striking ship created a big hole around the waterline area of the side shell of the LNG ship and also dented the upper deck structure, as reported by Zhang et al. (2018). It has been demonstrated that the present collision analysis procedure is capable of producing accurate results for this collision event.

For general risk assessments, the LNG carrier could be struck by ships of different dimensions. Then the energy absorption capability of the LNG carrier at the striking ship bow that just reaches the inner hull can be calculated in the same way as above. However, for easy and quick estimation, a series of numerical calculations has been carried out, and by curve fitting, the formula $E = 7 \cdot M^{0.25}$ for the energy absorbed by the LNG carrier (in MJ) has been developed, where M is the displacement of the striking ship (in tons). For midship perpendicular collision, the critical striking speed (in knots) can be estimated from $V_{cr} = 115 \cdot M^{-0.3}$.

4.4.4 Absorbed energy of oil tankers at initial rupture of the side shell in ship collisions

It is often important to know how much collision energy a ship can absorb before the rupture of the side shell in a collision. There is not a definite answer to this question, because it depends on the collision location, the striking ship size and bow shape, etc. However, from an engineering practical judgement, it is possible to give an indication of the energy absorbed before the side shell ruptures. To illustrate this, seven double-hull oil tankers with length from 170 to 320m are selected. These ships were built between the period 2002 and 2012 in high-tensile steels (yield stress 315 and 355 MPa).

The striking ships are assumed to be container ships of the same length as that of the struck oil tankers. The side impacts are assumed perpendicular (90 degrees) between two transverse frames of the struck oil tankers. Only the relatively sharp rigid bulbous bows of the striking container ships are assumed to be in contact with the tanker structures. That is, the upper part of the striking ship does not get into contact with the oil tanker.

Under such collision scenarios, the energy absorbed by the oil tankers at the initial rupture of the outer side shell is calculated using Eq. (4.1) and presented in Table 4.12. Here, the critical penetration at the initial rupture of the side shell is determined using the method given in Chapter 3, Section 3.4.6 ($\delta_f = {\varepsilon_f}^{0.4} b^{0.4} {R_b}^{0.6}$). It is seen that a larger ship will absorb considerably more energy than a smaller ship.

Table 4.12 Absorbed energy of double-hull oil tankers at initial rupture of side shell in ship collisions.

Oil tanker length (m)	Frame spacing (m)	Rupture strain (–)	Bulb radius (m)	Penetration at initial rupture (m)	Absorbed energy (MJ)
170	2.59	0.35	2.08	1.12	8
219	3.46	0.35	2.18	1.30	13
220	3.44	0.32	2.15	1.25	13
234	3.70	0.35	2.30	1.38	18
263	4.79	0.32	2.53	1.58	22
319	5.45	0.35	3.24	1.98	69
320	5.70	0.35	3.30	2.03	70

4.5 Prediction of bottom damage distributions for high-speed vessels

The International Maritime Organization (IMO, 1995a) adopted the International Code of Safety for High-Speed Craft (HSC), which states: "the longitudinal extent of damage should be $0.1L$ or $3\,\text{m}+0.03L$ or $11\,\text{m}$, whichever is the least, and the assumed damage should be increased by 50% in the case of damage in the forward $0.5L$ of the craft"; L is the length of the craft.

After several years of use, it was found that the code for defining grounding damage length was inadequate. In 1996, several European countries submitted a revision to the IMO, suggesting that the longitudinal extent of damage should be for the "full underwater length of the craft".

In 1997, Simonsen (1997a,b) proposed a procedure for the analysis of the bottom raking damage distribution for powered grounding events of a high-speed crafts. Simonsen used the existing bottom damage distribution of conventional ships (such as IMO damage statistics) to predict the damage distribution of new high-speed crafts. This prediction method is presented here together with application examples.

4.5.1 The prediction method

Simonsen (1997a,b) and Cerup-Simonsen et al. (2009) used a parameter to describe the bottom damage. The parameter is the ratio between the damage length and the ship length (L_{dam}/L). Based on a balance of energy approach, the following relation for two different ships can be established:

$$\frac{(L_{dam}/L)_1}{(L_{dam}/L)_2}=\frac{M_1}{M_2}\cdot\left(\frac{V_1}{V_2}\right)^2\cdot\frac{L_2}{L_1}\cdot\frac{P_2}{P_1}$$

where M is the ship mass, L is the ship length, V is the ship speed, and P is the horizontal grounding force. For comparison between various ship classes, the speed can be expressed by the Froude number ($F_n=V/\sqrt{gL}$), instead of the ship velocity. Then, the following equation can be derived:

$$\frac{(L_{dam}/L)_1}{(L_{dam}/L)_2}=\frac{M_1}{M_2}\cdot\left(\frac{F_{n1}}{F_{n2}}\right)^2\cdot\frac{P_2}{P_1} \tag{4.7}$$

Eq. (4.7) indicates that if we know the damage distribution $(L_{dam}/L)_1$ for one ship, then the damage distribution $(L_{dam}/L)_2$ for the other ship can be determined.

The major challenge of this procedure is to determine the horizontal grounding forces P_1 and P_2. These forces depend on many factors, such as the rock shape, bottom structural arrangement and material and indentation depth. Using Eq. (4.3), the energy balance for a ship grounded on a rock can be expressed as

$$\frac{1}{2}MV^2 = 3.21\left(\frac{t}{\lambda}\right)^{0.6}\sigma_0 R_t$$

The volume of the destroyed material in a bottom raking damage can be approximately calculated from $R_t = L_{dam} \cdot B_{dam} \cdot t_{eq}$, where L_{dam} is the damage length, B_{dam} is the damage width, and t_{eq} is the equivalent thickness of the whole bottom, including transverse and longitudinal webs and stiffeners. Note that t is the equivalent thickness of the outer bottom including longitudinal webs and longitudinal stiffeners only. As discussed previously, the critical tearing length is taken to be equal to the damage width ($\lambda = B_{dam}$). Then, the relationship for the relative damage between two different ships can be expressed as

$$\frac{(L_{dam}/L)_1}{(L_{dam}/L)_2} = \frac{M_1}{M_2} \cdot \left(\frac{F_{n1}}{F_{n2}}\right)^2 \left(\frac{\sigma_2}{\sigma_1} \cdot \frac{t_{eq2}}{t_{eq1}} \cdot \left(\frac{t_2}{t_1}\right)^{0.6} \cdot \left(\frac{B_{dam2}}{B_{dam1}}\right)^{0.4}\right) \tag{4.8}$$

All parameters of the right-hand side of Eq. (4.8) are known for the two given ships, except the damage width (B_{dam}). The damage width is related to the ship size and the colliding obstacle. Here is proposed that the vertical indentation of a rock into the ship bottom is proportional to the ship draught, implying that a ship with a larger draught will suffer larger vertical penetration and consequently, a larger damage width. The ratio of the damage width between two ships can be expressed as $\frac{B_{dam2}}{B_{dam1}} = \frac{T_2}{T_1}$, where T is the ship draught.

By simple substitution, the final expression for the relative damage length between the two ships is

$$\frac{(L_{dam}/L)_1}{(L_{dam}/L)_2} = \frac{M_1}{M_2} \cdot \left(\frac{F_{n1}}{F_{n2}}\right)^2 \left(\frac{\sigma_2}{\sigma_1} \cdot \frac{t_{eq2}}{t_{eq1}} \cdot \left(\frac{t_2}{t_1}\right)^{0.6} \cdot \left(\frac{T_2}{T_1}\right)^{0.4}\right) \tag{4.9}$$

If Eqs. (4.7) and (4.9) are compared, it is seen that the ratio of the horizontal grounding forces of the two ships can be expressed by the factor

$$K = \left(\frac{\sigma_2}{\sigma_1} \cdot \frac{t_{eq2}}{t_{eq1}} \cdot \left(\frac{t_2}{t_1}\right)^{0.6} \cdot \left(\frac{T_2}{T_1}\right)^{0.4}\right) = \frac{P_2}{P_1}$$

4.5.2 Application example to a high-speed craft

Four different ships are analysed here: a cargo ship, a midsize ferry, an oil tanker, and a high-speed craft (HSC). The main dimensions of the ships and the scantling of the bottom structures are presented in Table 4.13.

From Table 4.13, the equivalent thickness (t_{eq}) of the whole bottom, including transverse and longitudinal webs and stiffeners, and the equivalent thickness (t) of

Table 4.13 Main dimensions of the ships and scantling of the bottom structures.

	Cargo ship	**Ferry**	**Oil tanker**	**HSC**
Length BP (m)	122	128	219	63
Beam (m)	20.5	17.5	32.0	23.4
Depth (m)	12.2	12.4	21.6	10.7
Draught (m)	9.4	4.5	16.0	3.0
Mass (t)	18,000	6300	82,000	910
Service speed (m/s)	8.0	9.5	6.7	19.5
Froude no. (F_n)	0.23	0.27	0.15	0.79
Material	Steel $\sigma_0 = 260\,\text{MPa}$ Double bottom	Steel $\sigma_0 = 260\,\text{MPa}$ Double bottom	Steel $\sigma_0 = 260\,\text{MPa}$ Double bottom	Aluminium $\sigma_0 = 250\,\text{MPa}$ Single bottom
Outer shell thickness	19.0	12.0	18.5	8.0
Inner bottom thickness	19.5	12.0	18.5	NA
Double-bottom height	1500	1250	2150	NA
Spacing of trans. webs	750	2250	2700	1200
Equivalent thickness of floors	6.5	9.0	12.0	8.0
Spacing of girders	3400	3600	3400	NA
Equivalent thickness of girders	14.5	10.0	12.3	NA
Spacing of longitudinals	NA	610	850	240
Web height of longitudinals	NA	220	340	250
Web thickness of longitudinals	NA	10	12.0	8
Flange width of longitudinals	NA	70	NA	50
Flange thickness of longitudinals	NA	15	NA	10

NA, not applicable.

Table 4.14 The equivalent thickness t and t_{eq} of the bottom of the four ships.

	Cargo	Ferry	Tanker	HSC
t (mm)	22.2	19.1	27.2	18.4
t_{eq} (mm)	57.9	43.2	64.8	20.4

the outer bottom including longitudinal webs and longitudinal stiffeners are calculated and shown in Table 4.14.

The factor K representing the horizontal grounding force ratio between two ships is determined as

$$K_1 = \frac{Cargo}{Ferry} = 1.0 \times 1.34 \times 1.09 \times 1.34 = 1.96$$

$$K_2 = \frac{Cargo}{HSC} = 1.04 \times 2.84 \times 1.12 \times 1.58 = 5.23$$

$$K_3 = \frac{Ferry}{HSC} = 1.04 \times 2.12 \times 1.02 \times 1.18 = 2.65$$

$$K_4 = \frac{Tanker}{HSC} = 1.04 \times 3.18 \times 1.26 \times 1.95 = 8.15$$

The relative damage ratio between the two ships is given as (Eq. 4.9)

$$\frac{(L_{dam}/L)_{Ferry}}{(L_{dam}/L)_{Cargo}} = \frac{6300}{18{,}000} \times \left(\frac{0.27}{0.23}\right)^2 \times 1.96 = 0.95$$

$$\frac{(L_{dam}/L)_{HSC}}{(L_{dam}/L)_{Cargo}} = \frac{910}{18{,}000} \times \left(\frac{0.79}{0.23}\right)^2 \times 5.23 = 3.12$$

$$\frac{(L_{dam}/L)_{HSC}}{(L_{dam}/L)_{Ferry}} = \frac{910}{6300} \times \left(\frac{0.79}{0.27}\right)^2 \times 2.65 = 3.28$$

$$\frac{(L_{dam}/L)_{HSC}}{(L_{dam}/L)_{Tanker}} = \frac{910}{82{,}000} \times \left(\frac{0.79}{0.15}\right)^2 \times 8.15 = 2.51$$

It is seen that the damage ratios of the cargo ship and the ferry are similar. These vessels should be well represented by the IMO damage statistics. The results also show that the damage length ratio for the high-speed craft is more than twice the damage length ratio for the conventional ships. When the bottom damage distribution for conventional ships is known, the bottom damage distribution for a high-speed craft or other new type of ship can be established quite easily by those damage results.

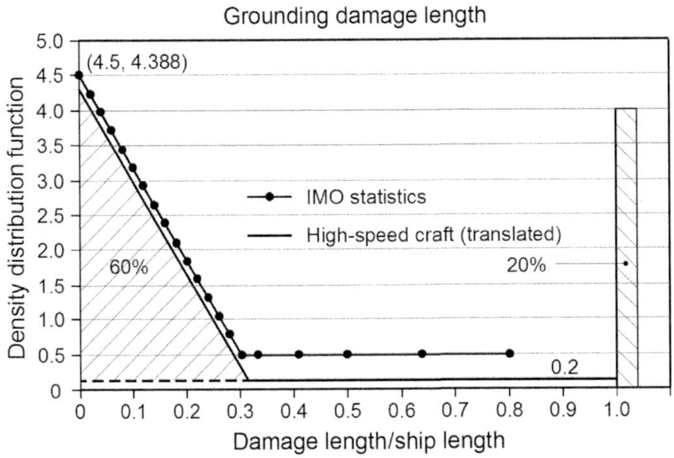

Fig. 4.41 The predicted bottom damage density distribution for high-speed craft translated from that of tankers (IMO, 1995b).

Let us assume that the damage length ratio between a high-speed craft and an oil tanker is approximately equal to 2.5. Based on this ratio, Fig. 4.41 shows the predicted damage density distribution function of the longitudinal extent for the high-speed craft converted from the IMO density distribution function for tankers (IMO, 1995b). In this translation, we have assumed that in the IMO damage distribution, the raking damage is represented by a constant density distribution equal to 0.5 for damage lengths between 0% and 80% of the ship length. The reason is that the raking damage is one of several types of grounding damages to ships. Other grounding damages are caused by grounding on soft bottoms, sideways stranding, etc. However, in the IMO damage distribution, the damage lengths above 30% of the ship length are all assumed to be caused by raking damage. With this assumption, the constant density distribution of raking damage for the high-speed craft is simply equal to $0.5/2.5 = 0.20$.

It is also seen from Fig. 4.41 that the probability of full ship length damage for the high-speed craft is 20% while the probability of full ship length damage for tankers in the IMO statistics is zero. This means that the high-speed craft suffers significantly higher probability of large damage lengths than that of the conventional ships.

4.5.3 Application example to a high-speed Ro-Ro ferry

In this section, investigations of the bottom damage to a fast new ferry are carried out. The damage in powered grounding events to the new ferry is compared with that to old conventional ships. The new fast ferry is shown in Fig. 4.42. The main particulars and the bottom structure of the new ferry and the old conventional ships are presented in Table 4.15. The equivalent thickness t_{eq} of the whole bottom, including transverse and longitudinal webs and stiffeners, and the equivalent thickness t of the outer bottom,

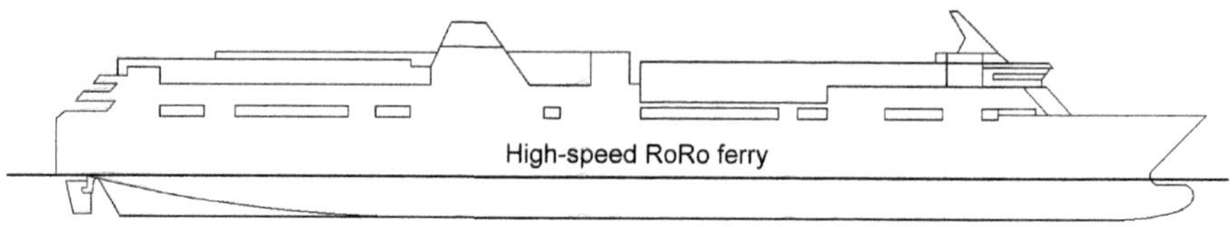

Fig. 4.42 A new high-speed Ro-Ro ferry ($V=27$ knots).

Table 4.15 Main particulars and bottom structure of the new and conventional ships.

	New ferry	Old ferry 1	Old ferry 2	Oil tanker
Length L_{pp} (m)	173	128	158.5	219
Beam (m)	26.0	17.5	24.0	32.0
Depth (m)	15.7	12.4	14.0	21.6
Draught (m)	6.5	4.5	6.1	16.0
Mass (t)	16,073	6300	15,000	82,000
Service speed (m/s)	14	9.5	11	6.7
Froude no. (F_n)	0.34	0.27	0.28	0.15
Material $\sigma_0=300$ MPa	Steel	Steel	Steel	Steel
	Double bottom	Double bottom	Double bottom	Double bottom
Outer shell thickness (mm)	13.0	12.0	14.0	18.5
Inner bottom thickness	12.0	12.0	12.0	18.5
Double-bottom height	1480	1250	2000	2150
Spacing of trans. webs	2400	2250	3000	2700
Equivalent thickness of floors	10.0	9.0	10.0	12.0
Spacing of girders	3900	3600	3200	3400
Equivalent thickness of girders	12.7	10.0	15.4	12.3
Spacing of longitudinals	650	610	800	850
Web height of longitudinals	180	220	260	340
Web thickness of longitudinals	10	10	11	12.0
Flange width of longitudinals	NA	70	NA	NA
Flange thickness of longitudinals	NA	15	NA	NA
Equivalent thickness t (mm)	18.2	19.1	22.3	27.2
Equivalent thickness t_{eq} (mm)	41.5	43.2	49.0	64.8

NA, not applicable.

only including longitudinal webs and longitudinal stiffeners, are calculated and also shown in Table 4.15.

The factor K representing the ratio of the horizontal grounding resistance between two ships is

$$K_1 = \frac{OldFerry(1)}{NewFerry} = 0.92;\; K_2 = \frac{OldFerry(2)}{NewFerry} = 1.30;\; K_3 = \frac{OilTanker}{NewFerry} = 2.85$$

The relative damage ratio between two different ships is.

$$\frac{(L_{dam}/L)_{New}}{(L_{dam}/L)_{Old1}} = 3.73;\; \frac{(L_{dam}/L)_{New}}{(L_{dam}/L)_{Old2}} = 2.05;\; \frac{(L_{dam}/L)_{New}}{(L_{dam}/L)_{Tanker}} = 2.87$$

It is seen that the "new" fast ferry suffers larger grounding damage in powered grounding events than the conventional ships. The nondimensional damage length to the new ferry is more than twice the nondimensional damage length to the conventional ships. Therefore, designers of new high-speed ferries may have to consider these results in order to improve safety in accidental situations.

4.6 Investigation of grounding statistical damage distributions

4.6.1 Introduction

The International Maritime Organization (IMO, 1995b) adopted Interim Guidelines for Approval of Alternative Methods of Design and Construction of Oil Tankers under Regulation 13F(5) of Annex I of MARPOL 73/78. These guidelines give a probabilistic procedure for assessing the oil outflow performance of an oil tanker design in a collision and/or grounding accident.

One of the important elements in the guidelines is the damage density distributions, which were derived from the actual damage data of 52 collisions and 63 grounding accidents of oil tankers, chemical tankers of 30,000 tons deadweight and above (Hysing, 1993). These data were collected in the period from 1980 to 1990 by different classification societies, such as American Bureau of Shipping (ABS), ClassNK, Det Norske Veritas (DNV), Lloyd's Register (LR), and Registro Italiano Navale (RINA). Fig. 4.43 shows the probability density distributions for the longitudinal length, the vertical penetration, and the transverse extent of the expected grounding damages according to the IMO guidelines. It is seen that the bottom grounding damages are assumed to scale linearly with the ship main dimensions. The same approach is assumed for the side collision damage distributions.

Since the publication of the IMO interim guidelines, many authors have used them to assess the environmental performance of oil tankers; see, for example, Bockenhauer and Jost (1995) and Michel et al. (1996). The Society of Naval Architects and Marine Engineers (SNAME) formed a special technical committee to make further

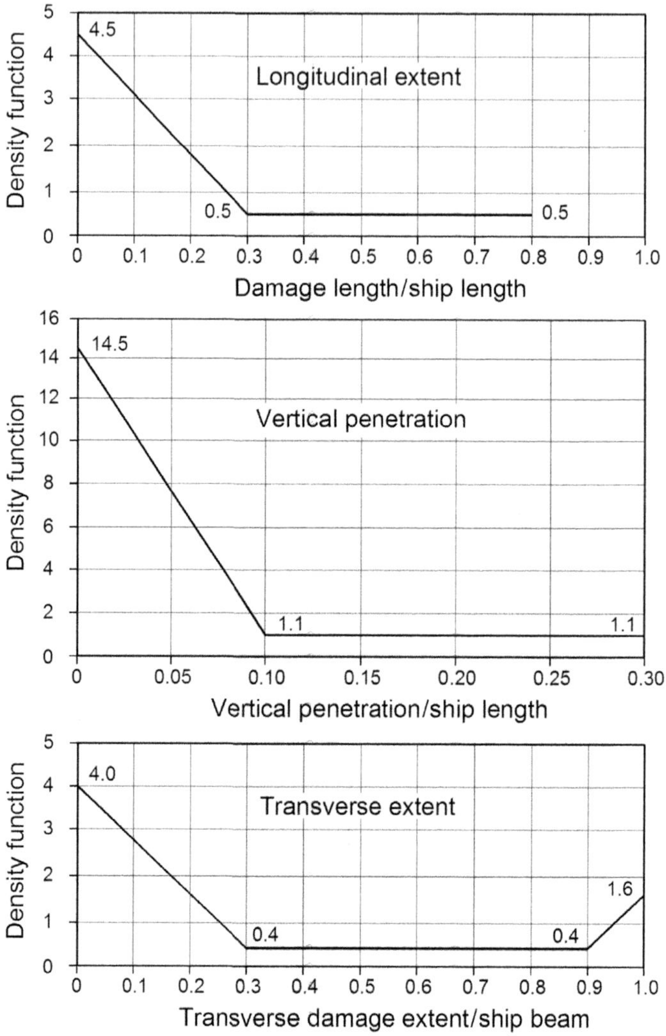

Fig. 4.43 Probabilistic density distributions for bottom grounding damages.
Based on International Maritime Organization (IMO), 1995b. Interim guidelines for approval of alternative methods of design and construction of oil tankers under regulation 13F(5) of annex I of MARPOL 73/78. Technical Report, Resolution MEPC. 66 (37), pp. 1–40.

assessment of the performance of oil tankers suffering collision and grounding accidents from 1995 to 1997 (Sirkar et al., 1997). As discussed by Sirkar et al. (1997) and Rawson et al. (1998), a major shortcoming of the IMO guidelines is that they do not consider the effect of the local structural design or the crashworthiness on the damage extent and that all tankers have the same nondimensional damage distributions. Sirkar et al. (1997), Simonsen (1997a,b), Cerup-Simonsen et al. (2009), and Rawson et al. (1998) performed theoretical grounding analyses and established damage density distributions given a grounding event for a specific ship. These calculations are based upon many assumptions, such as the distribution of grounding speeds and the distribution of rock shapes and rock elevations, and thus, the validity of the damage density distributions obtained by such theoretical calculation needs further verification. One way to validate these assumed distributions of speeds, ground shapes, etc. is to choose the speed and the rock shape distributions so that the calculated grounding damage distributions for old and traditional single-hull tankers become identical to the damage distributions represented by Fig. 4.43. The idea is then to use the same ground and speed distributions to construct damage distributions for the new generations of tankers. It is evident that the result of such direct calculation procedures depends strongly on the validity of the IMO damage distributions given in Fig. 4.43 for grounding damages.

Previous analyses of bottom damages due to grounding on plane, sloping sand, or rock bottoms have shown that larger ships suffer considerably larger bottom damages than smaller ships. In addition, larger ships are exposed to larger hull girder sectional forces due to grounding (Pedersen, 1994).

In this section, we shall first derive a procedure for analysis of the effect of the ship size and the building material on grounding on irregular rocks, that is, grounding scenarios resulting in raking damage to the ship's bottom. Thereafter, the results will be validated by a comparison with statistical grounding damage data. One purpose of this analysis is to investigate whether it is reasonable to assume that grounding damages scale linearly with ship dimensions, as assumed in the IMO guidelines.

4.6.2 Prediction of the relative damage extent in grounding

As discussed previously, the relative damage length (L_{dam}/L) between two ships can be determined from

$$\frac{(L_{dam}/L)_1}{(L_{dam}/L)_2} = \frac{M_1}{M_2} \cdot \left(\frac{V_1}{V_2}\right)^2 \cdot \frac{L_2}{L_1} \cdot \frac{P_2}{P_1} \tag{4.10a}$$

$$\frac{(L_{dam}/L)_1}{(L_{dam}/L)_2} = \frac{M_1}{M_2} \cdot \left(\frac{F_{n1}}{F_{n2}}\right)^2 \left(\frac{\sigma_2}{\sigma_1} \cdot \frac{t_{eq2}}{t_{eq1}} \cdot \left(\frac{t_2}{t_1}\right)^{0.6} \cdot \left(\frac{T_2}{T_1}\right)^{0.4}\right) \tag{4.10b}$$

In this formula, it has been assumed that the larger ship suffers larger vertical penetration and larger transverse damage extent. Obviously, the elevation of the rock above the baseline of the larger ship is greater than that of the smaller ship, as shown in Fig. 4.44. This also means that the damage distribution for the vertical penetration and the transverse damage extent are assumed to be independent of the local structure.

Based on statistics of existing ships and classification design rules, it is found that the equivalent thickness t and t_{eq} may be approximated as

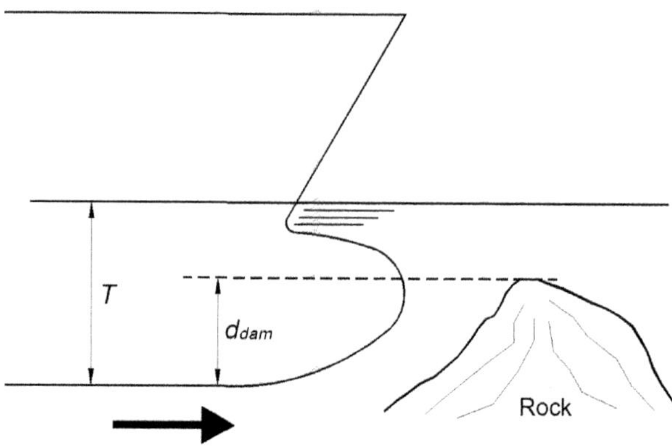

Fig. 4.44 The relation between the vertical penetration and the ship draught in a grounding event.

$$\begin{cases} t = k_1 \cdot L^{0.7}/\sqrt{f} \\ t_{eq} = k_2 \cdot L^{0.7}/\sqrt{f} \end{cases}$$

where k_1 and k_2 are constants, L is the ship length in m, and $f = \sigma_0/235$ is a material factor where σ_0 is the flow stress (MPa) of the material. The design draught of the ship is taken to be proportional to the ship length, that is, $T = k_3 \cdot L$ where k_3 is a constant. By simple substitution, the ratio between the horizontal grounding forces for two different tankers is obtained as

$$\frac{P_2}{P_1} = \left(\frac{\sigma_2}{\sigma_1}\right)^{0.2} \cdot \left(\frac{L_2}{L_1}\right)^{1.52}$$

It is seen that the grounding force increases with the ship length. Based on statistics and economic optimization criteria, a ship length can be related to the ship displacement by (Schneekluth, 1987)

$$L = C \cdot M^{0.3} V_0^{\,0.3} \tag{4.11}$$

where M is the ship displacement in tonnes, V_0 is the design speed in knots, $C = 3.2$ if the block coefficient has the approximate value of $C_B = 0.145/F_n$, and L is the ship length in metres. After simple derivations, we get the relationship between the relative damage length for two different ships:

$$\frac{(L_{dam}/L)_1}{(L_{dam}/L)_2} = \left(\frac{\sigma_2}{\sigma_1}\right)^{0.2} \cdot \left(\frac{L_1}{L_2}\right)^{0.813} \cdot \left(\frac{V_1}{V_2}\right)^{2} \cdot \left(\frac{V_{02}}{V_{01}}\right) \tag{4.12}$$

Investigating the actual oil tanker designs, their design speed V_0 is commonly around 14–15 knots, and the difference is not large. Therefore, it is assumed that the design speed for all oil tankers is similar. It is also reasonable to assume that the distribution of the grounding speed is similar for all oil tankers, that is, $V_1/V_2=1$ and $V_{01}/V_{02}=1$. Thus, an approximation for the ratio between the relative damage length for two different tankers is

$$\frac{(L_{dam}/L)_1}{(L_{dam}/L)_2}=\left(\frac{\sigma_2}{\sigma_1}\right)^{0.2}\cdot\left(\frac{L_1}{L_2}\right)^{0.813} \tag{4.13}$$

For general cargo vessels, where the speed of the vessel is determined mainly by the Froude number F_n, this expression takes the form

$$\frac{(L_{dam}/L)_1}{(L_{dam}/L)_2}=\left(\frac{\sigma_2}{\sigma_1}\right)^{0.2}\cdot\left(\frac{F_{n1}}{F_{n2}}\right)\cdot\left(\frac{L_1}{L_2}\right)^{1.313} \tag{4.14}$$

From Eqs. (4.13), (4.14), it is seen that the relative damage length in powered grounding events depends upon the ship size, that is, larger ships suffer larger relative damage length. This reflects the influence of the structural design or the ship size on the distribution of the grounding damage length. It is also seen that the flow stress of the material has a weak influence on the relative damage length.

4.6.3 Calculation example and comparison with grounding damage data

Fig. 4.45 shows how the calculated ratio of relative damage length of tankers due to powered grounding varies with lengths from 100 to 300 m. From the results, it is seen that for similar conditions, the relative damage length of a 235 m tanker is twice the damage of a 100 m tanker. This shows that the tanker size has a significant influence on the relative damage length in accidental grounding. For conventional vessels where the speed is mainly governed by the Froude number, the difference is even larger.

As mentioned previously, the damage density distributions in the IMO guidelines were derived from the actual damage data of ships of 30,000 tons deadweight and above. Here, we shall assume that the mean value of the displacement in the IMO data for grounded ships is 50,000 tons. Thus, the IMO bottom damage distribution is considered to represent a tanker with a length of about 185 m when the design speed is assumed to be 15 knots. Then, by the application of Eq. (4.13), the converted density distributions of the longitudinal extent of a grounding damage for a 300 m tanker (250,000 tons displacement) can be seen in Fig. 4.46; the corresponding cumulative probabilities are shown in Fig. 4.47.

In these transformations, we have assumed that in the IMO damage distribution for tankers, the raking damage is represented by a constant density distribution equal to 0.5 for damage lengths between 0% and 80% of the ship length. The reason for this assumption, as also discussed previously, is that the raking damage is only one of several types of grounding damages to ships. Note that soft grounding and sideways stranding cause other grounding damages.

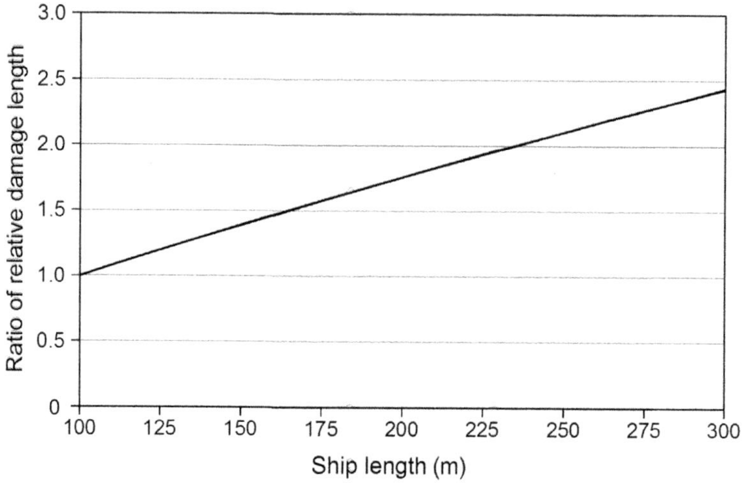

Fig. 4.45 Effect of the ship size on the relative damage length in ship grounding.

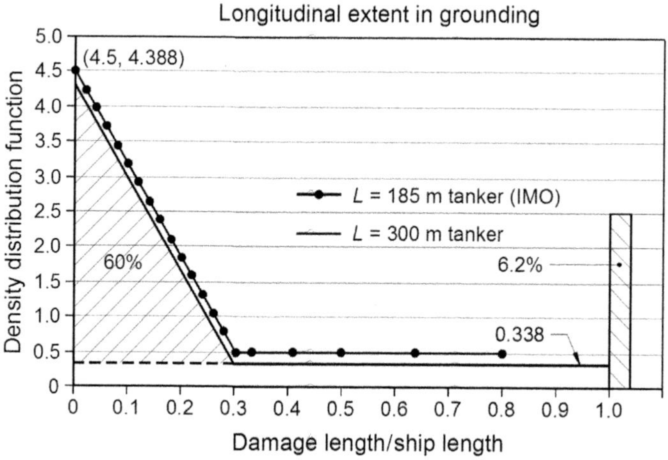

Fig. 4.46 Translated density distribution functions obtained by the present method for longitudinal extent with different ship sizes in grounding.

It is seen from Figs 4.46 to 4.47 that the larger tankers suffer a higher probability of large relative damage length than that of the smaller tankers. For damage lengths above 30% of the ship length, the probability is 30% for the 300 m tanker and 25% for the 185 m tanker.

Unfortunately, the IMO tanker database for grounding damages is not large enough to give a statistical verification of the calculated variations (Eq. 4.13) of expected grounding damages with ship size, but for merchant ships, the results (Eq. 4.14) obtained by the present method correlate well with statistical analyses of ship

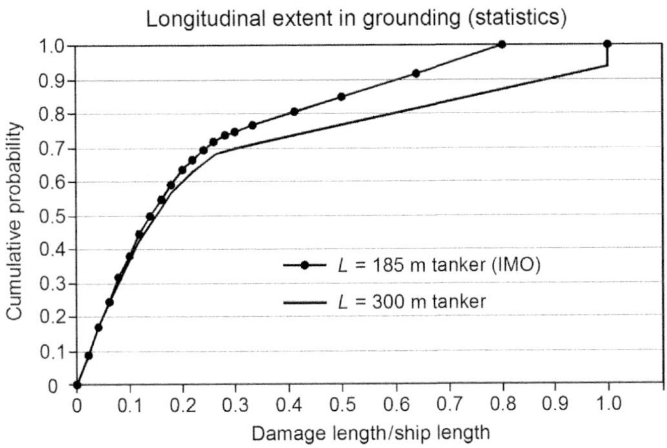

Fig. 4.47 Translated cumulative probabilities obtained by the present method for longitudinal extent with different ship sizes in grounding.

grounding accidents (Bjørneboe et al., 1999). These statistical results for bottom damage density functions of longitudinal extent are shown in Fig. 4.48, and the cumulative probabilities are shown in Fig. 4.49.

The results in Figs 4.48 and 4.49 are based on 128 grounding accidents that occurred during the period 1945–65 and mainly involved various cargo ships. In order to investigate the effect of the ship size on the damage distribution, the 128 grounding accidents are divided into two groups based on the ship size. One group represents ship lengths below 100m (the average length is 65m) and the other group ship lengths above 100m (the average length is 135m). The small ship group contains 65 grounding cases, and the large ship group includes 63 cases. The statistical results support the conclusions derived from the theory. It is clearly seen from the statistical results that the group with the larger ships has a higher probability of relatively large damage length. For damage lengths above 30% of the ship length, the probability is 25.4% for the large ship group, while it is only 9.2% for the small ship group.

A similar conclusion has also been reached by analysing the grounding damages that occurred during 1980–2008. For this analysis, for the damage lengths above 30% of the ship length, the probability is 25.0% for the large ship group (the average ship length of 126m), while it is 7.8% for the small ship group (the average ship length of 74m).

4.7 Concluding remarks

In this chapter, simple expressions for the relationship between the absorbed energy and the destroyed material volume are presented for analysis of ship collisions and grounding. Eqs (4.1)–(4.3) represent the energy absorption by plastic tension, crushing and folding, and tearing. It has been demonstrated that the structural damage

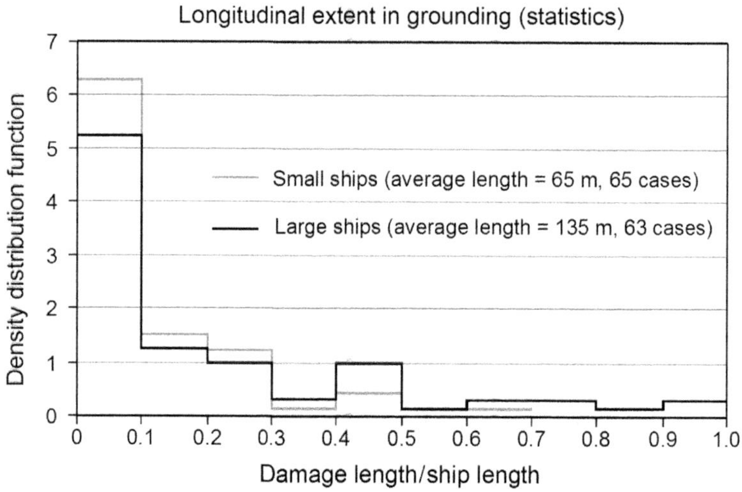

Fig. 4.48 Density functions for the longitudinal damage extent with different ship sizes in grounding obtained by statistical data in the period 1945–65.

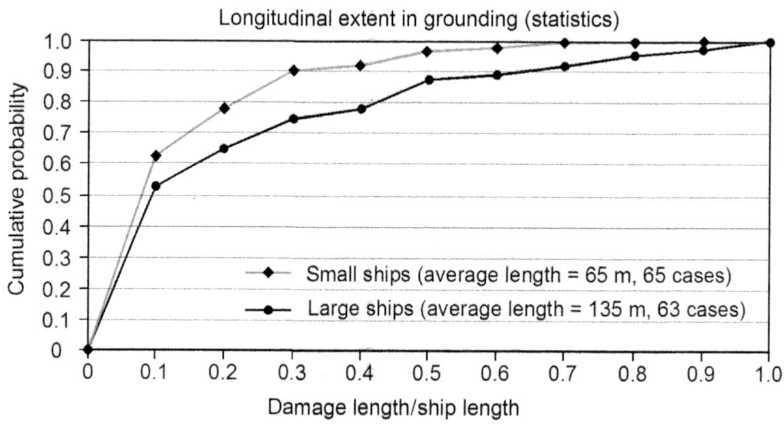

Fig. 4.49 Cumulative probabilities for the longitudinal damage extent with different ship sizes in grounding obtained by statistical data in the period 1945–65.

can be represented by the sum of these energy terms. The formulae are simple to use and have forms similar to Minorsky's empirical expression. The proposed method overcomes the major drawback of Minorsky's classical method as it takes into account the structural arrangement, the materials properties, and the damage pattern.

A large number of comparisons between the present method and experimental and numerical results indicate that the analytical method can give good predictions. Consequently, it can be considered as a rapid design tool/method for analysing ship collisions and grounding.

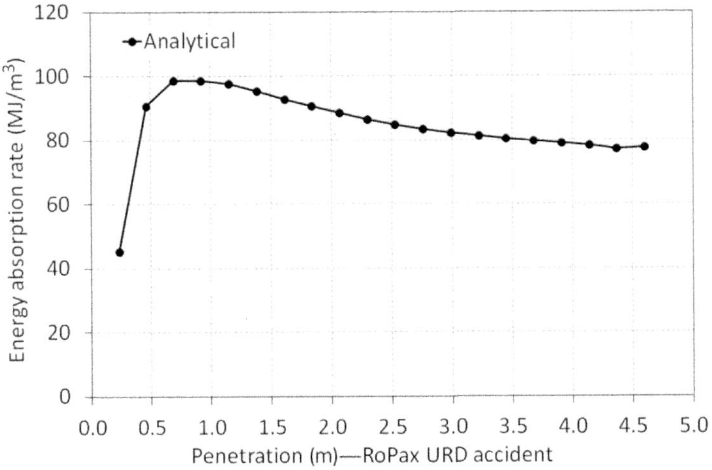

Fig. 4.50 Energy absorption efficiency of the RoPax URD collision accident at various damage penetrations.

The study shows that the energy absorption efficiency, which represents the absorbed energy per unit volume of destroyed material, varies strongly from structure to structure. It depends on the structural arrangement, the materials properties, and the failure mode. The range of the energy absorption efficiency, determined from the present examples, is 20–90 MJ/m^3 while that of Minorsky's results is always constant (47.2 MJ/m^3). Fig. 4.50 presents the energy absorption efficiency of the RoPax URD collision accident at various damage penetrations (example in Section 4.4.1). The results show that after the rupture of the side shell, the energy absorption rate decreases from 98 to 77 MJ/m^3.

It is demonstrated how the proposed method can be used to translate grounding damage distribution from conventional vessels to new types of ships, such as high-speed vessels. The examples show that the high-speed crafts suffer larger grounding damage length than the conventional ships.

Finally, an investigation of the effect of the ship size on the damage distributions in ship grounding is carried out. The results show that the damage density distribution for the longitudinal extent depends on the ship size in grounding. A larger ship has higher probability of a larger relative damage length than that of a smaller ship in grounding.

References

Amdahl, J., 1983. Energy Absorption in Ship-Platform Impact (PhD thesis). Norwegian Institute of Technology. Report No. UR-83-34.

Arita, K., Aoki, G., 1985. Strength of ship bottom in grounding (1st report)—an investigation into the case of a ship stranded on a rock (in Japanese). J. Soc. Nav. Archit. Jpn. 158, 359–367.

ASIS, 1998. Research on the Methodology for the Prediction of Accidental Damage to Tanker Structure in Case of Collision or Grounding and Development of New Hull Design With Improved Crashworthiness (1991-1997). The Association for Structural Improvement of the Shipbuilding Industry (ASIS), Japan.

Bjørneboe, N., Simonsen, B.C., Hansen, P.F., 1999. Statistical and Theoretical Analysis of Ship Grounding Accidents. Department of Naval Architecture and Offshore Engineering, Technical University of Denmark.

Bockenhauer, M., Jost, A., 1995. Guidelines for the assessing of alternative tanker designs. In: Lecture Note: Accidental Loading on Marine Structures—Risk and Response. 22nd WEGEMT Graduate School, Technical University of Denmark.

Cerup-Simonsen, B., Törnqvist, R., Lützen, M., 2009. A simplified grounding damage prediction method and its application in modern damage stability requirements. Mar. Struct. 22, 62–83.

Hysing, T., 1993. Oil spill from tankers in collisions and grounding—Damage statistics. Technical Report No. 93-0518, Det Norske Veritas.

International Maritime Organization (IMO), 1995a. International Code of Safety for High-Speed Craft (HSC Code). Resolution MSC. 36 (63), London.

International Maritime Organization (IMO), 1995b. Interim guidelines for approval of alternative methods of design and construction of oil tankers under regulation 13F(5) of annex I of MARPOL 73/78. Technical Report, Resolution MEPC. 66 (37), pp. 1–40.

Ito, H., Kondo, K., Yoshimura, N., Kawashima, M., 1984. A simplified method to analysis the strength of double hulled structures in collision. 1st report. J. Nav. Archit. Jpn. 156, 283–296.

Kuroiwa, T., 1996. Numerical simulation of actual collision and grounding accidents. In: International Conference on Design and Methodologies for Collision and Grounding Protection of Ships, San Francisco, California, USA, August 22-23.

Lehmann, E., Peschmann, J., 2002. Energy absorption by the steel structure of ships in the event of collisions. Mar. Struct. 15, 429–441.

Lloyd's Register, 2016. Guidance Notes for Collision Assessment for the Location of Low-Flashpoint Fuel Tanks.

Martens, I., 2014. Konstruktive Aspekte beim Entwurf von Bugwülsten zur Verbesserung des Energieaufnahmevermogens bei Schiffskollisionen. (PhD thesis), Schriftenreihe Schiffbau, Bericht 679, Technical University of Hamburg, Germany. ISBN: 978-3-89220-679-8 (in German).

Michel, K., Moore, C., Tagg, R., 1996. A simplified methodology for evaluating alternative tanker configurations. J. Mar. Sci. Technol. 1 (4), 209–219.

Minorsky, V.U., 1959. An analysis of ship collision with reference to protection of nuclear power ships. J. Ship Res. 3 (2), 1–4.

Paik, J.K., Chung, J.Y., Choe, I.H., Thayamballi, A.K., Pedersen, P.T., Wang, G., 1999. On rational design of double hull tanker structures against collision. In: The Society of Naval Architects and Marine Engineers. Annual Meeting, SNAME. vol. 107.

Pedersen, P.T., 1994. Ship grounding and hull-girder strength. Mar. Struct. 7, 1–29.

Pedersen, P.T., Zhang, S., 2000. Absorbed energy in ship collisions and grounding—revising Minorsky's empirical method. J. Ship Res. 44 (2), 140–154.

Peschmann, J., 2001. Berechnung der Energieabsorption der Stahlstruktur von Schien bei Kollisionen und Grundberuhrungen. (in German). (PhD thesis). Technical University of Hamburg, Germany.

Rawson, C., Crake, K., Brown, A., 1998. Assessing the environmental performance of tankers in accidental grounding and collision. In: SNAME Annual Meeting, San Diego, USA.

Rodd, J.L., 1997. Frame design effects in the rupture of oil tanks during grounding accidents. In: Int. Conf. on Advances in Marine Structures III. DERA, ROSYTH, UK.

Schneekluth, H., 1987. Ship Design for Efficiency and Economy. Butterworth & Co. (Publishers) Ltd.

Simonsen, B.C., 1997a. Ship grounding on rock—I. Theory. Mar. Struct. 10 (7), 519–562.

Simonsen, B.C., 1997b. Ship grounding on rock - II. Validation and application. Mar. Struct. 10 (7), 563–584.

Sirkar, J., Ameer, P., Brown, A., Goss, P., Michel, K., Nicastro, F., Willis, W., 1997. A framework for assessing the environmental performance of tankers in accidental groundings and collisions. Report of SNAME T&R Ad Hoc Panel on the Environmental Performance of Tankers, also presented at SNAME Annual Meeting 1997.

Storheim, M., Amdahl, J., Martens, I., 2015. On the accuracy of fracture estimation in collision analysis of ship and offshore structures. Mar. Struct. 44, 254–287.

Tautz, I., Schöttelndreyer, M., Lehmann, E., Fricke, W., 2013. Collision tests with rigid and deformable bulbous bows driven against double hull side structures. In: Amdahl, Ehlers, Leira, (Eds.), Proceedings of the 6th Conference on Collision and Grounding of Ships and Offshore Structures. Taylor & Francis Group, London. ISBN: 978-1-138-00059-9.

Tautz, I., Fricke, W., Schöttelndreyer, M., 2014. Validierung von Kollisionsberechnungen durch Großversuche an Konstruktionsvarianten von Seitenhüllen (in German). Technical Report, Technische Universität Hamburg. ISBN: 978-3-89220-676-7.

Wang, G., Ohtsubo, H., Liu, D., 1997. A simple method for predicting the grounding strength of ships. J. Ship Res. 41 (3), 241–247.

Wang, G., Arita, K., Liu, D., 2000. Behavior of a double hull in a variety of stranding or collision scenarios. Mar. Struct. 13, 147–187.

Woisin, G., 1979. Design Against Collision. vol. 31 (2). Schiff & Hafen, Germany, pp. 1059–1069.

Zhang, S., 1999. The Mechanics of Ship Collisions (PhD thesis). Technical University of Denmark, Denmark. ISBN: 87-89502-05-1.

Zhang, S., Pedersen, P.T., 2016. A method for ship collision damage and energy absorption analysis and its validation. Ship Offshore Struct. https://doi.org/10.1080/17445302.2016.1254584 (Special Issue).

Zhang, S., Villavicencio, R., Zhu, L., 2018. Effect of friction on the impact response and failure mode of double hull structures in ship collisions. In: 3rd International Conference on Safety and Reliability of Ships and Offshore & Subsea Structures (SAROSS 2018). Wuhan University of Technology, Wuhan, China.

Collision damage assessment by nonlinear finite element simulations

5

5.1 Introduction

The external dynamics and the internal mechanics of ship collisions have been discussed in Chapters 2 and 3 by means of simplified analytical methods. Chapter 4 introduced the damage material volume method that, by using Eq. (4.4), manages to predict the energy–displacement response of typical ship structures struck by a rigid indenter, as summarized in Table 4.5 and Fig. 4.26. The present chapter discusses nonlinear finite element simulations as a method to predict the impact response of typical ship structures. In particular, this chapter provides insight in different input parameters needed to conduct finite element simulations for ship collision assessments. The most important input parameters are the mesh size of the finite elements and the failure criterion selected to initiate rupture of the material.

The sensitivity to calculation results of the mesh size and the failure criterion was demonstrated when Ehlers et al. (2008) conducted a failure criteria benchmark study with three of the four large-scale ship collision tests carried out in the Netherlands (see Fig. 3.3 as reference). Fig. 5.1 shows the large deviations between the experimental results and the calculated nonlinear finite element force–displacement responses of the collision experiments when using three different mesh sizes (25, 50, and 100 mm) and three different failure criteria named as GL, PES, and RTCL; these failure criteria are introduced in Section 5.5. It should be mentioned that the differences observed in the benchmark study were not only in terms of the structural impact response but also in terms of the deformation mode (see Fig. 5.2). Note that test model 3 is the same as the one used in Section 4.3.1 (large-scale model test by Lehmann and Peschmann, 2002) to validate the simplified analytical method (Eq. 4.4). Test model 1 was built with a plate of 20 mm stiffened by symmetric profiles (T $440 \times 12 + 200 \times 20$), and test model 2 was built as a sandwich panel of height 300 mm and plates of thickness 12.5 mm (Ehlers et al., 2008). One of the conclusions of the benchmark study was that the mesh size sensitivity might be more important than the failure criterion itself.

By analysing the force–displacement responses in Fig. 5.1, it is difficult to quantify the differences between the experiments and the respective finite element simulations. Therefore, the energy absorbed at a certain penetration is calculated for the experiments E_{exp} and for the finite element simulations E_{sim}, to evaluate the resulting energy ratio E_{sim}/E_{exp} (see Table 5.1 and Fig. 5.3). It can be observed that independently of the mesh size and failure criterion, the energy ratio varies from 0.35 to 0.83 for test model 1, from 0.44 to 1.06 for test model 2, and from 0.78 to 1.69 for test model 3.

Probability and Mechanics of Ship Collision and Grounding. https://doi.org/10.1016/B978-0-12-815022-1.00005-0

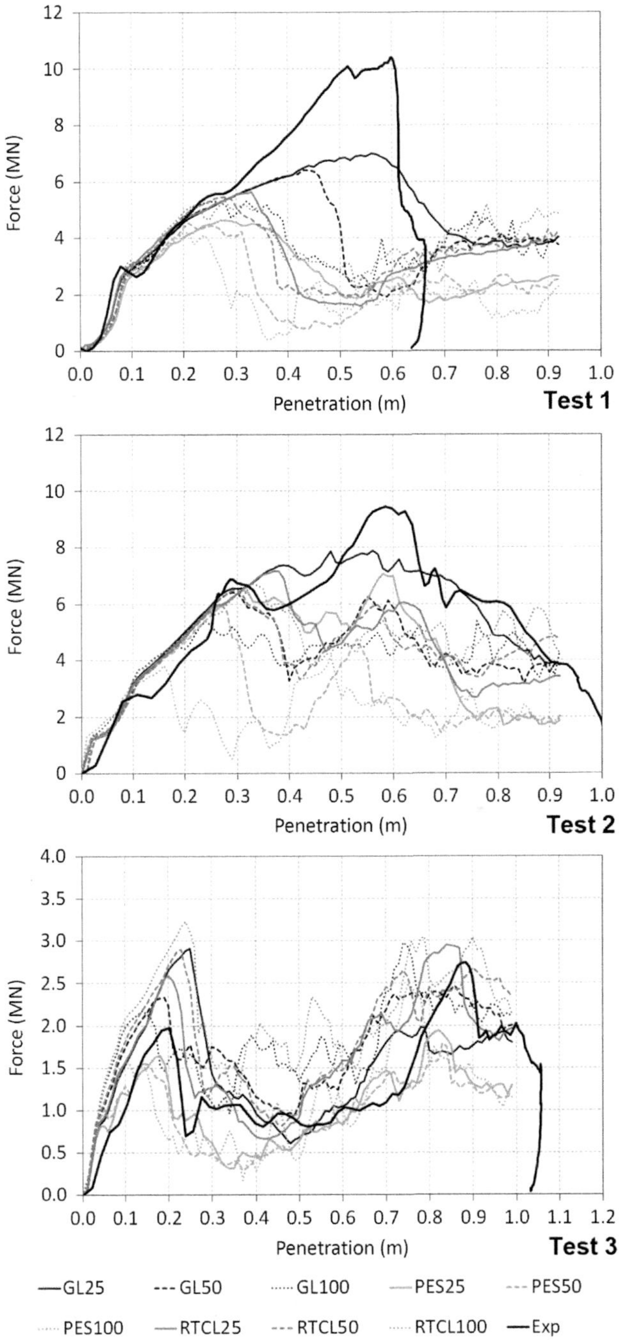

Fig. 5.1 Force–displacement responses for test model 1, test model 2, and test model 3 with different mesh sizes and fracture criteria reported in the ship collision benchmark study. GL25 means GL failure criterion with a mesh size of 25 mm (the notation is similar for the other simulation results).
cf. Ehlers, S., Broekhuijsen, J., Alsos, H.S., Biehl, F., Tabri, K., 2008. Simulating the collision response of ship structures: a failure criteria benchmark study. Int. Shipbuild. Prog. 55, 127–144.

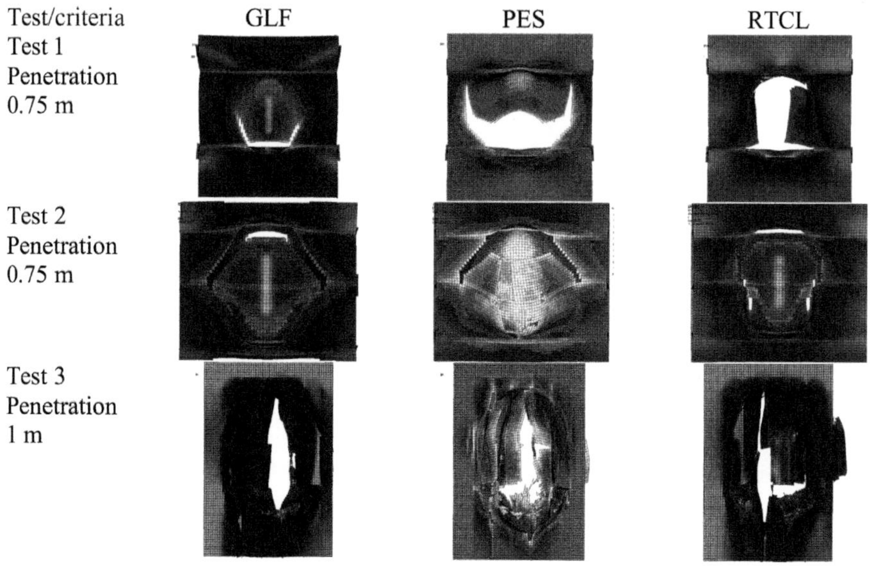

Fig. 5.2 Comparative fracture propagation plots of test model 1, test model 2, and test model 3 with the 25 mm mesh size. The views are thru the indenter.
cf. Ehlers, S., Broekhuijsen, J., Alsos, H.S., Biehl, F., Tabri, K., 2008. Simulating the collision response of ship structures: a failure criteria benchmark study. Int. Shipbuild. Prog. 55, 127–144.

This simple analysis highlights the importance of the mesh size and failure criterion when assessing potential ship collision events.

Following the benchmark study of Ehlers et al. (2008), a number of nonlinear finite element simulations of experiments evaluating the effect of different mesh sizes and failure criteria have been presented. However, the findings remain unchanged since the scatter between baseline experimental results and the respective finite element simulations is still relatively large. For example, Storheim et al. (2015a) found that the mean of the normalized energy (E_{sim}/E_{exp}) from various simulations of three experiments of stiffened plates, supported between webs and using different mesh sizes and failure criteria, varies from 1.02 to 1.52 at peak force and from 0.74 to 1.15 at the end of the quasistatic impact event. It should be mentioned that these tests were quasistatic laboratory experiments and thus less complex than the actual ship collision tests carried out in the Netherlands. Storheim et al. also found that by using two mesh sizes ($5t$ and $10t$ where t is the plate thickness) and eight failure criteria, the energy ratio E_{sim}/E_{exp} for test model 3 varies from approximately 0.25 to 2.00 at peak force and from 0.60 to 1.50 at the end, independently of these two parameters.

Similarly, Ringsberg et al. (2018) found that the energy ratio E_{sim}/E_{exp} at the end of the impact varies from 0.39 to 1.19 when summarizing the results of the 15 participants in the MARSTRUCT benchmark study on nonlinear finite element simulation of an experiment where an indenter impacts a ship side shell structure. The results of the

Table 5.1 Energy ratio E_{sim}/E_{exp} for the test models of the ship collision benchmark study reported by Ehlers et al. (2008).

Criterion/mesh size	25 mm	50 mm	100 mm
Energy ratio (E_{sim}/E_{exp}) for test model 1—penetration 0.88 m			
GL	0.83	0.67	0.59
PES	0.50	0.39	0.35
RTCL	0.53	0.52	0.58
Energy ratio (E_{sim}/E_{exp}) for test model 2—penetration 0.60 m			
GL	1.06	0.84	0.76
PES	0.91	0.60	0.44
RTCL	0.90	0.82	0.85
Energy ratio (E_{sim}/E_{exp}) for test model 3—penetration 0.60 m			
GL	1.21	1.38	1.44
PES	0.82	0.78	0.90
RTCL	1.32	1.44	1.69

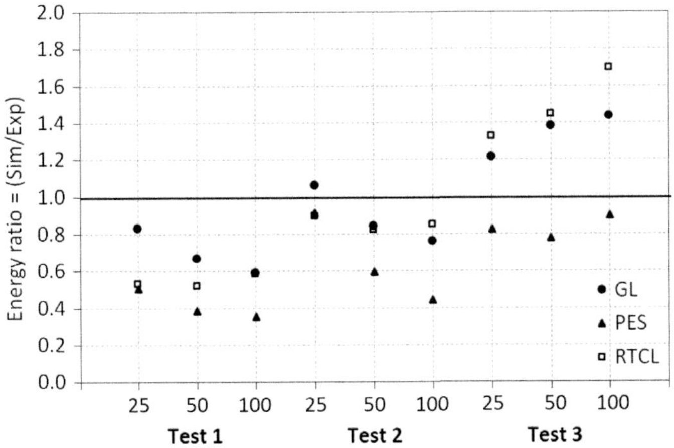

Fig. 5.3 Energy ratio E_{sim}/E_{exp} for the test models of the ship collision benchmark study reported by Ehlers et al. (2008). Penetration for test model 1 is 0.88 m, for test model 2 is 0.60 m, and for test model 3 is 0.60 m. The abscissa indicates the mesh sizes 25, 50, and 100 mm for each model test.

15 simulations can be seen in Fig. 5.4 together with a sketch of the specimen's geometry. It should be mentioned that the participants of the MARSTRUCT benchmark study were familiar with the experimental results but performed the analysis independently. Thus, apart from the mesh discretization and the failure criterion, other input parameters could also have influenced the calculations. Details of the experiment and the different mesh and failure criteria and other input parameters can be found in Ringsberg et al. (2018).

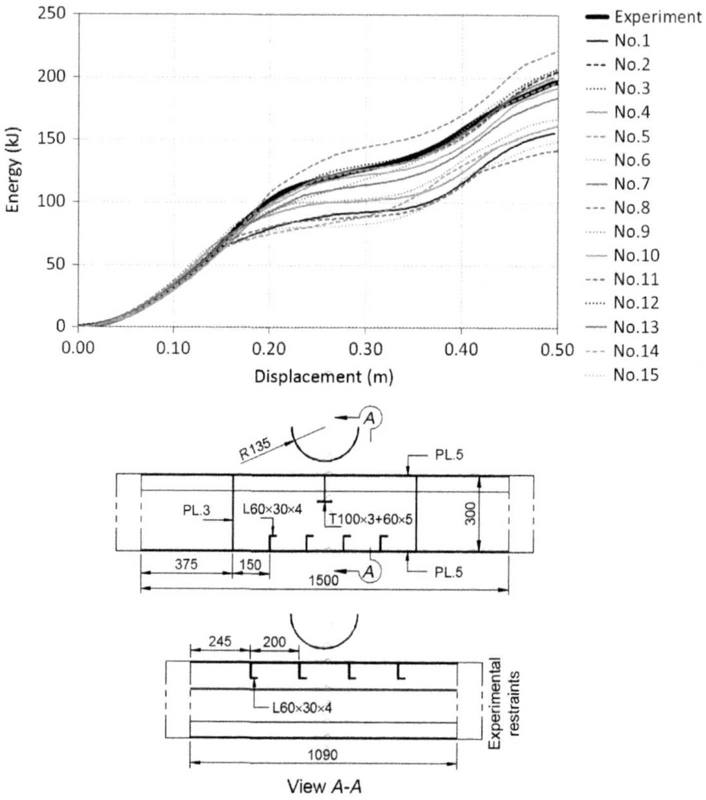

Fig. 5.4 Energy–displacement responses reported in the MARSTRUCT benchmark study and the geometry of the specimen.
cf. Ringsberg, J., et al., 2018. MARSTRUCT benchmark study on non-linear FE simulation of an experiment of an indenter impact with a ship side-shell structure. Mar. Struct. 59, 142–157.

The review of the three benchmark studies mentioned above shows that a nonlinear finite element ship-to-ship collision simulation relies on an accurate definition of the material nonlinearities, and most importantly, on the size of the finite elements. However, other input parameters also influence the finite element predictions, such as approximations of the external dynamic loadings and modelling of structural details. After reviewing the views of experienced analysts about the issues that arise in practice when conducting dynamic finite element simulations, Jones (2013) discussed how those and other factors influence the credibility of finite element predictions for structures subjected to large dynamic loadings. While the discussion of Jones was for a general dynamic design application and the evaluated dynamic loadings can be larger than those expected in an actual ship collision scenario, the idea behind this study and the conclusions presented also apply to ship collision analyses.

The present chapter summarizes the most important input parameters needed to conduct nonlinear finite element simulations of ship collisions. In particular, the discussion is concentrated on the definition of the material, including the characteristic true stress–true strain behaviour and representative failure criteria to estimate the critical failure strain of the finite elements. Comments are also offered on the type of finite elements used for collision simulations, the mesh size, the boundary conditions, the definition of the contact, and the associated static friction. It should be mentioned that the information to be presented about finite element formulations, available approaches to define failure, and contact-impact algorithm, among others, follows the recommendations given by the explicit finite element package LS-DYNA (Hallquist, 2006).

The idea is that with the information provided here, it will be possible for an analyst to set up a nonlinear finite element model to conduct collision simulations required in industry, for example, to assess the structure of the compartment of the LNG (as fuel) tank used in modern green ships (Lloyds Register, 2016). For such analysis, the output required is the critical absorbed energy until a predefined displacement, while other more complicated effects, such as evolution of the stress and strain during the impact process and an accurate representation of the deformation mechanisms, are omitted.

After introducing the input parameters required for the ship collision assessments, simulation results of a representative set of double-hull specimens subjected to quasistatic impact loads are presented. A calculation example of a collision between two container ships is also presented as a practical application of collision simulations where the results are compared with those estimated by the damage material volume method (Eq. 4.4).

5.2 Finite elements and mesh size

5.2.1 Finite elements

For finite element analyses, four-node shell elements are preferred and recommended to evaluate the response of ships and offshore structures subjected to impact loads. Usually, the number of integration points throughout the thickness is set to five to improve the accuracy of problems involving plasticity.

The Hughes–Liu shell element formulation (a four-node shell with uniformly reduced integration) was the first shell element implemented in LS-DYNA, essentially because (1) it has the quality that rigid body rotations do not generate strains, thus allowing for the treatment of finite strains that occur in many practical applications, (2) it has computational efficiency, and (3) it is based on a degenerated brick element formulation.

Later, the Belytschko–Lin–Tsay shell element was implemented as a computationally efficient alternative to the Hughes–Liu shell element. For that reason, it is usually the shell element formulation of choice and therefore the default formulation for explicit calculations. The efficiency of this element is obtained from the mathematical simplifications that result from two kinematical assumptions: (1) a coordinate system is embedded in the element (corotational formulation), and (2) the rate of deformation is restricted to small strains (velocity–strain formulation).

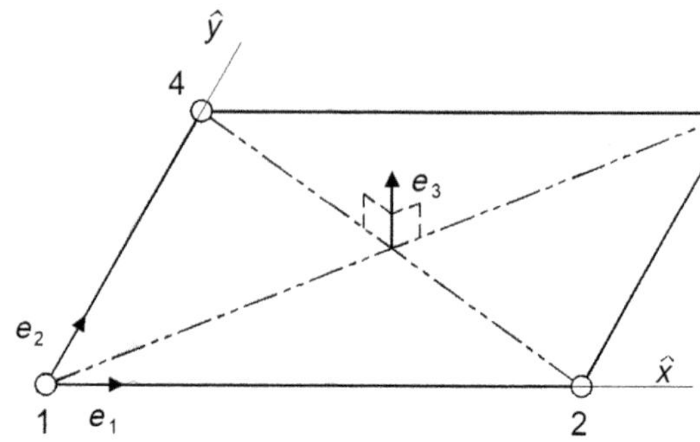

Fig. 5.5 Construction of the coordinate system of shell elements.

To construct the corotational coordinate system, it is necessary to calculate a unit vector normal to the main diagonal of the element (see Fig. 5.5). If the nodes are coplanar, the unit vectors e_1 and e_2 are tangent to the midplane of the shell element, and e_3 is in the thickness direction. As the element deforms, an angle may develop between the actual thickness direction and the normal e_3. The size of this angle depends on the magnitude of the strains, and the acceptable values are in the order of 0.01. In addition, the difference between the rotation of the corotational coordinates e and the material rotation should be small. This small rotation condition does not restrict the magnitude of the element rigid body rotations. The restriction is on the out-of-plane deformations (element strain), and consequently, the rate of deformation relations is also limited to small strains evaluated at the element quadrature.

The computational efficiency of the Belytschko–Lin–Tsay element is essentially derived from the use of one-point quadrature in the plane of the element. However, this simplification needs to control the zero-energy modes (hourglassing modes) that are characterized by a trapezoidal deformation of the elements (see Fig. 5.6). Hourglass modes are often observed to be oscillatory as they have periods that are typically

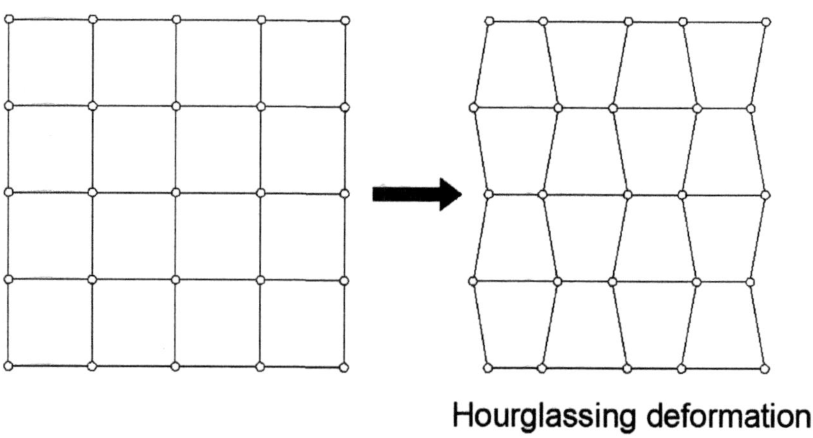

Fig. 5.6 Typical hourglassing deformation mode observed in shell elements.

much shorter than the periods of the structural response. If the periods are comparable with those of the structural response, the kinematic component of the global deformation modes may be stable, and thus, they are admissible. Hourglassing is often controlled with a viscous damping or a small elastic stiffness capable of stopping the formation of those anomalous modes but having a negligible effect on the stable global modes. Since the hourglass deformation modes are orthogonal to the strain calculations, the work done by the hourglass resistance is neglected leading to a slight loss of energy. For structures subjected to large deformations, mainly involving crushing modes, the hourglassing energy should be less than 10%. In general, the hourglassing energy is found to be less important in tension modes.

To conclude, the Belytschko–Lin–Tsay shell element is at present a relevant choice for collision simulations of ship structures. For reference, in the MARSTRUCT benchmark study (Ringsberg et al., 2018), 8 of the 11 participants using LS-DYNA selected this element formulation, only one selected the Hughes–Liu formulation, and the other two chose a fully integrated shell element formulation (shell element type 16 in LS-DYNA). Details of the various finite element formulations can be found in the LS-DYNA's theory manual (Hallquist, 2006).

5.2.2 Mesh size

The selection of the mesh size essentially depends on the structure to be evaluated and the scope of the analysis. For example, solid elements or a very fine mesh, probably in the order of the plate thickness, is needed to simulate the response of a tensile test or a small-scale plate forming experiment to capture all local phenomena such as necking, rupture, and crack propagation; see, for example, Ehlers and Varsta (2009), Ehlers (2010), and Liu et al. (2013). On the other hand, for practical engineering applications, such as approval of alternative arrangements of cargo vessels (Zhang et al., 2004), the mesh configuration, in general, has to be much coarser since the representative finite element model of the struck ship should have a length of at least one cargo hold/compartment.

For ship collision analyses, it is common practice to design a finer mesh in the area to be impacted by the striking ship, while the surrounding areas can be modelled much coarser, and even with linear material properties. This is mainly because the damage to the struck ship is usually confined around the rigid bow of the striking ship (see Fig. 4.1 for reference). Ideally, the mesh should be able to capture the local folding of the web frames and stringers and the tension and subsequent rupture of the outer and inner shell. From calculation experience, it has been stated that the length of the finite elements should not be more than 200mm in the collision area (Zhang et al., 2004). This size sounds reasonable, as for large tanker vessels, LNG carriers, or container ships, 4–5 elements would be modelled between the shell longitudinal stiffeners and the size of the elements will be about 10 times the plate thickness ($10t$).

Liu et al. (2017a) showed that the force–displacement response until necking of plates punched by a spherical indenter is essentially the same when the size of the elements is 2, 5, or 10 times the plate thickness. However, when the size increases to $15t$ or $20t$, the response is stiffer and deviates at the early stages of the impact

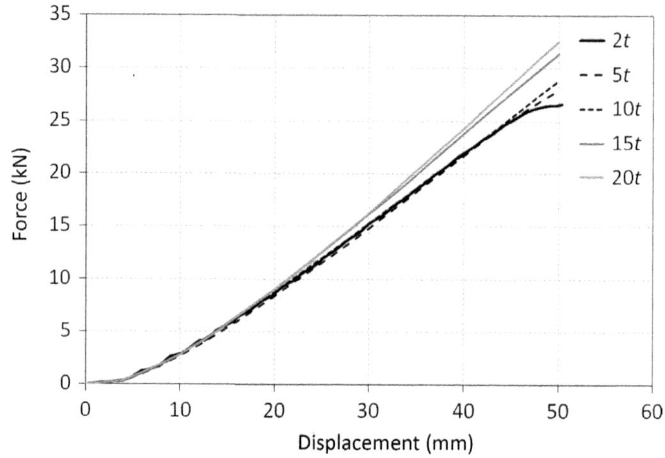

Fig. 5.7 Force–displacement response using different mesh sizes of a square steel plate model ($320 \times 320 \times 1.0$ mm) punched by a rigid indenter of radius 25 mm.
cf. Liu, B., Villavicencio, R., Zhang, S., Guedes Soares, C., 2017a. A simple criterion to evaluate the rupture of materials in ship collision simulations. Mar. Struct. 54, 92–111.

(see Fig. 5.7). These observations should justify the choice of a relatively coarse mesh for ship collision analysis, aiming at predicting the energy to be absorbed at large penetrations.

Certainly, a coarse mesh does not represent the shape of the deformation as accurately as when a very fine mesh is used, in particular with regard to folding of primary supporting members and propagation of cracks. However, the energy absorption can be predicted with a certain degree of accuracy, as it is later shown in Section 5.7. An analogy with respect to the mesh size and the deformation modes of structures can be made with a simpler linear modal analysis: while the determined natural frequencies are essentially the same for a fine or a coarsely meshed structure, the natural deformation modes of the structure normally vary for different mesh sizes/configurations.

The mesh size determines the critical failure strain to be used as an input for the finite element calculation (see Sections 5.4.2 and 5.5) since most simplified expressions to estimate that strain depends on the length and the thickness of the elements. This mesh dependence is critical for structures failing by tension such as the outer and inner shell, but it is less relevant for crushing and folding mechanisms of primary supporting members as only few of the elements connected to the outer shell may fail (reach the critical failure strain) for ductile steel structures. Further details about the mesh dependency when evaluating the critical failure strain are given in Sections 5.4.2 and 5.5.

A relatively coarse mesh has to be applied for ship collision assessments in order to achieve a balance between computational time and reasonable results. However, to evaluate the behaviour of local structural components under impact loading, fine meshes are required. For example, crushing analysis requires a very fine mesh to reflect the folding configuration of web frames. Paik (2007) recommended that the mesh size to capture the crushing behaviour can be estimated by the expected half-length of one structural fold; analytical formulae to predict the length of one fold can be found in Fig. 3.32. Plate forming analysis also requires a very fine mesh, and studies have shown that the ratio between the element size and the indenter radius should preferably be approximately 1/5 in order to satisfactorily define the shape of the deformation (Simonsen and Lauridsen, 2000; Villavicencio et al., 2012). If the interest is to determine the local necking region and/or the initiation of rupture, the size of the elements has to be less than twice the plate thickness (Liu et al., 2013).

For large ship structures, the calculation examples to be presented in Section 5.7 indicate that a mesh of size $10t$ should be sufficient for practical estimation of the energy of ship structures under impact loads. This can be also seen in the finite element results of test model 3 reported by Storheim et al. (2015a), where similar energies at the end of the simulations were obtained when the mesh size is $5t$ or $10t$ as it will be presented in Section 5.5 (Fig. 5.20). Note that this was only observed for three of the eight failure criteria evaluated (GL, SHEAR, and PES), while significant differences were found when using the other five failure criteria.

5.3 Boundary conditions

The in-plane displacements at the supports affect considerably the impact response of structures. For example, a fully clamped beam laterally impacted by a mass exhibits a much larger impact force and a much smaller deflection than a beam having no axial restraints at the supports. This is because the development of geometry changes can introduce axial membrane forces along the centreline of the beam. Certainly, the in-plane restraints at the boundaries can be modelled in detail if the aim is to provide accurate predictions. This approach is commonly used to evaluate the dynamic impact response of small-scale specimens, such as beams and plates. However, when the structure is much larger and complex, such as ship structures, the restraints at the supports might be less important since the penetration and associated damage is only confined to the area of the rigid indenter (see Fig. 4.1).

In Section 3.4.6, it is pointed out that Eqs (3.63), (3.64), and (3.65) manage to predict the experimental deflection of the plates that were used to derive these expressions, but they overestimate the deflection of the plates attached to the webs of the double-hull structures (see Fig. 3.27). It is also mentioned that one possible reason is that the experimental axial restraints at the supports of the plates were not adequately designed and that this may be the reason why Eqs (3.63), (3.64), and (3.65) predict a larger deflection than that occurring in a plate better constrained by actual webs.

In the experiment shown in Fig. 4.7, it is noticed that the inner hull suffered small deflections before the indenter got into contact with it. Therefore, the penetrations are slightly adjusted to reflect this in the analytical calculations. This phenomenon is probably due to inadequate experimental restraints since the double-hull specimen should have been reinforced by strong brackets at the boundaries of the outer and inner plates (top and bottom plates). See, for example, the experimental boundary conditions shown in Fig. 5.8, which were used in the experiments carried out by

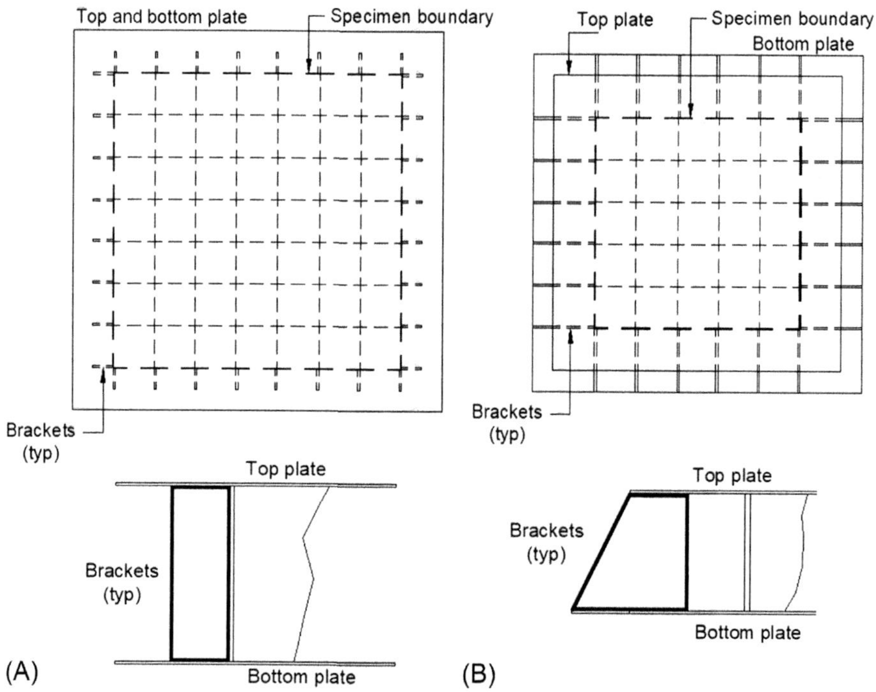

Fig. 5.8 Illustration of the experimental restraints designed for small-scale double-hull specimens punched by a rigid indenter. (A) Experiments conducted by Arita and Aoki (1985). (B) Experiments conducted by Wang et al. (2000).

Arita and Aoki (1985) and Wang et al. (2000); such experimental restraints represent better a fully clamped condition.

The representation of the boundary conditions of the ship collision experiment presented in Section 4.3.1 (large-scale model test by Lehmann and Peschmann, 2002) is a real challenge, considering that the clamping system in the experiment was not critically assessed, and therefore, the simulations conducted by Ehlers et al. (2008) simply assumed a fully rigid clamped condition. This idealized restraint can be considered responsible for some of the deviations observed in the finite element results (see Fig. 5.1 and Table 5.1). It should be noted that the simplified method (Eq. 4.4) also relies on the assumption of a fully clamped condition.

When strain rate effects can be neglected, it has been demonstrated that the energy predicted in a perpendicular ship collision situation, in which the cargo hold structure is considered as fully clamped at the watertight bulkheads and the striking bow is subjected to a constant speed (displacement-controlled simulation), is essentially the same as the energy predicted by a dynamic simulation that accounts for the ship motions in the water plane and starts with an initial velocity of the striking bow (Pill and Tabri, 2011). For reference, an illustration of a dynamic simulation model is provided in Fig. 5.9 (Liu et al., 2017b). The comparison of the force–displacement responses predicted by a displacement-controlled and a dynamic simulation is shown

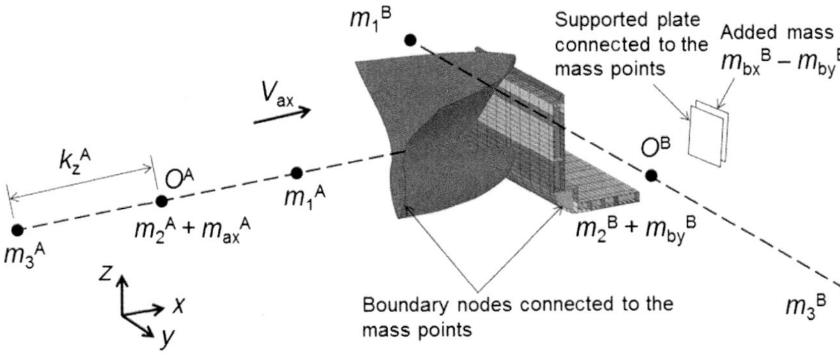

Fig. 5.9 Loading and boundary conditions concept for a coupled dynamic finite element simulation that accounts for the ship motions in the water plane area and that starts with an initial velocity of the striking bow. The discrete mass points m_1, m_2, and m_3 are located at the longitudinal centre of gravity of the ship to represent the total mass of the striking and struck ship. The added mass due to ship motions is included by adjusting the positions of the mass points and by modelling a rigid block positioned at the opposite side of the struck structure. cf. Liu, B., Villavicencio, R., Zhang, S., Guedes Soares, C., 2017b. Assessment of external dynamics and internal mechanics in ship collisions. Ocean Eng. 141, 326–336.

in Fig. 5.10 (Liu et al., 2017b). Certainly, the same agreement is not achieved for oblique collisions. Then, the dynamic simulation would be more appropriate. However, restraining the displacements of the structure at the ends of the cargo hold/compartment (location of watertight bulkheads) is a reasonable approach to adopt when the interest is in design assessments.

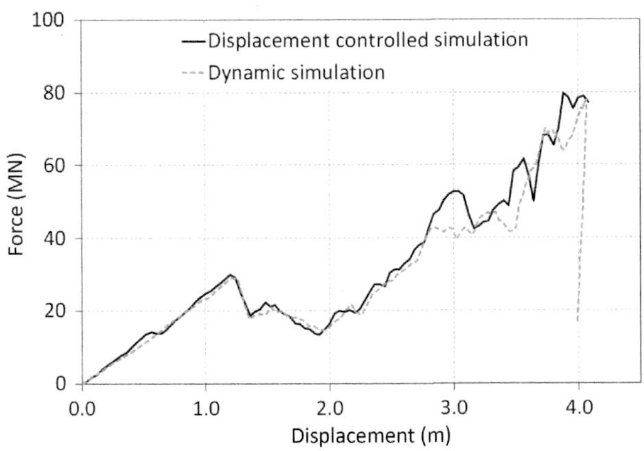

Fig. 5.10 Force–displacement response. Comparison between a perpendicular ship collision displacement-controlled simulation (the struck structure is fully clamped at the ends) and a dynamic simulation (the struck structure is free to move in the water plane area).
cf. Liu, B., Villavicencio, R., Zhang, S., Guedes Soares, C., 2017b. Assessment of external dynamics and internal mechanics in ship collisions. Ocean Eng. 141, 326–336.

5.4 Material characterization

The mechanical behaviour of materials is generally described by their deformation and fracture characteristics under applied tensile loads by means of tension tests. The tensile properties of steel materials and the temperature and strain rate effects have already been discussed in Section 3.2. The concept of the true stress and true strain, or strain hardening, has also been introduced in Section 3.2 indicating that it is needed for conducting nonlinear finite element simulations of structures subjected to large plastic deformations.

Here, additional details about the derivation of the true stress–true strain curve from standard tension tests are summarized, and simplified expressions to represent the strain hardening behaviour of materials are provided. Most importantly, the concept of the critical failure strain is introduced as a key parameter to evaluate the rupture and eventual fracture of structures. A brief discussion of strain rate effects when conducting collision simulations is also presented here.

5.4.1 *True stress–true strain curve*

The engineering stress–strain curve (see Fig. 5.11) does not give a true indication of the deformation of a tensile piece, because it is entirely based on the original length L_0 and original cross-sectional area S_0 of the tensile specimen. Since the cross-sectional area decreases rapidly after necking, the engineering stress–strain curve falls off until fracture occurs. Necking, or localized deformation, usually begins at maximum load where the increase in stress due to the decrease in cross-sectional area of the specimen becomes greater than the increase in the load-carrying capacity of the material.

The true stress is based on the actual cross-sectional area of the specimen, and thus, the stress–strain curve increases continuously until fracture. If instantaneous measurements of the strain are also available, the obtained curve is known as true stress–true strain (see Fig. 5.11); this curve is also known as flow curve since it represents the basic plastic flow characteristics of the material.

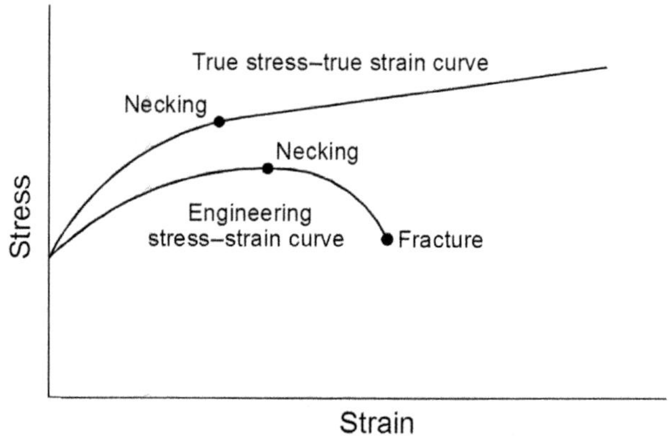

Fig. 5.11 Representation of an engineering stress–engineering strain curve and a true stress–true strain curve.

The true stress σ_t and the true strain ε_t can be expressed in terms of the engineering stress σ and the engineering strain ε by (see also Section 3.2.1)

$$\sigma_t = \sigma(\varepsilon + 1) \tag{5.1}$$

$$\varepsilon_t = \ln(\varepsilon + 1) = \ln\frac{L}{L_0} \tag{5.2}$$

These expressions are only valid until necking since they assume both constancy of volume and a homogeneous distribution of the strain along the specimens gauge length. Beyond the maximum load, the true stress and the true strain are to be determined from actual measurements of the load F and the cross-sectional area S:

$$\sigma_t = F/S \tag{5.3}$$

$$\varepsilon_t = \ln\frac{S_0}{S} \tag{5.4}$$

The true stress–true strain curve is always to the left of the engineering curve until the maximum load, as illustrated in Fig. 5.11 (see Eqs 5.1, 5.2). However, beyond necking, the high and localized strains in the necked region exceed largely the engineering strain (Eq. 5.4). Frequently, the flow curve is linear from maximum load to fracture (see Fig. 5.11), while in other cases, its slope continuously decreases until fracture (Dieter, 1986).

The true stress at maximum load corresponds to the true tensile strength. If it is considered that σ_{tu} and ε_{tu} denote the true stress and the true strain at the maximum load F_{max} when the cross-sectional area of the specimen is S_u, the engineering ultimate tensile strength σ_u can be defined as

$$\sigma_u = \frac{F_{max}}{S_0} \tag{5.5}$$

and

$$\sigma_{tu} = \frac{F_{max}}{S_u} \tag{5.6}$$

By eliminating F_{max}, it is obtained

$$\sigma_{tu} = \sigma_u \frac{S_0}{S_u} \tag{5.7}$$

and by using Eq. (5.4) in the form $\varepsilon_{tu} = \ln(S_0/S_u)$ and rewriting Eq. (5.7) using the natural logarithmic constant $e\,(e^{\varepsilon_{tu}} = S_0/S_u)$, the true stress at maximum load can be expressed as

$$\sigma_{tu} = \sigma_u e^{\varepsilon_{tu}} \tag{5.8}$$

The formation of a necked region introduces triaxial stresses that make it difficult to determine accurately the longitudinal tensile stress from the onset of necking until fracture. This is because the necked region behaves as a mild notch that under tension produces stresses in the transverse and thickness directions that raise the value of the longitudinal stress required to cause plastic flow compared with the case where simple tension prevailed. This state of instability in tension, leading to localized deformation, is defined by the condition $dF=0$. Considering that

$$F=\sigma_t S \tag{5.9}$$

the above condition becomes

$$dF=\sigma_t dS+S d\sigma_t=0 \tag{5.10}$$

From the constancy of volume relationship

$$\frac{dL}{L}=-\frac{dS}{S}=d\varepsilon_t \tag{5.11}$$

and using the instability condition (Eq. 5.10) rewritten in the form

$$-\frac{dS}{S}=\frac{d\sigma_t}{\sigma_t} \tag{5.12}$$

it is possible to obtain the following relationship at a point of tensile instability (at the point of necking) as

$$\frac{d\sigma_t}{d\varepsilon_t}=\sigma_t \tag{5.13}$$

To obtain the true stress–true strain curve beyond the maximum load, accurate optical measurement systems are required to get actual records of the cross-sectional area during the tension tests; a study to obtain the true stress–strain relation until failure for mild steel specimens under quasistatic tension loads can be found in Ehlers and Varsta (2009). In practice, however, such detailed measurements are rarely available for practical structural design assessments. Therefore, simplified mathematical expressions to determine the true plastic characteristics of metals have been derived.

In Section 3.2.1, the basic mathematical expression for the flow curve has been presented (power curve relation) as well as equations to estimate its parameters. Here, the derivation of those parameters is briefly summarized, and the expressions commonly used for ship collision analysis are reviewed.

The simple power curve relation is the most commonly used expression for the representation of the true stress–true strain curve:

$$\sigma_t=K\varepsilon_t^n \tag{5.14}$$

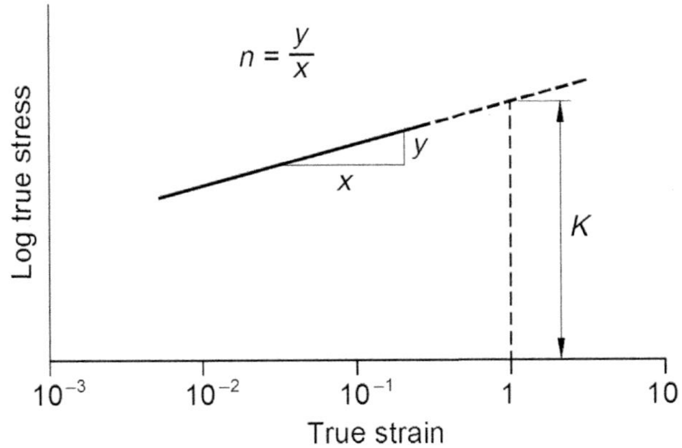

Fig. 5.12 Log–log plot of the true stress and true strain to determine the strength coefficient K and the strain hardening exponent n.

where n is the strain hardening exponent and K is the strength coefficient. These parameters are obtained from a log–log plot of the true stress and true strain until necking (Eqs 5.1 and 5.2), commonly resulting in a straight line; the linear slope of this line is n, and K is the true stress at $\varepsilon_t = 1.0$ (see Fig. 5.12).

It should be noticed that the rate of strain hardening $d\sigma_t/d\varepsilon_t$ is not the same as the strain hardening exponent

$$n = \frac{d(\log \sigma_t)}{d(\log \varepsilon_t)} = \frac{d(\ln \sigma_t)}{d(\ln \varepsilon_t)} = \frac{\varepsilon_t \, d\sigma_t}{\sigma_t \, d\varepsilon_t} \tag{5.15}$$

or

$$\frac{d\sigma_t}{d\varepsilon_t} = n\frac{\sigma_t}{\varepsilon_t} \tag{5.16}$$

By substituting the necking criterion given in Eq. (5.13) into Eq. (5.16), a simple relationship for the strain at which necking occurs is obtained:

$$\varepsilon_{tu} = n \tag{5.17}$$

In the benchmark study of the ship collision tests carried out in the Netherlands (Ehlers et al., 2008), the simple power law expression was selected for all simulations where the strength coefficient and the strength hardening exponent were obtained by reversing the true stress–true strain curve reported by Peschmann (2001). The expressions summarized below have also been used for finite element simulations of panel indentation experiments and/or to perform actual ship collision analysis.

Zhang et al. (2004) recommended Eq. (5.14) to characterize the flow curve, where the strain hardening exponent can be obtained from Eqs (5.2), (5.17). In Eq. (5.2), the engineering strain is the maximal uniform strain related to the ultimate tensile stress σ_u; thus, the obtained true strain corresponds to the strain hardening exponent (see Eq. 5.17). The strength coefficient is estimated with Eq. (5.14) by using the true stress and the true strain at maximum load. Thus, K is expressed in terms of the natural logarithmic constant e with the help of Eqs (5.7), (5.8) as

$$K = \frac{\sigma_{tu}}{\varepsilon_{tu}^n} = \frac{\sigma_u (S_0/S_u)}{n^n} = \sigma_u \left(\frac{e^n}{n^n}\right) = \sigma_u \left(\frac{e}{n}\right)^n \tag{5.18}$$

Eq. (5.18) still needs the data from the tension test since the strain hardening exponent is determined from the engineering maximal uniform strain ($n = \ln(\varepsilon_u + 1)$). In practice, however, the material properties are commonly obtained from standards which do not include such maximal strain. Therefore, Zhang et al. (2004) proposed to estimate the maximal uniform strain of steels with yield stress up to 355 N/mm^2 by the following expression:

$$\varepsilon_u = \frac{1}{0.24 + 0.01395\sigma_u} \tag{5.19}$$

Villavicencio and Guedes Soares (2012) proposed a material relation where the true stress–strain curve is divided in two parts with respect to the onset of necking: Eqs (5.1), (5.2) define the process before necking and Eq. (5.14) after necking. Following the recommendation of Zhang et al. (2004), Eq. (5.19) is selected to estimate the maximum uniform strain to derive the strain hardening exponent (Eqs 5.2, 5.17). In general, this selection provides good continuity of the logarithmic stress and logarithmic strain (Eqs 5.1, 5.2) beyond the maximum load. In practice, the last point of the first part of the curve is defined at the maximum load. However, the first point of the second part of the curve is selected at a true strain a few percent larger than that at the maximum load. Consequently, the two curves are simply connected by a straight line tangent at the last and the first point of the two parts of the flow curve. This material relation is named combined material since it combines two approximations for the flow curve.

Marinatos and Samuelides (2015) proposed two approaches to determine the true stress–true strain curve used in collision simulations of various experiments: (1) the power law type and (2) the tangent type. For both approaches, the true curve until the maximum load is characterized by Eqs (5.1), (5.2). Similarly to the combined material model, the power law type uses Eq. (5.14) beyond localization; however, the engineering stress σ_u and engineering strain ε_u at maximum load are used to estimate the strain hardening exponent (Eqs 5.2, 5.17) and the strength coefficient (Eqs 5.1, 5.14, and 5.17). Thus, the power-law-type curve can be written as

$$\sigma_t = \frac{\sigma_{tu}}{\varepsilon_{tu}^{\varepsilon_{tu}}} \varepsilon_t^{\varepsilon_{tu}} \tag{5.20}$$

On the other hand, the tangent type assumes that the true curve beyond necking is linear and tangent to the point (ε_{tu}, σ_{tu}), having the following form:

$$\sigma_t = c\varepsilon_t + d \tag{5.21}$$

where the coefficient $c = \sigma_{tu}$ is determined from Eq. (5.13) and then $d = \sigma_{tu}\,(1 - \varepsilon_{tu})$. Finally, the tangent-type curve takes the form

$$\sigma_t = \sigma_{tu}(1 - \varepsilon_{tu} + \varepsilon_t) \tag{5.22}$$

This linear/tangent-type assumption agrees with observations from tensile tests since frequently the flow curve determined from test data is linear from maximum load to fracture (Dieter, 1986).

The power law type (Eq. 5.20) and the tangent type (Eq. 5.22) represent the respective lower bound ($w = 0$) and upper bound ($w = 1$) of the following relationship for the true curve beyond localization (Ling, 1996):

$$\sigma_t = \sigma_{tu}\left[w(1 - \varepsilon_{tu} + \varepsilon_t) + (1 - w)\left(\frac{\varepsilon_t}{\varepsilon_{tu}}\right)^{\varepsilon_{tu}}\right] \tag{5.23}$$

where the variable w is a weight constant that varies between 0 and 1.0.

It should be noticed that all these mathematical expressions for the flow curve require input data from the tension tests for their definition. The approach of Zhang et al. (2004) could be considered as an exception since Eq. (5.19) allows representing the characteristic plastic flow of the material without the tensile test data. In practice, most often, the engineering stress–strain curve is not available to determine the strain hardening parameters K and n, and thus, analytical expressions are essential for the estimation of these parameters only with the information provided in the standards of materials (e.g. ASTM, 2010; Lloyds Register, 2018): yield tensile strength, ultimate tensile strength, and elongation at break (engineering fracture strain).

For steel, Liu et al. (2017a) proposed to estimate the strain hardening exponent and the strength coefficient by using only the yield tensile strength and the ultimate tensile strength together with the following expressions:

$$\sigma_0 = K \cdot 0.006^n \tag{5.24}$$

$$\sigma_u(n+1) = K[\,\ln(n+1)]^n \tag{5.25}$$

where the assumptions and justifications behind these expressions are as follows:

(1) The true strain at the origin of the flow curve is $\varepsilon_t = 0.006$, which is the beginning of plasticity for steels without/with negligible yield plateau; this magnitude is recommended by Sever et al. (2011).

(2) The engineering strain ε_u at the ultimate tensile strength is equal to the strain hardening exponent n; theoretically, this is not true as it is usually assumed that $\varepsilon_{tu} = n$ (Eq. 5.17). In practice, however, the difference between ε_u and ε_{tu} is not significant. This assumption allows using the strain hardening exponent to estimate the true stress $\sigma_{tu} = \sigma_u\ (n+1)$ and the true strain $\varepsilon_{tu} = \ln(n+1)$ at the maximum load in Eq. (5.25).

(3) The yield stress and the origin of the flow curve are simply connected by a line (negligible yield plateau). The yield strain is determined with the modulus of elasticity which is also provided in the standards.

Then, to estimate the variables K and n, the equivalent value of K from Eq. (5.25) is substituted into Eq. (5.24), and then, Eq. (5.24) is solved by a numerical method. Once n is calculated, Eq. (5.24) or (5.25) can be used to calculate K. Eventually, the true stress–true strain curve is simply characterized by the power law relation (Eq. 5.14) when $\varepsilon_t > 0.006$. Eqs (5.24), (5.25) are valid for mild and high-tensile-strength steels (yield tensile strength between 235 and 355 N/mm^2, as defined in Lloyds Register, 2018), since the flow stress curve of these steels can be described by simple power law relations (Dieter, 1986).

Fig. 5.13 shows an example of an engineering stress–strain curve and the simplified expressions to characterize the plastic flow of the material.

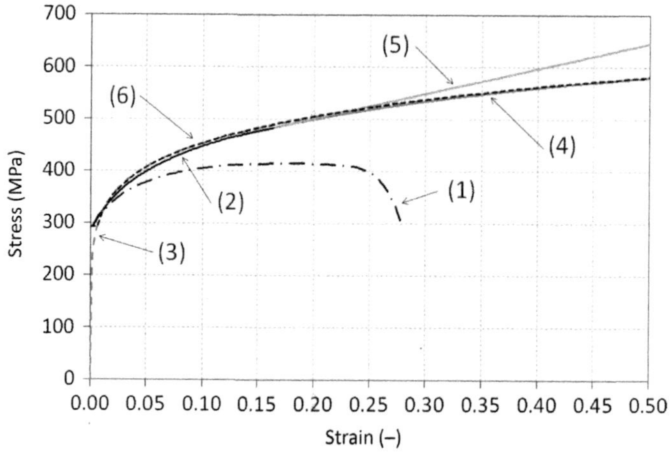

Fig. 5.13 Comparison between different approximations of the true stress–true strain curve: (1) baseline engineering stress–strain curve, (2) logarithmic true stress and strain until maximum load, (3) simple power law relation using the approach of Zhang et al. (2004), (4) power law type beyond maximum load as defined by Marinatos and Samuelides (2015), (5) tangent type beyond maximum load as defined by Marinatos and Samuelides (2015), and (6) true stress–true strain curve using the approach of Liu et al. (2017a). Curve (2) and curve (3) beyond necking characterize the combined material model (Villavicencio and Guedes Soares, 2012).

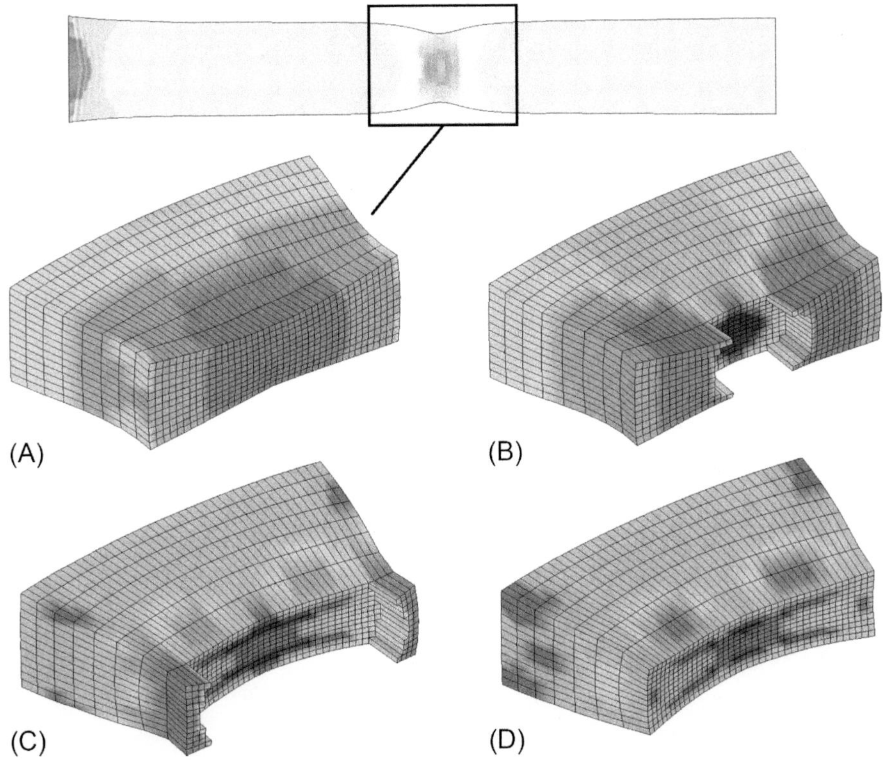

Fig. 5.14 Representation of the initiation of rupture and eventual fracture of a tensile piece using solid elements. The axial deformation of the elements is scaled for illustration. (A) The finite elements reach the maximum strain before rupture initiates. (B) Rupture has initiated at the central axis of the tensile piece. (C) The cracks propagate perpendicularly to the four edges of the specimen. (D) Eventually, fracture occurs as the tension piece is completely separated.

5.4.2 Critical failure strain

The formulation presented in the previous section for estimating the true stress–true strain beyond the maximum load (Eqs 5.3, 5.4) does not indicate whether the initial rupture of the tensile piece has occurred or not, since these expressions are only based on actual measurements of the cross-sectional area. The initial rupture and eventual fracture of a tensile specimen are illustrated in Fig. 5.14. Such phenomenon cannot be easily captured in a tensile test since the initial failure starts in a microscale at the central axis of the transverse sectional area and then propagates perpendicularly to the surfaces of the specimen until fracture occurs. Thus, Eqs (5.3), (5.4) are valid only to estimate the true fracture stress and the true fracture strain by measuring the cross-sectional area at fracture, although such measurements are quite difficult; in particular, for test pieces machined from a plate (flat bars).

For a nonlinear finite element analysis, the input needed is the true strain at rupture, which is usually named as critical failure strain. This means that only few elements fail first as show in step (B) of Fig. 5.14, while fracture occurs at step (D). It is considered that rupture occurs when the following failure criterion is satisfied:

$$\varepsilon_{eq} \geq \varepsilon_f \tag{5.26}$$

where ε_{eq} is the maximum equivalent strain at a deformed finite element and ε_f is the critical failure strain used as input for the calculations.

Normally, the critical failure strain is based on effective plastic strain (or equivalent plastic strain). However, it can be also based on plastic thinning, the major principal in the plane strain component, or a minimum time step (Hallquist, 2006). Expressions to estimate the critical failure strains commonly used for ship collision simulations are summarized in Section 5.5.

In nonlinear finite element simulations, the elements that reach the maximum equivalent strain are removed from the structure at the next incremental time step, decreasing the impact loading strength of the structure. Fig. 5.15 shows the deletion

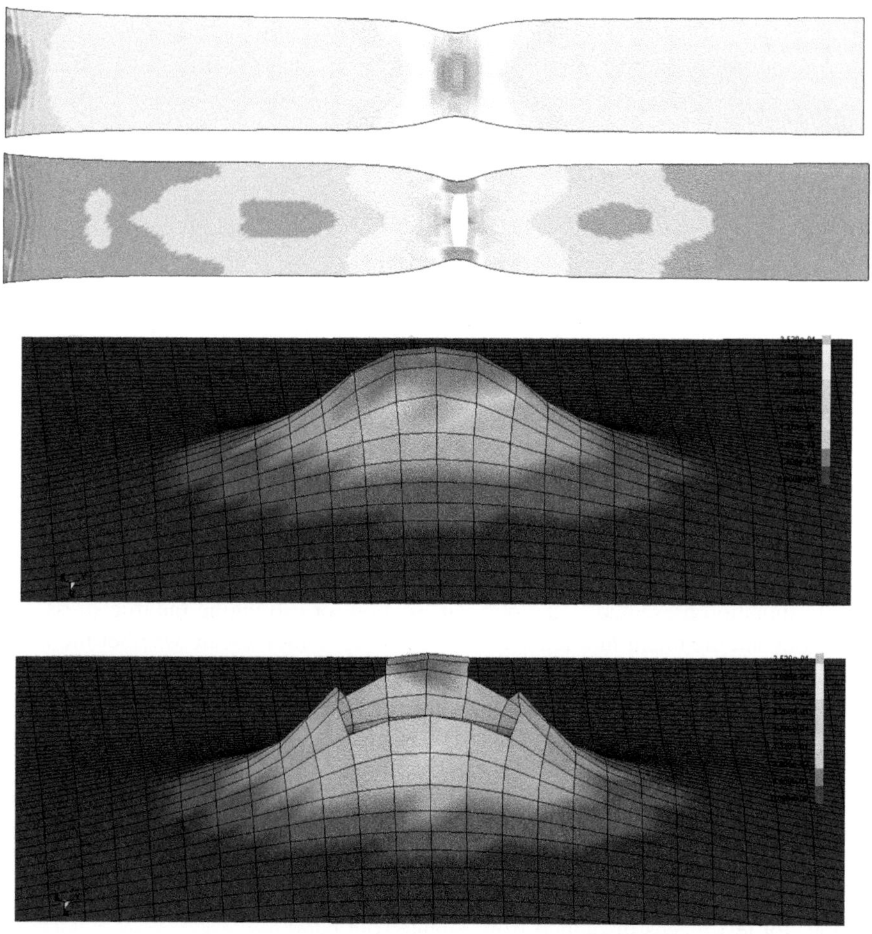

Fig. 5.15 Material rupture in a tension test piece and a plate specimen punched by a rigid indenter.

of the first failing element in the simulation of a tensile test and in a simulation of a plate forming experiment. Normally, more than one element reaches the critical failure strain at a particular time step of the simulation. Thus, in order to observe in detail the onset of rupture and the actual propagation of the cracks, the number of time steps is to be increased. This is not always practical as it results in an increase of the computational time.

Based on the equations for isotropic von Mises plasticity, the equivalent (or effective) plastic strain ε_{eq} is the integral over time of the equivalent plastic strain rate ε_{eq}^{r}:

$$\varepsilon_{eq} = \int_0^t \varepsilon_{eq}^{r} dt \tag{5.27}$$

and the equivalent plastic strain rate is defined in terms of the plastic strain rate:

$$\varepsilon_{eq}^{r} = \sqrt{\frac{2}{3}\varepsilon_{ij}^{r}\varepsilon_{ij}^{r}} \tag{5.28}$$

The volume of the material is unchanged by plastic flow. This corresponds to a Poisson's ratio of 0.5. For uniaxial stress, the plastic strain rate has the form

$$\varepsilon_{11}^{r} = \varepsilon_{p}^{r}\ ,\ \ \varepsilon_{22}^{r} = -\frac{1}{2}\varepsilon_{p}^{r}\ ,\ \ \varepsilon_{33}^{r} = -\frac{1}{2}\varepsilon_{p}^{r} \tag{5.29}$$

The terms ε_{ii}^{r} sum to zero, and all shear strain rates are zero. Then, the equivalent plastic strain rate for uniaxial tension can be expressed as

$$\begin{aligned} \varepsilon_{eq}^{r} &= \sqrt{\frac{2}{3}\left[\left(\varepsilon_{11}^{r}\right)^2 + \left(\varepsilon_{22}^{r}\right)^2 + \left(\varepsilon_{33}^{r}\right)^2\right]} \\ &= \sqrt{\frac{2}{3}\left[\left(\varepsilon_{p}^{r}\right)^2 + \left(-\frac{1}{2}\varepsilon_{p}^{r}\right)^2 + \left(-\frac{1}{2}\varepsilon_{p}^{r}\right)^2\right]} = \left|\varepsilon_{p}^{r}\right| \end{aligned} \tag{5.30}$$

The same concept is used to evaluate the critical failure strain in the simulations (Hallquist, 2006).

The critical failure strain is highly affected by the size of the elements, and therefore, the failure criteria are often defined to be mesh-dependent; failure criteria commonly used for ship collision simulations are given in Section 5.5. Fig. 5.16 illustrates the dependency of the failure strain on the mesh size of a tensile specimen of thickness 2.0 mm; note that the critical failure strain is normalized by the nominal fracture strain obtained by the tension test. This curve is based on an expression derived by Paik (2007) to estimate the critical failure strain. It is observed that for a mesh size larger than 10 mm (5t), the tendency of the curve is almost linear where the slope is relatively small. However, for mesh sizes smaller than 10 mm, the failure strain increases exponentially. It should be noticed that the critical failure strain is also affected by the true stress–true strain curve selected for the analysis.

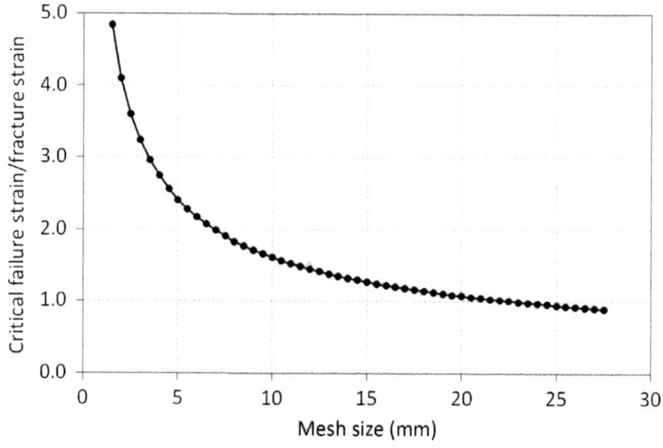

Fig. 5.16 Variation of the critical failure strain with the mesh size for a tensile specimen of thickness 2.0mm. The ordinate in the plot can be obtained by $4.1(t/l)^{0.58}$ where t is the plate thickness and l is the length of the finite elements.

5.4.3 *Strain rate*

Since the strain rate effect most often is omitted in ship collision simulations, this section just offers a brief discussion on this highly nonlinear phenomenon.

The plastic strain rate is often determined by the material constants D and q of the Cowper–Symonds equation (see Section 3.2.3 in Chapter 3). The recommended material constants to account for the strain rate characteristics of mild steels are $D=40.4\,s^{-1}$ and $q=5$ (Jones, 1989), and on the other hand, for high-tensile steels, they are $D=3200\,s^{-1}$ and $q=5$ (Paik, 2007). While these values have been used in various dynamic analyses of marine structures, the actual D and q values for mild steels can vary widely. For example, the values given by Hsu and Jones (2004) are $D=114.27\,s^{-1}$ and $q=5.56$ at the yield strain and $D=841\,s^{-1}$ and $q=4.27$ at the ultimate strain. Also, Jones and Jones (2002) found that the material constants at the yield and ultimate tensile stress of mild steels change not only with the increase of strain but also with the plate thickness, as shown in Table 5.2 (Jones, 2013). This abundance of D and q values suggests that experiments on the actual material selected for a particular structural design may be required. This was also highlighted by Jones (2013) when discussing the credibility of predictions for structural designs subjected to dynamic loads.

Table 5.2 Cowper–Symonds coefficients for mild steel reported by Jones (2013).

t (mm)	D_y (s^{-1})	q_y (–)	D_u (s^{-1})	q_u (–)
1.6	0.76	7.2	190,000	0.10
3.0	24.0	6.1	116,000	0.30
6.0	44.3	5.0	230,000	0.23

t is the plate thickness, and the subscripts y and u identify the values at the yield and ultimate tensile stresses.

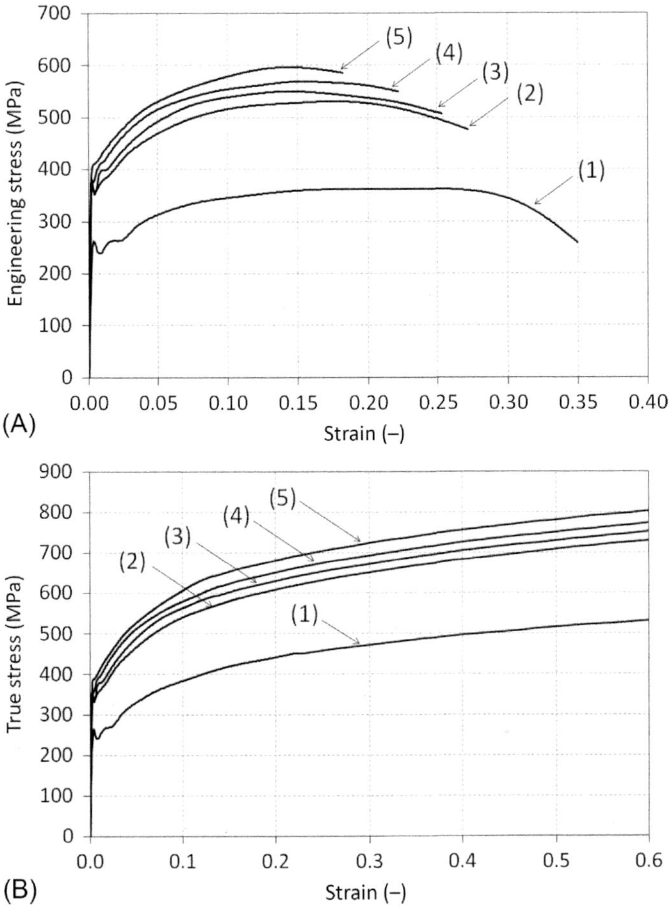

Fig. 5.17 (A) Quasistatic engineering stress–strain curve and dynamic engineering stress–strain curves using different strain rates: (1) quasistatic tension test, (2) strain rate $1.0\ s^{-1}$, (3) strain rate $10\ s^{-1}$, (4) strain rate $100\ s^{-1}$, and (5) strain rate $500\ s^{-1}$. (B) True stress–true strain curves of the quasistatic and dynamic engineering curves: (1) quasistatic tension test, (2) strain rate $1.0\ s^{-1}$, (3) strain rate $10\ s^{-1}$, (4) strain rate $100\ s^{-1}$, and (5) strain rate $500\ s^{-1}$.
cf. Liu, K., Liu, B., Villavicencio, R., Wang, Z., Guedes Soares, C., 2018b. Assessment of material strain rate effects on square steel plates under lateral dynamic impact loads. Ships Offshore Struct. 13 (2), 217–225.

In nonlinear finite element simulations, two options are available to account for strain rate effects. The first and most common practice is to input the D and q variables of the Cowper–Symonds equation to scale the already defined true stress–true strain curve. The second is to input a set of true stress–true strain curves, commonly determined from quasistatic and dynamic tension tests performed at different strain rates. An example of these curves is shown in Fig. 5.17. In Section 3.2.3, the estimated range of strain rates for actual ship collisions is about $0.5–5.0\,s^{-1}$ that corresponds to an increase of the dynamic flow stress by 55% for mild steels according to

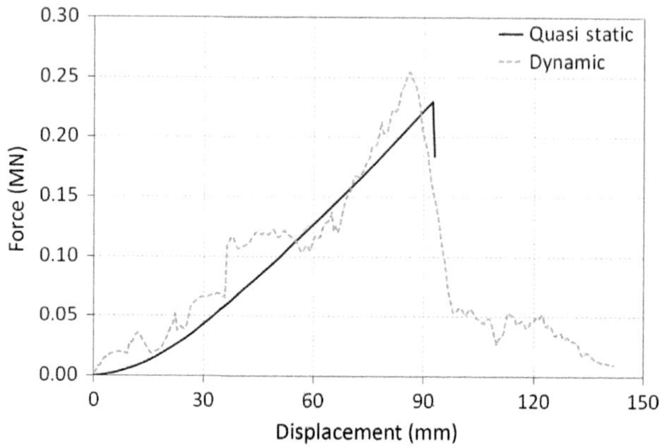

Fig. 5.18 Force–displacement response. Comparison between a quasistatic and a dynamic impact test of a square plate (dimensions $1000 \times 1000 \times 4.0$ mm) impacted by a rigid indenter (diameter 150 mm and mass 1350 kg) at an initial velocity of 6.2 m/s.
cf. Liu, K., Liu, B., Villavicencio, R., Wang, Z., Guedes Soares, C., 2018b. Assessment of material strain rate effects on square steel plates under lateral dynamic impact loads. Ships Offshore Struct. 13 (2), 217–225.

the Cowper–Symonds theoretical model. Fig. 5.17 shows that for the strain rate $1.0\,s^{-1}$, the ratio between the dynamic and the static stress is about 1.40 and for the rate $10\,s^{-1}$ is in the order of 1.60, which also agrees with the theoretical observations.

Fig. 5.18 shows the comparison of the quasistatic and dynamic experimental responses of an unstiffened square plate (dimensions $1000 \times 1000 \times 4.0$ mm) impacted by a rigid indenter with a diameter of 150 mm (Liu et al., 2018b). It is observed that the dynamic forces are larger than the quasistatic until a displacement of about 60 mm. The fact that this difference is more evident at the initial stages of the impact (deflection < 15 mm) agrees with the basis of the Cowper–Symonds theoretical model (Eq. 3.16). In that model, the constants are valid for strains with a maximum value of only a few percents, implying that at these strains, the strain rate effect is more important. This result also agrees with the data in Table 5.2, where the scale of the flow stress is larger at the yield stresses than at the ultimate tensile stresses (that scale is about twice for a strain rate of $100\,s^{-1}$).

While during the latest stages of the impact (few millimetres before failure), the dynamic material effect decreases, as the indenter has been deaccelerated, the dynamic effect is still manifested in a reduction of the failure strain. The ratio between the dynamic and the quasistatic displacement at failure is about 0.93 in Fig. 5.18. This effect can be also seen in Fig. 5.17 where the fracture strain decreases with the strain rate. In Fig. 5.17, the ratio between the dynamic and the quasistatic fracture strain is approximately 0.87 for the strain rate $1.0\,s^{-1}$, but a tensile test at a smaller strain rate should result in a larger fracture strain, which would agree with that observed in the impact response since at few instants before failure, the strain rate should be smaller than $1.0\,s^{-1}$.

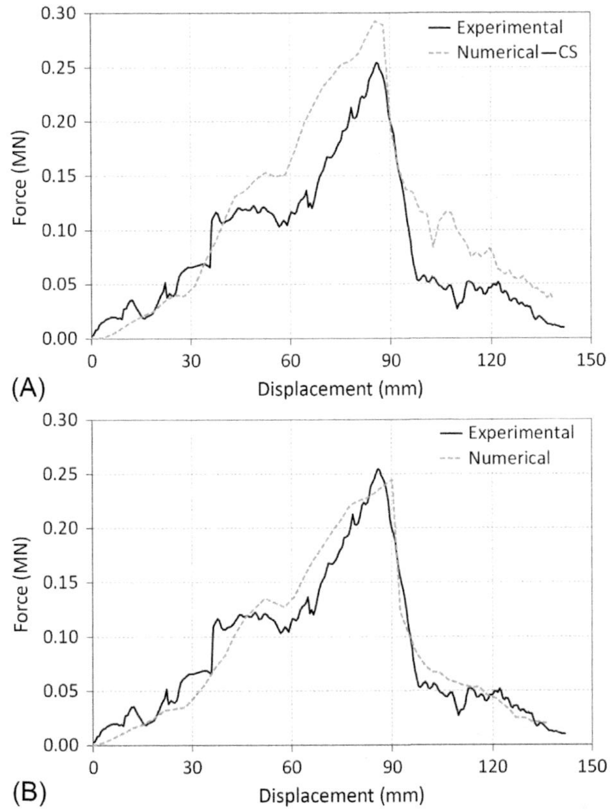

Fig. 5.19 Force–displacement response. (A) The finite element simulation uses the strain rate material parameters $D=40.4\,s^{-1}$ and $q=5$. (B) The simulation uses the actual set of true stress–true strain curves shown in Fig. 5.17.
cf. Liu, K., Liu, B., Villavicencio, R., Wang, Z., Guedes Soares, C., 2018b. Assessment of material strain rate effects on square steel plates under lateral dynamic impact loads. Ships Offshore Struct. 13 (2), 217–225.

Fig. 5.19 presents the results of the finite element simulations using the two approaches mentioned above to evaluate the strain rate effect. These simulations were performed by Liu et al. (2018b) using the solver of ABAQUS. For the first approach (Fig. 5.19A), the material constants commonly recommended for mild steels were used ($D=40.4\,s^{-1}$ and $q=5$). For the second approach (Fig. 5.19B), the set of true dynamic stress–true strain curves plotted in Fig. 5.17 was used. It is observed that the Cowper–Symonds model describes reasonably the experimental response until a displacement of about 45 mm, but beyond that point, the forces are largely overestimated. On the other hand, the input of the actual strain rates helps to predict better the magnitude of the instantaneous forces until failure.

These results are expected since the Cowper–Symonds material constants are only valid for small strains (Jones, 1989, 2013) (see Table 5.2) and thus, the scaling of the flow stress at larger strains is overestimated. Therefore, care is to be taken when analysing numerical predictions that use the Cowper–Symonds model, since the results can be misunderstood if the coefficients D and q are kept constant during the entire impact simulation. Both impact simulations used quasistatic tensile test simulations to predict the critical failure strain. Consequently, the fact that the Cowper–Symonds simulation predicts better the deflection at failure should not be considered reliable since for consistency a dynamic failure strain should have been used in the numerical analysis. As discussed in Section 3.2.3, the dynamic engineering fracture strain can be roughly estimated with the quasistatic fracture strain and the inverse of the Cowper–Symonds equation. The inclusion of a dynamic failure strain could have improved the prediction of failure of the model based on the actual strain rate material curves.

Finally, it should also be mentioned that in the absence of complete data on the strain rate and dynamic rupture behaviour, which is the usual situation, an effort should be made to demonstrate the sensitivity of finite element results to changes in the constitutive equations for the material (Jones, 2013).

5.5 Failure criteria for ship collision assessments

The finite element simulations for the ship collision tests conducted in the Netherlands used the GL, PES, and RTCL failure criteria for the benchmark study (Ehlers et al., 2008). The results of these simulations are given in Fig. 5.1 to highlight the importance of the mesh size and the associated failure criterion for ship collision assessments. Ehlers et al. (2008) summarized the background of these criteria; the GL criterion was proposed by Scharrer et al. (2002) and Zhang et al. (2004), the PES criterion – by Peschmann (2001) and the RTCL criterion – by Törnqvist (2003). These three criteria continue to be used as a reference when evaluating the response of ship structures under impact loading; see, for example, Villavicencio et al. (2014), Marinatos and Samuelides (2015), Storheim et al. (2015a), and Liu et al. (2017a).

Storheim et al. (2015a) not only selected the GL, PES, and RTCL criteria to assess the impact response of typical ship structures but also evaluated other criteria reported in the literature (see Table 5.3). As shown in Table 5.3, the failure criteria were named and grouped as simple strain-state-independent and advanced strain-state-dependent where the strain state is defined as the ratio of the minor and major principal strain rates ($\varepsilon_2^{\text{rate}}/\varepsilon_1^{\text{rate}}$). Similarly, the stress state can be defined as the ratio between the minor and major principal stresses (σ_2/σ_1) or in terms of the stress triaxiality T which is the ratio of the mean stress (through the principal stresses σ_i) and the effective von Mises stress (equivalent stress), $T=(\sigma_1+\sigma_2+\sigma_3)/(3\sigma_{\text{eq}})$. These three parameters can be also used to characterize multiaxial loading states. The simple criteria assume a constant critical equivalent strain often defined as mesh-dependent.

Storheim et al. (2015a) repeated the collision simulations of test model 3 (see Fig. 5.20) using two mesh sizes ($5t$ and $10t$) and most failure criteria listed in Table 5.3. It can be seen that the GL, PES, and SHEAR criteria were the best options to predict the absorbed energy at the end of the impact for both mesh sizes, while the results using the other criteria deviated largely from the experimental energy, in

Table 5.3 Different failure criteria reported in the literature.

Criterion	Reference
Simple strain-state-independent	
DNV RP-C204	Det Norske Veritas (2010)
PES	Peschmann (2001)
GL	Scharrer et al. (2002)/Zhang et al. (2004)
SHEAR	Used by Marinatos and Samuelides (2015)
Damage	ABAQUS (2012)
Advanced strain-state-dependent	
Swift criterion for diffuse instability	Swift (1952)/Altmeyer (2013)
Hills criterion for local necking	Hill (1952)
Bressan–Williams shear instability	Bressan and Williams (1983)
BWH: Bressan–Williams–Hill instability criterion	Alsos et al. (2008)
BWH: Bressan–Williams–Hill instability criterion with post-necking damage	Storheim et al. (2015b)
RTCL: Rice–Tracey Cockroft–Latham damage criterion	Törnqvist (2003)

Data from Storheim, M., Amdahl, J., Martens, I., 2015a. On the accuracy of fracture estimation in collision analysis of ship and offshore structures. Mar. Struct. 44, 254–287.

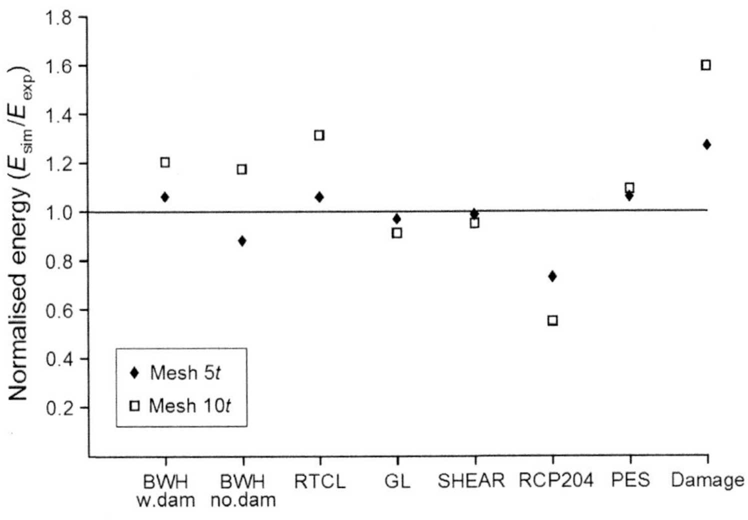

Fig. 5.20 Normalized energy E_{sim}/E_{exp} at the end of test model 3 using various simple strain-state-independent and advanced strain-state-dependent failure criteria.
cf. Storheim, M., Amdahl, J., Martens, I., 2015a. On the accuracy of fracture estimation in collision analysis of ship and offshore structures. Mar. Struct. 44, 254–287.

particular for the mesh $10t$. It was concluded that at the end of the impact when several large fractures occur, the criteria neglecting the state of stress in the definition of the threshold value for rupture, that is, the simple criteria, are able to represent the impact process better than those that account for the state of stress. This conclusion was based not only on the results of test model 3 but also on the response of a total of 47 different simulations of various collision experiments.

In a practical industrial application, such as collision assessment for the appraisal of ship structural arrangements (Lloyds Register, 2016), the interest is on the critical energy at a prescribed maximum penetration. Thus, a simple criterion represents advantages as a rapid and reasonably reliable design tool. Here, the formulations of the GL and PES criteria for steel structures are provided for easy reference. In addition, the failure criterion recently reported in Liu et al. (2017a) is reviewed as it is also used in the calculation examples presented in Section 5.7.

5.5.1 GL criterion

The GL criterion is based on the thinning plastic strain (strain in the thickness direction) at failure which for uniaxial tension can be evaluated from Eq. (5.29). The following expression was proposed to estimate the critical thinning at rupture:

$$\varepsilon_f = 0.056 + 0.54 \cdot \frac{t}{l} \tag{5.31}$$

where t is the plate thickness and l is the individual length of the elements. The value 0.056 represents the uniform strain, and 0.54 is the necking strain for shell elements. It is commonly recommended that the ratio l/t is not less than five for shell elements. The expression of the GL criterion can be found in Zhang et al. (2004), where an approval procedure for alternative arrangements of cargo ships that need to comply with damage stability requirements after severe collision damage was presented. Eq. (5.31) was derived by measuring the thickness from prototype damaged structural components, such as shell plating and stiffeners, and determining the uniform strain, the necking, and the necking length. This work was carried out by Scharrer et al. (2002).

5.5.2 PES criterion

The PES criterion is based on the equivalent plastic strain at failure that can be estimated by the following expression:

$$\varepsilon_f = \varepsilon_g + \alpha \cdot \frac{t}{l} \tag{5.32}$$

where t is the plate thickness, l is the individual length of the elements, ε_g is the uniform strain, and α is a factor that depends on the necking strain and the length of the neck. Peschmann (2001) obtained experimentally these variables from forming limit diagrams of the plates used for test model 3. For the plate thickness of 5.0mm, $\varepsilon_g = 0.10$, and $\alpha = 0.80$. For the plate thickness of 12.5–20mm, $\varepsilon_g = 0.08$, and $\alpha = 0.65$.

5.5.3 LIU criterion

The LIU criterion (Liu et al., 2017a) assumes a constant effective (equivalent) plastic strain at failure determined by

$$\varepsilon_{\mathrm{f}} = 0.5 - 0.01 \cdot \frac{l}{t} \tag{5.33}$$

where t is the plate thickness and l is the length of the finite elements. The expression indicates that the necking strain is 0.50 and that the failure strain decreases linearly with the mesh size. The ratio l/t should range between 5 and 20. This expression was derived from finite element simulations of coarsely meshed square plates ($l/t = 5$–20), punched by a hemispherical indenter until the onset of necking. The critical necking point was first determined by using the same plates but modelled with a finer mesh (ratio $l/t = 2$). Thus, Eq. (5.33) may be used to estimate the failure strain of coarsely meshed ship structures simulating a ship-to-ship collision scenario where the bulbous bow penetrates the side shell structure between web frames. The criterion is valid for mild and high-strength steel materials (yield stress between 235 and 355 MPa).

5.6 Contact definition and friction

In general, the contact interface between two colliding objects requires that one side of the interface is denoted as the slave side and the other as the master side; the nodes lying in those surfaces are respectively referred as slave and master nodes. In practice, automatic contact types are commonly preferred for ship collision analyses where the slave and master surfaces are conveniently generated internally by the software. In this case, the contact forces between the struck and striking structure can be postprocessed by simply indicating which part is acting as the rigid striker.

Automatic contacts always consider that the thickness of the shell elements is projected to both the master and the slave surfaces and they are based on mid surface normal projection vectors (they do not have orientation). Thus, during the modelling stage, the most important aspect is to account for the thickness of the elements, implying that the surfaces must be offset by at least one-half of the thickness (see Fig. 5.21). This modelling technique may contribute to unwanted initial contact interpenetrations that can cause numerical problems when the analysis begins.

In ship collision analyses, the indenter (striking structure) is often modelled as a rigid body. This is a preferred modelling technique not only for the effective cost of computations but also because it is a fairly good approximation for ship collision calculations (see Fig. 4.1). A rigid body is defined by associating a rigid material model to it, where realistic properties (Young's modulus and Poisson's ratio) are to be defined since they are used to determine sliding interface parameters in the contact interaction. Any unrealistic value may contribute to numerical problems. As illustrated in Fig. 5.21, a realistic thickness is to be specified for a rigid body (striker) modelled with shell elements to avoid loss of contact or excessive interpenetrations.

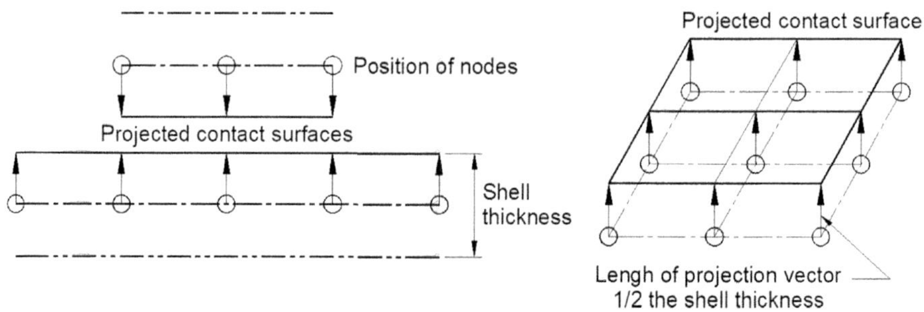

Fig. 5.21 Definition of the projected contact surfaces in automatic contacts used for collision simulations.

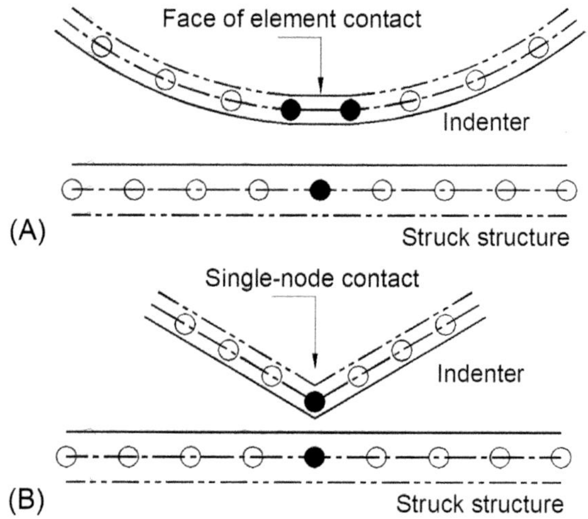

Fig. 5.22 (A) Face of the indenter element in contact with the struck structure. (B) Single-node contact between the striking and struck structure.

From a practical point of view, it is recommended that the indenter is meshed to ensure that the face of an element gets initial contact with the struck structure, thus providing a more realistic simulation of the contact area (see Fig. 5.22A). Also, it has been found that when a single-node corner of the striker first contacts a node of the struck structure (see Fig. 5.22B), the postprocessed force and absorbed energy values experience unrealistic peaks at the beginning of the response.

To initiate the contact between the struck structure and the striker, there are two possibilities: (1) an initial impact velocity is assigned in the direction of the displacement of the rigid striker, or (2) a prescribed displacement-time curve (constant velocity) is assigned to the striker. The use of an initial impact velocity requires that an artificially large density is used for the striking body to represent its actual mass and consequently the initial kinetic energy of the body. This approach allows the striker to deaccelerate during the simulation, and depending on the impact velocity, it may be possible to take into account strain rate effects. The use of a prescribed constant velocity during the entire simulation does not require increasing of the density of the striker, and it is valid as far as strain rate effects are omitted. Care is to be taken when assigning a high constant velocity to decrease the computational time as it can provoke unrealistic fluctuations during the early stages of the force–displacement response.

The selection of an automatic contact requires correct modelling/offset of the structures, as most of the remaining input parameters are set as default by the software. The exception is the friction between the interacting structures. Friction is based on the Coulomb formulation ($F_f = \mu_s \cdot F_n$), meaning that the force of friction F_f exerted between the surfaces is equal to the normal force F_n exerted by each surface on the other times the coefficient of friction μ_s. The coefficient of friction is a property of the contacting materials.

Typical values found in engineering handbooks for the static coefficient of friction for steel-steel contact surfaces differ considerably from those that have been actually used for ship collision analysis or even simpler simulations of metal forming experiments (see Table 5.4). It can be seen that in most of the cases, the value of static friction is assumed as 0.23 or 0.30.

The fact that the effect of friction is relatively small when evaluating perpendicular ship collisions has been already pointed out by Paik (2007). Of course, this effect will be significant for side collisions of raking type or for ship grounding events where a relative velocity between the struck structure and the indenter exists. A demand for investigation of the energy dissipated by friction was highlighted by Liu et al. (2018a) when they reviewed the 'lack-of-knowledge topics in ship collision analysis' for calculation procedures.

The influence of the friction coefficient on the impact response of double-hull structures can be observed by performing parametric studies varying this parameter; see, for example, Zhang et al. (2018). Fig. 4.2 shows the results for specimen S-II where the effect of the friction is evaluated by plotting the ratio between the energy absorbed by friction and the energy absorbed by the structural deformation of the double-hull specimens. The geometry of specimen S-II and other characteristics of the impact event can be found in Section 4.3.1 (three model tests by Arita and Aoki, 1985). It should be mentioned that the tendency of the results for the other two specimens S-I and S-III is similar, as reported by Zhang et al. (2018).

Table 5.4 Static coefficient of frictions that have been used to conduct ship collision simulations.

Reference	Friction coefficient
Lehmann and Peschmann (2002)	0.10
Ehlers et al. (2008)	0.30
Karlsson et al. (2009)	0.23
Marinatos and Samuelides (2015)	0.30
Storheim et al. (2015a)	0.20/0.23/0.30
Liu et al. (2017a)	0.30
Zhang et al. (2019)	0.23

From Fig. 4.2, the following can be observed. (1) The absorbed energy increases with the friction coefficient. (2) The energy absorbed by friction increases with the penetration. This may be due to an increase of the contact surface areas and the relative displacements. (3) Before the outer shell plate ruptures (at a penetration of about 70mm), the friction contribution is very small, less than 3.0%. (4) Overall, the energy absorbed by friction is not very significant. Even for the largest friction coefficient, $\mu_s = 0.45$, the average absorbed energy is only about 10%.

Finally, it should be mentioned that the selected friction coefficient also influences the crack initiation and growth directions. This is mainly because friction influences the contact area between the struck structure and the indenter, thus affecting the deformations and strains. Moreover, if the friction coefficient is set to zero, the friction energy would be almost equal to zero, too, with only very small negative/positive contact energies due to minor numerical problems commonly caused by undetected penetrations.

5.7 Calculation examples

The most relevant parameters required to conduct nonlinear finite element simulations of typical ship structures subjected to impact loads have been introduced in the previous sections. This section presents calculation examples using those input parameters. The results of the collision simulations are compared with those recorded experimentally and those predicted by the simplified method (Eq. 4.4). Various double-hull specimens presented in Chapter 4, Section 4.3.1, are selected for the simulations. In addition, a collision simulation of two identical container ships is presented as a practical example. These results are compared with the predictions of the simplified method.

5.7.1 Double hull models

The selected double-hull specimens are listed in Table 5.5, indicating the geometric characteristics and material properties. Additional details of the specimens can be found in Section 4.3.1. All finite element models are analysed with the same input parameters: (1) The simple power law relation is used to derive the true stress–true strain curve (Eq. 5.14); (2) the strain hardening parameters K and n are estimated with Eqs (5.24), (5.25); and (3) the critical failure strain is evaluated with the GL, PES, and LIU criteria given in Section 5.5. The mesh size, strain hardening parameters, and friction coefficient are given in Table 5.6. The finite element model of specimen S-II is shown in Fig. 5.23 for reference.

For specimens S-I, S-II, and S-III, the force- and energy–displacement responses are presented in Figs 5.24–5.26. These collision simulations were performed by Zhang et al. (2019). As an example, the numerically determined shape of the deformation of specimen S-I is shown in Fig. 5.27. This shape is similar to the experimentally observed. For reference, the deformation pattern can be considered comparable to that

Table 5.5 Dimensions and scantlings of the selected double-hull specimens.

Specimen	L (mm)	s (mm)	h (mm)	t (mm)	R (mm)	σ_y (MPa)	σ_u (MPa)
S-I	1400	200	300	3.2	80	284.5	440
S-II	1400	280	300	4.5	80	294.3	447
S-III	1400	200	300	4.5	80	294.3	447
ST-3-BW	1410	282	350	2.8	80	245.3	337.8
ST-3-OW	1410	235	350	2.8	80	245.3	337.8
P-50	1000	200	200	2.3	50	282	438
P-100	1000	200	200	2.3	100	282	438
P-200	1000	200	200	2.3	200	282	438
P-300	1000	200	200	2.3	300	282	438

L is the length of the square panel, s is the spacing between webs, h is the width of the double hull (distance between the outer and inner hull), t is the thickness of the plating, R is the end radius of the conical indenter, σ_y is the yield stress, and σ_u is the maximum tensile strength of the material.

Table 5.6 Inputs for the finite element calculations of the double-hull models.

Specimen	Mesh sizes (target: $5t/10t$)	Strength coefficient K (MPa)	Strain hardening exponent (–)	Friction coefficient (–)
S-I	$6.25t/12.5t$	650	0.17	0.23
S-II	$4.44t/8.88t$	670	0.17	0.23
S-III	$4.44t/8.88t$	670	0.17	0.23
ST-3-BW	$4.57t/9.14t$	518	0.15	0.30
ST-3-OW	$4.66t/9.32t$	518	0.15	0.30
P-50	$4.35t/8.70t$	623	0.16	0.30
P-100	$4.35t/8.70t$	623	0.16	0.30
P-200	$4.35t/8.70t$	623	0.16	0.30
P-300	$4.35t/8.70t$	623	0.16	0.30

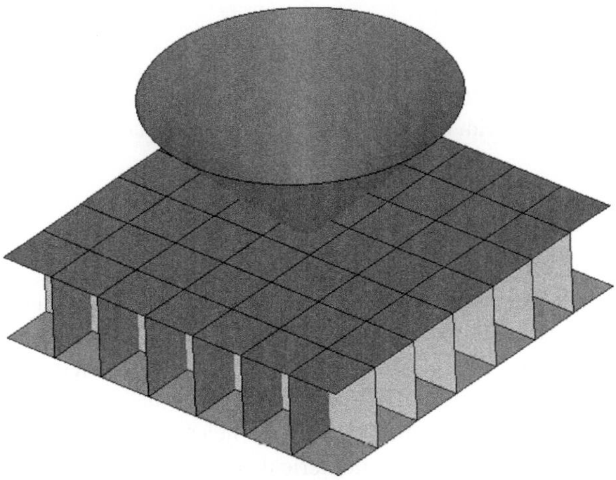

Fig. 5.23 Finite element model of specimen S-II.
cf. Zhang, S., Villavicencio, R., Zhu, L., Pedersen, P.T., 2019. Ship collision damage assessment and validation with experiments and numerical simulations. Mar. Struct. 63, 239–256.

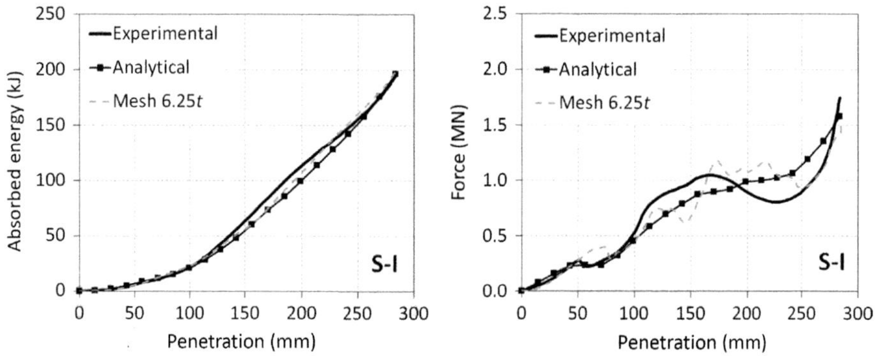

Fig. 5.24 Energy–penetration and force–penetration curves: Specimen S-I. cf. Zhang, S., Villavicencio, R., Zhu, L., Pedersen, P.T., 2019. Ship collision damage assessment and validation with experiments and numerical simulations. Mar. Struct. 63, 239–256.

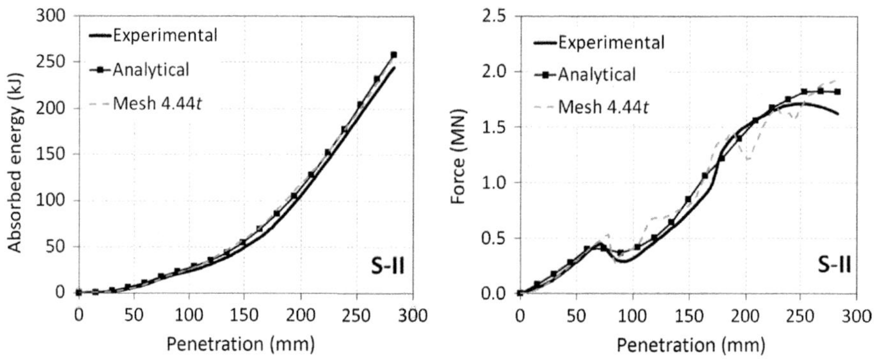

Fig. 5.25 Energy–penetration and force–penetration curves: Specimen S-II. cf. Zhang, S., Villavicencio, R., Zhu, L., Pedersen, P.T., 2019. Ship collision damage assessment and validation with experiments and numerical simulations. Mar. Struct. 63, 239–256.

illustrated in Fig. 4.12 for specimen P-100, as the structural arrangement of the specimen and the geometry of the indenter are similar.

From the comparison with the experiments, it can be observed that the energy and impact force responses are not particularly improved by the finite element results. In fact, the energy- and force–displacement curves are better predicted by the analytical method for specimens S-II and S-III. For specimen S-I, the results from the analytical method and the finite element simulations have nearly the same agreement with the experimental results. However, the numerical prediction of specimen S-I overestimates the force for the initial rupture of the outer plate. The prediction of this point

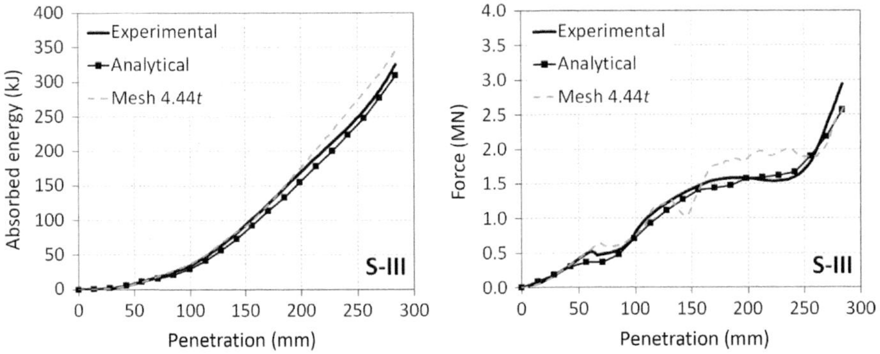

Fig. 5.26 Energy–penetration and force–penetration curves: Specimen S-III. cf. Zhang, S., Villavicencio, R., Zhu, L., Pedersen, P.T., 2019. Ship collision damage assessment and validation with experiments and numerical simulations. Mar. Struct. 63, 239–256.

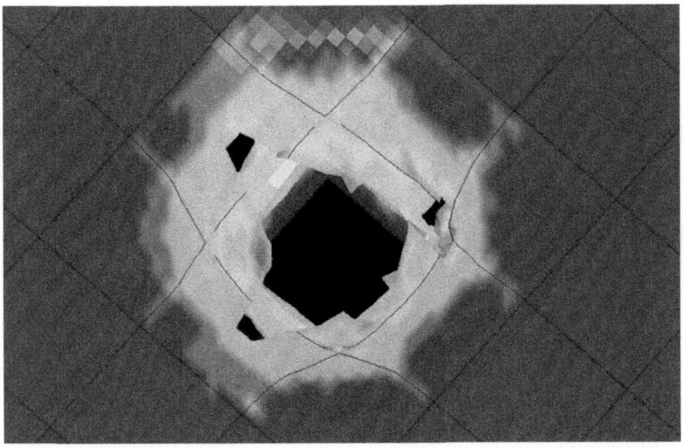

Fig. 5.27 Shape of the deformation of specimen S-I at a deflection of about 200 mm. cf. Zhang, S., Villavicencio, R., Zhu, L., Pedersen, P.T., 2019. Ship collision damage assessment and validation with experiments and numerical simulations. Mar. Struct. 63, 239–256.

is reasonably good for the finite element models of S-II and S-III. Beyond this initial peak force, the cracks continue propagating in the outer plate, and the forces increase continuously as a consequence of the larger contact area with the conical indenter and the resistance offered by the crushed web frames.

Fig. 5.28 compares the energy–displacement curves for the LIU, PES, and GL failure criteria for specimens P-300 (indenter radius = 300 mm) and P-100 (indenter radius = 100 mm). These results were reported by Liu et al. (2017a). It is observed

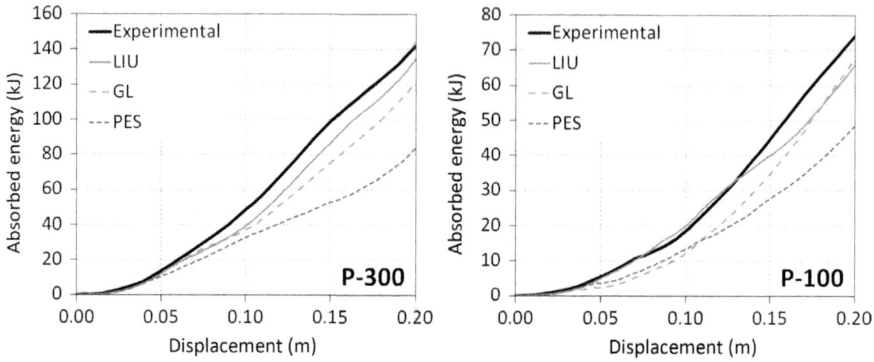

Fig. 5.28 Energy–displacement curves for specimen P-300 (indenter radius = 300 mm) and specimen P-100 (indenter radius = 100) using the LIU, GL, and PES criteria. The mesh size is 8.7*t* as indicated in Table 5.6.
cf. Liu, B., Villavicencio, R., Zhang, S., Guedes Soares, C., 2017a. A simple criterion to evaluate the rupture of materials in ship collision simulations. Mar. Struct. 54, 92–111.

that the LIU criterion gives a more reasonable prediction when the smooth and larger indenter diameter is used (as in a bulbous bow collision), although the response is still lower than the experimental results. For the sharper indenter (P-100), LIU predicts the energy well until a displacement of about 0.13 m, but beyond this point, the LIU response is underestimated and equals the predictions of GL criterion at the end of the impact. While the PES criterion improves the predictions for the sharper indenter, the energy is largely underestimated for these two cases. The reason is that the effective plastic strain estimated by PES criterion is much smaller than that estimated by LIU. On the other hand, the GL criterion cannot be compared directly with LIU and PES since GL evaluates the critical through thickness strain at the moment of fracture (plastic thinning). It should be highlighted that for these two cases, the predictions of the simplified method are far better than those given by the nonlinear finite element simulations (see for comparison Figs 4.15 and 4.17).

The ratio between the simulated and the experimental energy (E_{sim}/E_{exp}) at the end of the impact for the meshes 5*t* and 10*t* (see Table 5.6) is presented in Fig. 5.29. Only the GL and the LIU criteria are used for the simulations and the energy predicted by the simplified method is included for reference. The results from the simulations were presented in Villavicencio et al. (2018). Fig. 5.29 is an illustration of the agreement between the finite element simulations and all of the selected experiments. It should be recalled here that the energy absorbed at the end of the impact (or until a prescribed penetration) represents the most important parameter to assess the crashworthiness of ship structures, since the actual failure modes and/or the initial rupture and subsequent crack propagation of the material are normally not required for practical applications.

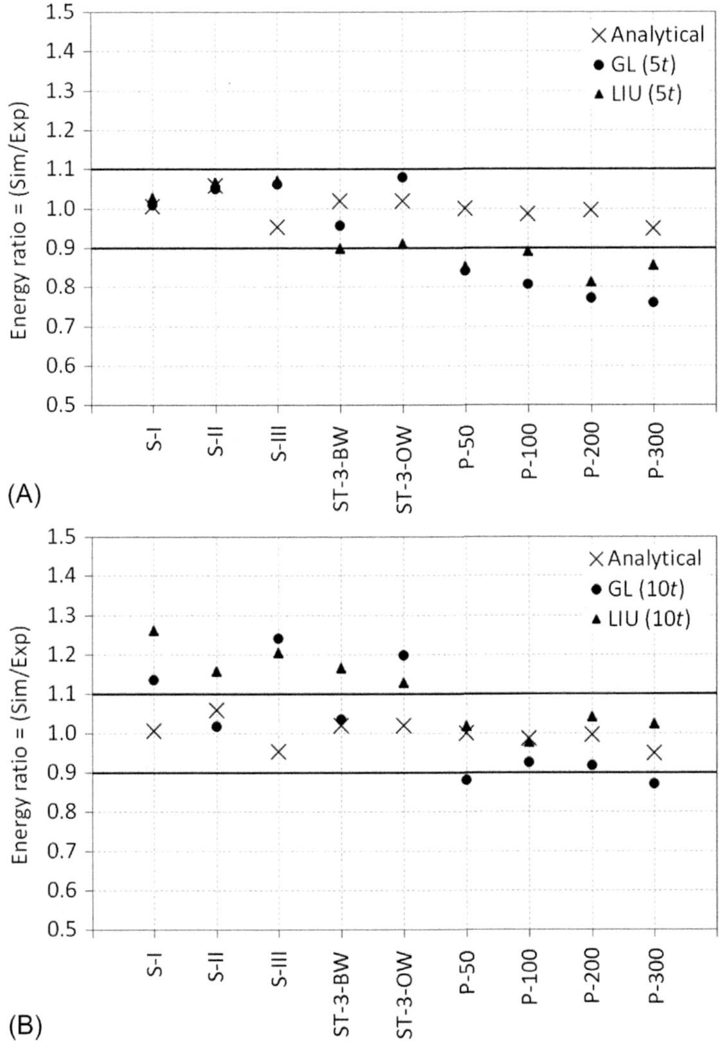

Fig. 5.29 Energy ratio E_{sim}/E_{exp} for all double-hull specimens. The calculations using the simplified method (Eq. 4.4 in Chapter 4) are also included. (A) Mesh size 5*t*. (B) Mesh size 10*t*. The results of the simulations were presented in Villavicencio et al. (2018).

Assuming that a ±10% deviation is acceptable, Fig. 5.29 shows that the energy is predicted satisfactorily for the first five specimens when using the mesh 5*t* and the failure criteria GL and LIU. On the other hand, the energy of the last four specimens is better predicted when the mesh is 10*t*. Note that the GL and LIU criteria predict a similar energy when the mesh is 5*t*.

It should be highlighted here that the predictions of the simplified method agree well with all experimental results. Considering this agreement plus that with other experiments, as presented in Chapter 4, Section 4.3.1, the damage material volume method has been chosen as a reference for the ship collision calculation in Section 5.7.2.

5.7.2 Container ship model

The Guidance Notes for Collision Assessment for the Location of Low-flashpoint Fuel Tanks (Lloyds Register, 2016) recommend the simplified analytical method (Eq. 4.4) to estimate the energy absorbed by destroying ship structural members in tension and crushing mode, as the scatter between experimental results and other alternative methods, such as nonlinear finite element simulations, is still relatively large. Here, the structure of a 4200 TEU container ship is selected for comparison between the simplified analytical method and the collision simulations using the three criteria presented in Section 5.5 (PES, GL, and LIU). This example illustrates a practical application of finite element simulations. The finite element model of the container ship and the simulation results presented here were reported by Liu (2015).

The main particulars of the 4200 TEU container ship are given in Table 5.7, and the scantlings of the side structure are shown in Fig. 5.30. The length of the side structure is 28.6 m, which corresponds to the distance between two consecutive watertight bulkheads (one cargo hold). The spacing of the web frames is 3.2 m, and those supporting the cross decks is 1.5 m. Fig. 5.31 shows the arrangement of the web frames and stringers as well as the impact point.

The collision scenario can be described as follows: the striking ship (rigid bulbous bow of the container ship) sails with a forward speed and collides with a stationary ship (side structure of the hold of the container ship) at a right angle (see Fig. 5.32). The draught of the striking and the struck ship is the same, and the striking ship is subjected to a constant impact velocity.

The mesh size is chosen as 215 mm (four elements between the longitudinal stiffeners). This mesh is equal to $15t$ for the shell plating (thickness 14 mm). The material of the struck structure is HT36 steel where the mechanical properties are summarized in Table 5.8. The nonlinear characterization of the material uses the simple power law relation for true strains larger than 0.006 where the parameters $K = 924$ MPa and $n = 0.187$ are estimated with Eqs (5.25), (5.26). The strain rate sensitivity of the material is omitted.

The energy- and force–penetration curves are plotted in Fig. 5.33. The simplified method (Eq. 4.4) uses a rupture strain of 0.275 estimated with Eq. (3.18a). The energy response shows that the LIU criterion follows the analytical prediction until a

Table 5.7 Main particulars of the container ship as reported in Liu (2015).

Overall length	Length between perpendiculars	Moulded breadth	Depth	Design draught
261.20 m	248.00 m	32.25 m	19.20 m	11.00 m

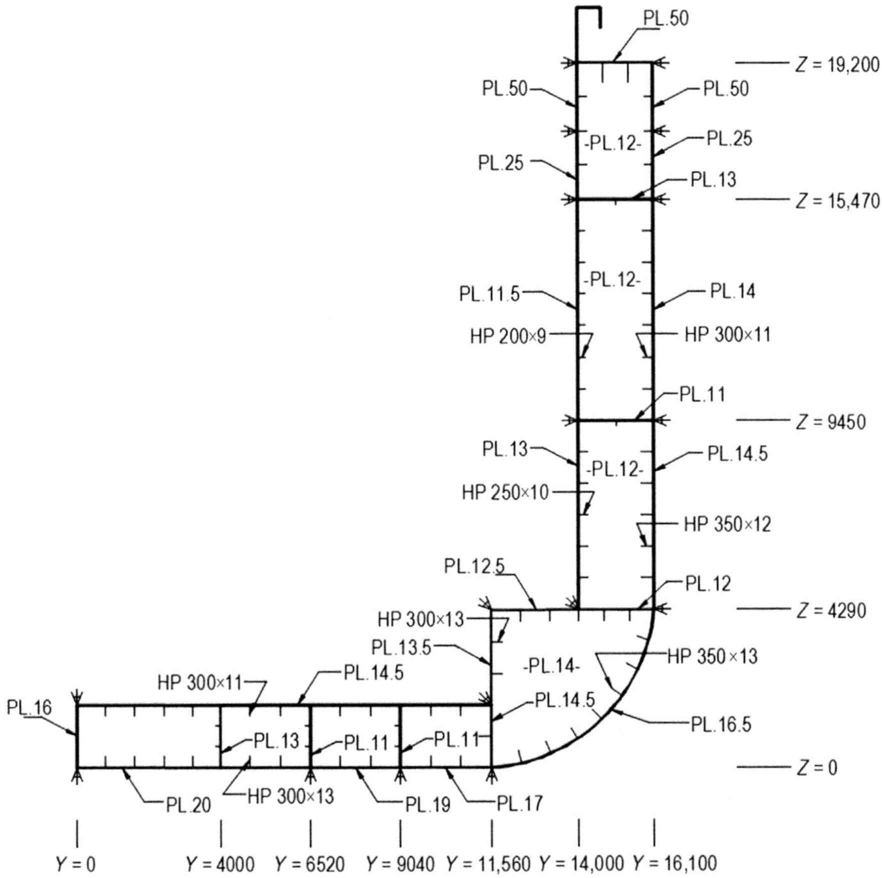

Fig. 5.30 Scantling of the side structure of the container ship.
cf. Liu, B., 2015. Energy Absorption of Ship Structural Components Under Impact Loading. (PhD thesis), Instituto Superior Técnico, University of Lisbon.

penetration of 2.5 m. Beyond that point, the energy is overestimated. The GL criterion, on the other hand, underestimates the energy until the penetration of 4.0 m, while at the latest stages of the impact event, the results using the GL criteria equal the simplified prediction. The PES criterion underestimates the entire response. The shifting in the predictions by the LIU and GL criteria is also clearly observed in the force–penetration response, where LIU follows the instantaneous forces at the early stages, while later, the forces are overestimated. It should be noticed, however, that the magnitude of the forces at the end are similar for both the LIU and GL criteria. Such similitude in the forces is not reflected in the energy response at the end, since the energy curve is a result of the integration of the instantaneous forces and it should be considered that the GL and LIU responses deviate at the initial stages of the impact event.

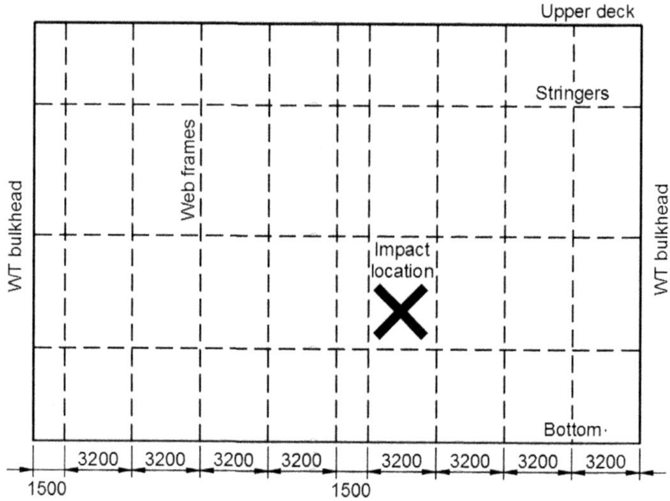

Fig. 5.31 Structural arrangement of the web frames and stringers in the hold compartment of the container ship.
cf. Liu, B., 2015. Energy Absorption of Ship Structural Components Under Impact Loading. (PhD thesis), Instituto Superior Técnico, University of Lisbon.

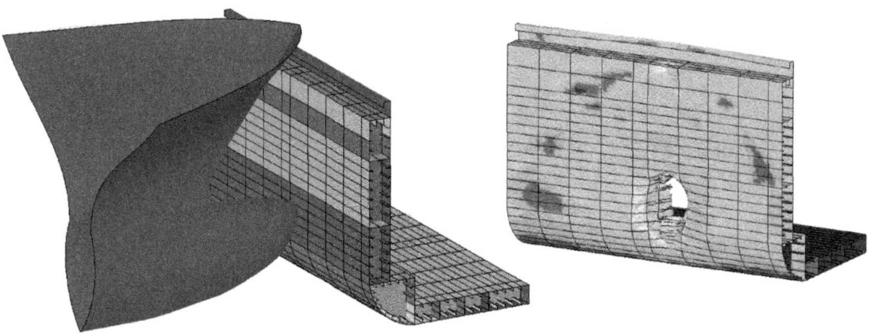

Fig. 5.32 Finite element model representing the collision scenario and the eventual damage to the struck structure.
cf. Liu, B., 2015. Energy Absorption of Ship Structural Components Under Impact Loading. (PhD thesis), Instituto Superior Técnico, University of Lisbon.

Table 5.8 Mechanical properties of the material of the container ship.

Property[a]	**Units**	**Plates and stiffeners**
Yield stress	MPa	355
Ultimate tensile strength	MPa	460–630[b]

[a]The mass density is 7850 kg/m^3, the Young's modulus is 206 GPa, and the Poisson's ratio is 0.3.
[b]The mean value 560 MPa is used in the material definition.

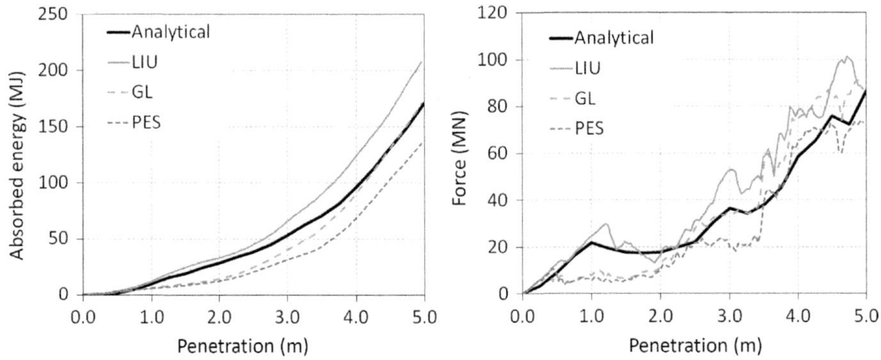

Fig. 5.33 Energy–penetration and force–penetration curves. Comparison between the analytical method and simulations using LIU, GL, and PES criteria. The finite element simulation results were reported by Liu (2015).

5.8 Concluding remarks

It has been shown that the results from nonlinear finite element simulations of ship–ship collisions are highly dependent on the size of the elements and the associated critical failure strain and that care is to be taken when evaluating numerical nonlinear results since these can vary significantly from one approach to another.

Input parameters required for conducting the finite element simulations have been summarized and discussed. Besides the needed mesh sizes, the critical failure strain that can be used for ship collision assessments has been reviewed and used for the calculation examples presented in Section 5.7. It has been concluded that the existing more complex failure criteria do not predict a better response than the simple ones (refer to Fig. 5.1 for example).

The true stress–true strain curves that can be used for the characterization of the strain hardening have been also reviewed. Since the information from a tensile test is required to obtain the strength coefficient and the strain hardening exponent, alternative expressions have been suggested to determine these parameters when only the yield stress and the ultimate tensile stress are available; see, for example, Eqs (5.24), (5.25).

Other parameters reviewed in this chapter, such as boundary conditions and contact definition, are commonly considered as standard/default for conducting collision simulations, although a correct definition is a must when setting up the finite element model.

The fact that nowadays finite element analyses are so popular is mainly because they allow studying very complex structures, as well as visualizing the deformation modes of a structure to identify the areas that could be better optimized/designed. For linear response analyses, the finite element analysis is a robust and practical tool. However, the application of nonlinear finite element software for large-scale damage analysis requires experienced analysts to obtain reliable results. For this reason, there is at present significant research efforts devoted to improve the nonlinear calculation procedures and to reduce the uncertainties of the results.

Examples of input files have not been provided here since they are easily found in the manual/tutorials of the selected/available finite element codes. Instead, the calculation examples can be used to replicate collision simulations.

Finally, it should be mentioned that while the calculation examples and definitions reviewed here are applicable for ship–ship collision cases, they can be also used to evaluate certain grounding scenarios.

References

ABAQUS, 2012. ABAQUS/Standard 6.12 Users Manual. Providence, RI, USA.

Alsos, H.S., Hopperstad, O.S., Törnqvist, R., Amdahl, J., 2008. Analytical and numerical analysis of sheet metal instability using a stress based criterion. Int. J. Solids Struct. 45 (78), 2042–2055.

Altmeyer, G., 2013. Theoretical and numerical comparison of limit point bifurcation and maximum force criteria. Application to the prediction of diffuse necking.

American Society for Testing and Materials (ASTM), 2010. Standard Methods of Tension Testing of Metallic Materials, E 8.

Arita, K., Aoki, G., 1985. Strength of ship bottom in grounding (1st report)—an investigation into the case of a ship stranded on a rock. (in Japanese), J. Soc. Nav. Archit. Jpn. 158, 359–367.

Bressan, J.D., Williams, J.A., 1983. The use of a shear instability criterion to predict local necking in sheet metal deformation. Int. J. Mech. Sci. 25 (3), 155–168.

Det Norske Veritas, 2010. DNV-RP-C204: Design Against Accidental Loads.

Dieter, G.E., 1986. Mechanical behavior under tensile and compressive loads. In: Mechanical Metallurgy, third ed. McGraw-Hill, New York, pp. 275–295.

Ehlers, S., 2010. Strain and stress relation until fracture for finite element simulations of a thin circular plate. Thin-Walled Struct. 48 (1), 1–8.

Ehlers, S., Varsta, P., 2009. Strain and stress relation for non-linear finite element simulations. Thin-Walled Struct. 47 (11), 1203–1217.

Ehlers, S., Broekhuijsen, J., Alsos, H.S., Biehl, F., Tabri, K., 2008. Simulating the collision response of ship structures: a failure criteria benchmark study. Int. Shipbuild. Prog. 55, 127–144.

Hallquist, J.O., 2006. LS-DYNA Theory Manual. Livermore Software Technology Corporation.

Hill, R., 1952. On discontinuous plastic states, with special reference to localized necking in thin sheets. J. Mech. Phys. Solids 1 (1), 19–30.

Hsu, S.S., Jones, N., 2004. Quasi-static and dynamic axial crushing of thin walled circular stainless steel, mild steel and aluminium alloy tubes. Int. J. Crashworthiness 9 (2), 195–217.

Jones, N., 1989. Structural Impact. Cambridge University Press.

Jones, N., 2013. The credibility of predictions for structural designs subjected to large dynamic loadings causing inelastic behaviour. Int. J. Impact Eng. 53, 106–114.

Jones, N., Jones, C., 2002. Inelastic failure of fully clamped beams and circular plates under impact loading. Proc. Inst. Mech. Eng. C J. Mech. Eng. Sci. 216 (2), 133–149.

Karlsson, U., Ringsberg, J.W., Johnson, E., Hosseini, M., Ulfvarson, U., 2009. Experimental and numerical investigation of bulb impact with a ship side-shell structure. Mar. Technol. 46 (1), 16–26.

Lehmann, E., Peschmann, J., 2002. Energy absorption by the steel structure of ships in the event of collisions. Mar. Struct. 15, 429–441.

Ling, Z., 1996. Uniaxial true stress-strain after necking. AMP. J. Technol. 5, 37–48.

Liu, B., 2015. Energy Absorption of Ship Structural Components Under Impact Loading (PhD thesis). Instituto Superior Técnico, University of Lisbon.

Liu, B., Villavicencio, R., Guedes Soares, C., 2013. Experimental and numerical plastic response and failure of laterally impacted rectangular plates. J. Offshore Mech. Arct. Eng. 135 (4) 041602.

Liu, B., Villavicencio, R., Zhang, S., Guedes Soares, C., 2017a. A simple criterion to evaluate the rupture of materials in ship collision simulations. Mar. Struct. 54, 92–111.

Liu, B., Villavicencio, R., Zhang, S., Guedes Soares, C., 2017b. Assessment of external dynamics and internal mechanics in ship collisions. Ocean Eng. 141, 326–336.

Liu, B., Pedersen, P.T., Zhu, L., Zhang, S., 2018a. Review of experiments and calculation procedures for ship collision and grounding damage. Mar. Struct. 59, 105–121.

Liu, K., Liu, B., Villavicencio, R., Wang, Z., Guedes Soares, C., 2018b. Assessment of material strain rate effects on square steel plates under lateral dynamic impact loads. Ships Offshore Struct. 13 (2), 217–225.

Lloyds Register (LR), 2016. Guidance Notes for Collision Assessment for the Location of Low-Flashpoint Fuel Tanks. Lloyds Register, London.

Lloyds Register (LR), 2018. Rules for the Manufacture Testing and Certification of Materials. Lloyds Register, London.

Marinatos, J.N., Samuelides, M.S., 2015. Towards a unified methodology for the simulation of rupture in collision and grounding of ships. Mar. Struct. 42, 1–32.

Paik, J.K., 2007. Practical techniques for finite element modeling to simulate structural crashworthiness in ship collision and grounding (Part I: Theory). Ship Offshore Struct. 2 (1), 69–80.

Peschmann, J., 2001. Berechnung der Energieabsorption der Stahlstruktur von Schien bei Kollisionen und Grundberuhrungen. (in German). (PhD thesis). Technical University of Hamburg, Germany.

Pill, I., Tabri, K., 2011. Finite element simulations of ship collisions: a coupled approach to external dynamics and inner mechanics. Ships Offshore Struct. 6 (1), 59–66.

Ringsberg, J., et al., 2018. MARSTRUCT benchmark study on non-linear FE simulation of an experiment of an indenter impact with a ship side-shell structure. Mar. Struct. 59, 142–157.

Scharrer, M., Zhang, L., Egge, E.D., 2002. Final report n0614, Collision calculations in naval design systems. (in German), Report Nr. ESS 2002.183, Version 1/2002-11-22, Germanischer Lloyd, Hamburg.

Sever, N.K., Choi, C., Yang, X., Altan, T., 2011. Determining the flow stress curve with yield and ultimate tensile strengths. Part II. Stamp. J. (July/August), 14–15.

Simonsen, B.C., Lauridsen, L.P., 2000. Energy absorption and ductile failure in metal sheets under lateral indentation by a sphere. Int. J. Impact Eng. 24, 1017–1039.

Storheim, M., Amdahl, J., Martens, I., 2015a. On the accuracy of fracture estimation in collision analysis of ship and offshore structures. Mar. Struct. 44, 254–287.

Storheim, M., Alsos, H., Hopperstad, O.S., Amdahl, J., 2015b. A damage-based failure model for coarsely meshed shell structures. Int. J. Impact Eng. 83, 59–75.

Swift, H., 1952. Plastic instability under plane stress. J. Mech. Phys. Solids 1 (1), 1–18.

Törnqvist, R., 2003. Design of Crashworthy Ship Structures (PhD thesis). Technical University of Denmark, Denmark.

Villavicencio, R., Guedes Soares, C., 2012. Numerical plastic response and failure of a pre-notched transversely impacted beam. Ships Offshore Struct. 7 (4), 417–429.

Villavicencio, R., Sutherland, L.S., Guedes Soares, C., 2012. Numerical simulation of transversely impacted, clamped circular aluminium plates. Ships Offshore Struct. 7 (1), 31–45.

Villavicencio, R., Liu, B., Guedes Soares, C., 2014. Experimental and numerical analysis of a tanker side panel laterally punched by a knife edge indenter. Mar. Struct. 37, 173–202.

Villavicencio, R., Liu, B., Liu, K., 2018. Accuracy of failure criteria commonly used for ship collision simulations. In: Proceedings of the 37th International Conference on Ocean, Off-shore and Artic Engineering (OMAE 2018), Madrid, Spain. Paper OMAE2018-77377.

Wang, G., Arita, K., Liu, D., 2000. Behavior of a double hull in a variety of stranding or collision scenarios. Mar. Struct. 13, 147–187.

Zhang, L., Egge, E.D., Bruhns, H., 2004. Approval procedure concept for alternative arrangements. In: Proceedings of the 3rd International Conference on Collision and Grounding of Ships, Izu, Japan, pp. 87–96.

Zhang, S., Villavicencio, R., Zhu, L., 2018. Effect of friction on the impact response and failure mode of double hull structures in ship collisions. In: Proceedings of the 3rd International Conference on Safety and Reliability of Ships Off-shore SAROSS. Wuhan University of Technology, Wuhan, PR China.

Zhang, S., Villavicencio, R., Zhu, L., Pedersen, P.T., 2019. Ship collision damage assessment and validation with experiments and numerical simulations. Mar. Struct. 63, 239–256.

Index

Note: Page numbers followed by *f* indicate figures, *t* indicate tables, and *b* indicate boxes.

L

M

N

O

9780128150221